HAZARD IDENTIFICATION
AND RISK ASSESSMENT

HAZARD IDENTIFICATION AND RISK ASSESSMENT

Geoff Wells

INSTITUTION OF CHEMICAL ENGINEERS

The information in this book is given in good faith and belief in its accuracy, but does not imply the acceptance of any legal liability or responsibility whatsoever, by the Institution, or by the author, for the consequences of its use or misuse in any particular circumstances.

All rights reserved. No part of this publication may be reproduced, stored in a retrieval system, or transmitted, in any form or by any means, electronic, mechanical, photocopying, recording or otherwise, without the prior permission of the publisher.

Published by
Institution of Chemical Engineers,
Davis Building,
165–189 Railway Terrace,
Rugby, Warwickshire CV21 3HQ, UK.
IChemE is a Registered Charity

© 1996 Geoff Wells

ISBN 0 85295 353 4

Printed in the United Kingdom by Redwood Books, Trowbridge, Wiltshire, UK

PREFACE

The aim of this text is to demonstrate various ways of carrying out hazard analysis in the process industries, with reference to the way in which major incidents develop. The relationship between hazard and risk is defined and illustrated.

The methodology generally applied in the process industries shows that hazard identification methods can be used in different ways to model part of the incident scenario leading to a possible accident. So an effective model of the incident scenario is the key to a successful application. Such a model has been developed. It follows the structure of a generic fault tree up to the release of material, and of an event tree from this point to the impact of the release on people, plant and the environment.

A process goes through many stages during its life, and safety reviews can be carried out at many of them. The techniques described in this book can be applied at any stage in the life cycle of a process. But it is most cost effective to carry out studies early in a project when there is the greatest potential for reducing the hazardous characteristics of a planned process. Such techniques are often classified as 'pre-Hazop' studies.

Concept Hazard Analysis is primarily concerned with identifying the main hazards associated with a new process design. Several versions of this technique are presented, ranging from an initial project review at a very early stage of a project to a more detailed study recommending hazard controls which might be accomplished late in concept engineering or early in basic design.

Preliminary Process Hazard Analysis has been refined from a 'top down approach' based on Fault Tree Analysis into a technique in which the analysis is commenced by examining the dangerous disturbances immediately preceding the significant release of material by rupture or discharge. This method, when extended by Consequence Analysis, enables the examination of the key incident scenarios affecting the plant. It is applied during basic engineering or for special studies of an existing plant.

In order to reduce the tendency to add on safety, an old technique — Critical Examination — has been revived with better methods for its documentation. The aim is to introduce a study which, instead of asking what can go wrong in process safety, asks what can be done better. The change in focus encourages the use of inherent safety.

Hazop remains the prime method for the study of the safety of a detailed design. This book provides an update of Hazop on its own and when combined with part of a Task Analysis. The latter is described with the main aim of emphasizing further the important role of the operator as an agent of hazard control. It also allows a limited description of the root causes of incidents.

Fault tree and event tree construction and evaluation are discussed. The *pro forma* used for Preliminary Process Hazard Analysis has proved particularly useful for the development of fault trees. The use of Boolean algebra for the evaluation of fault trees is described.

What constitutes a tolerable risk of damage and harm stemming from a process plant is an emotive subject. The topic is discussed and suggestions made about appropriate risk criteria which companies might consider. The elements of risk assessment are described. A specific shortcut method for risk estimation is recommended.

The various techniques are illustrated by appropriate case studies.

In a text of this size, it is only possible to consider selected hazard identification methods in depth. Information on major hazards, Consequence Analysis and the study of root causes of incidents with appropriate case studies will be considered in a companion volume in preparation. It was also decided to exclude the study of reliability and availability from this book.

Please note that many of the diagrams given in this book have been simplified for presentation purposes. Also the safety exercises have not been subjected to the extensive review of team exercises. Consequently it is possible that a number of minor errors may have been unwittingly introduced.

The bulk of this material was originally used as part of the first module of a Masters course in process safety and loss prevention. Several chapters contain material which has been influenced by material presented in other modules of this Masters course; special thanks to David Embrey for material on Task Analysis and human factors, Dave Rochford for supplying Appendix 2, Colin Bullock and Steve Whitty for some of the material on Hazop, Keith Cassidy and Steve Allum for information on risk assessment, Rob James on Concept Hazard Analysis, Steve Kirby on Critical Examination, and Alan Reeves on Fault Tree Analysis. Christina Phang and Mike Wardman have provided research assistance in the development of hazard analysis techniques. Also thanks to all the Masters course students in process safety and loss prevention who have contributed to the development of this book during the teaching of the first module. Many thanks to Gillian Nelson and colleagues at IChemE and Alan Jones at the Health & Safety Executive without whose help the process safety and loss prevention course would not have been possible. Particular thanks for their great contribution at the editing stage to Peter Varey and Audra Morgan at IChemE.

Whilst associated with this course I was able to note how both experienced and inexperienced safety practitioners coped with different techniques. When difficulties were encountered appropriate methods were developed or refined. I hope that these changes are of general interest for a wider number of people; that is why I have written this book. The book provides a grounding in hazard analysis and risk assessment for newcomers to the field and experienced practitioners to pick up useful suggestions to help them in their work.

CONTENTS

		PAGE
PREFACE		iii
1.	AN INTRODUCTION TO HAZARDS AND RISK	1
2.	CONCEPT HAZARD ANALYSIS	19
3.	PRELIMINARY PROCESS HAZARD ANALYSIS	46
4.	CRITICAL EXAMINATION OF SYSTEM SAFETY	74
5.	HAZARD AND OPERABILITY STUDIES (HAZOP)	90
6.	FAULT TREE ANALYSIS	124
7.	TASK ANALYSIS	152
8.	TASK ANALYSIS AND HAZARD IDENTIFICATION	176
9.	RISK CRITERIA	199
10.	RISK ASSESSMENT	210
APPENDIX 1 — THE DEVELOPMENT OF P&I DIAGRAMS		242
APPENDIX 2 — HAZCHECK LISTING		257
APPENDIX 3 — A HAZOP STUDY		267
APPENDIX 4 — FURTHER STUDIES		288
APPENDIX 5 — LIST OF ACRONYMS		296
REFERENCES		297
INDEX		298

1. AN INTRODUCTION TO HAZARDS AND RISKS

PREAMBLE

In this chapter the aim is to introduce the different topics which will be dealt with in the book. The main hazard analysis methods described are Concept Hazard Analysis (CHA) and Preliminary Process Hazard Analysis (PPHA), which are recommended for use during conceptual and basic engineering as precursors to Hazop, which is preferred for use during detailed engineering. The use of Critical Examination to generate alternative schemes is recommended. Other techniques described include the use of fault trees, event trees and Task Analysis. Risk assessment is increasingly being required for major hazard plants and this is discussed with respect to estimating relative and absolute values of risk.

Particular attention is given to the importance of selecting hazard analysis methods which are appropriate to the project stage and which relate to the process information currently available, and the development of a valid model of the main incident scenarios which might occur on the plant.

HAZARDS

A *hazard* has the potential to cause harm. This can take the form of death, ill health and injury to people; damage to property, plant, products or the environment; production losses; business harm and increased liabilities. Ill health includes acute and chronic ill health caused by physical, chemical or biological agents as well as adverse effects on mental health.

A major accident might be defined as one having the potential to kill three or more people or damage a specific area of the environment or cause property damage and loss in excess of a particular sum. Such a definition means that most major accidents in the process industries involve a large accidental release of chemicals or the energy from their reaction, in such a way as to cause appreciable damage. A minor accident might arise when the release of a substance can cause ill health due to the physico-chemical properties of the substance and the way these affect people. Similar lesser incidents occur when people suffer cuts or burns or when a smell or smuts cause nuisance to the public. A near miss often causes hostile reaction, particularly if an appropriate authority must be informed of an incident which had the potential to cause major harm.

Many of the hazards which give rise to such incidents are readily identified. In the process industries such hazards include those falling into one or more of the following categories:
- *chemical hazards*, which include acidity, alkalinity, corrosivity, explosiveness, flammability, reactivity, toxicity and asphyxiation;
- *thermodynamic hazards*, which include high pressure and vacuum, heat transfer, high and low temperature and fluid jets;
- *electrical and electromagnetic hazards*, which include high voltage, radiation, static electricity and electrical current;

- *mechanical hazards*, which include mechanical energy, stresses, forces and impact blows and contact laceration;
- *health hazards*, which include noise, pollutants, chemicals, vibration, radioactivity and temperature extremes.

External threats include accidental impact damage by missiles and vehicles, act of God and natural causes, abnormal environmental extremes, external interference by a variety of causes including action by humans and chemicals, instability of structures or foundations, force majeure and external releases of energy or toxins.

Any unplanned change might constitute a threat to plant, personnel and environment. For example, it might introduce a weakness into the system affecting the defences of the system against loss of containment. This not only includes internal changes to pipework and vessels, such as increase in flow rate and pressure, but also changes which arise in design, operation

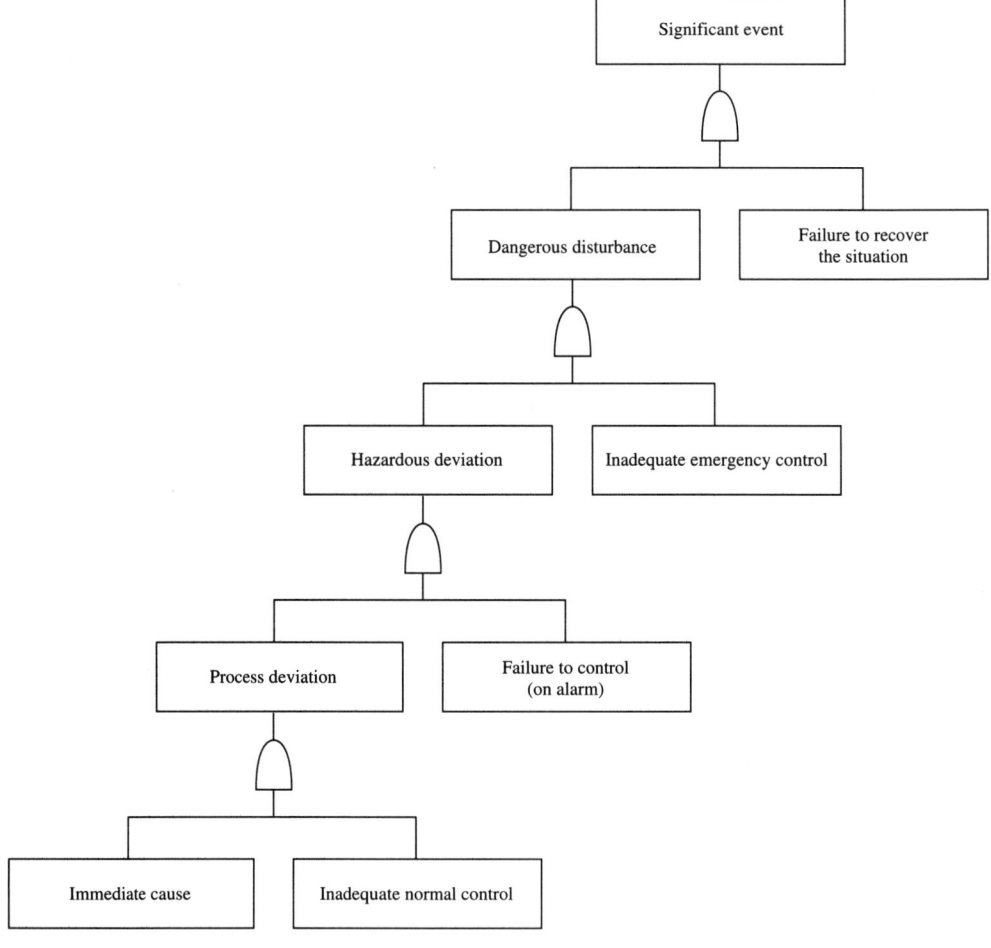

Figure 1.1 Fault tree representation of the general incident scenario.

and maintenance which affect performance of plant and equipment, personnel and procedures, communication and information processing. Furthermore such changes may not be immediately recognized as undesired events, but only appear at a much later time to affect behaviour.

So the position is more complicated than it first seems. The analyst cannot simply aim at any particular time to study all hazards but must restrict the study to the area of interest at that time. This can be done by paying particular attention to the way incidents develop.

INCIDENT SCENARIOS

An *incident* includes all undesired circumstances and near misses which have the potential to cause accidents. An *accident* includes any undesired circumstances which give rise to ill health or injury; damage to property, plant, products or the environment; production losses or increased liabilities. Some people believe that an accident is by definition an accidental random event which cannot, therefore, be anticipated or predicted. So the term 'incident' is preferred in this text as it is quite obvious from the study of major accidents that many of these could have been prevented.

A simplified incident scenario is indicated in Figure 1.1. This shows an event path or chain leading from the initiating event(s) specific to certain consequences, via undesired events and in a particular propagation time.

An undesired event is defined as an event which is either itself inherently unsafe if somebody or something is exposed directly to it, or part of a chain of events which lead to an inherently unsafe event.

Figure 1.1 takes the form of a simple fault tree. The symbols linking the events (or more correctly faults on the tree) are gates, which in this case simply means that both events must occur before the next state of the system is reached.

A sequence of hazardous events can often be stopped by appropriate action to control the situation. Hazard controls should be designed to minimize the probability of occurrence of initiating events and their propagation into undesired events. Any features which make the plant safe by design should be given priority. Capabilities for the implementation of corrective functions or contingency functions, or both, should be built into the design in order to prevent the propagation of these events. Special procedures must also be developed to counter undesired events which cannot be satisfactorily controlled by design or by safety interlock systems (emergency control systems).

The immediate causes of an incident include factors such as:
- action by plant personnel inadequate;
- defects directly cause loss of plant integrity;
- plant or equipment inadequate or inoperable;
- control system or emergency control inadequate;
- change from design intent at any point in the plant life cycle;
- environmental and external causes of disturbance.

Consequently operating parameters change from those specified by design either by alteration in operating conditions within the system, or because the construction is defective or deteriorating in service, or because an abnormal opening is left or made in equipment, or due to a planned discharge or vent changes.

The changes in operating parameters which occur in the incident scenario as the danger increases have been distinguished by the following terms, as illustrated for an uncontrolled increase in temperature of a stream:
- a *process deviation* occurs when the stream temperature increases to higher than normal design and an alarm might signify the need for intervention by the operator;
- a *hazardous disturbance* occurs when the stream temperature is reported as high-high temperature and emergency control might be effected;
- a *dangerous disturbance* occurs when the temperature has passed the point of overtemperature and a trip system has failed to correct the situation. There is still time to recover the situation before significant release on rupture or discharge.

The different changes are defined solely by the designer or analyst and usually not noted by any change of system state. In some cases — such as when an operator opens a valve to release material to atmosphere — the time interval in passing from process deviation to dangerous disturbance can be almost instantaneous.

Attempts to control the situation corresponding to these deviations are defined as follows:
- *inadequate normal control* corresponds to the failure of the normal control system to correct the trend for any reason, including the failure to provide such a control system. This may be effected automatically or by action by operators, either maintenance or production. Clearly detection of the change is important;
- *failure to control (on alarm)* implies that corrective systems applied already fail to prevent the trend and that further steps taken by the operator, probably alerted by alarm, to correct the situation have also failed;
- *failure of emergency control* means that any further emergency control systems — such as automatic shutdown systems, process relief or action by the operators following a set procedure — have failed to control the situation.

When a dangerous disturbance has been reached there may still be time to recover the situation by further and immediate recovery action. If this fails then the plant will either rupture on exceeding mechanical design limits, fail under design conditions due to a critical defect or deterioration in construction, lose significant material through an abnormal opening to atmosphere or discharge incorrect specification materials out of the plant.

The *significant event* here is the release of material by rupture or discharge, when the release creates a hazard or hazardous condition and countermeasures fail to recover the situation immediately. This in turn can lead to *harm* and *damage*. Such impact can be reduced by appropriate *emergency planning* involving avoiding escalation by toxic release, fire or explosion and measures to mitigate against the release and energetic and toxic events.

But immediate causes are only the perceived direct cause of the incident. Underlying these are the *root causes* which are not shown on Figure 1.1. These latent errors can affect any part of the incident scenario. They include and involve consideration of external systems such as:
- industrial bodies, contractors and public;
- the system climate, including legislation regulations, business climate and culture;
- organization and higher management, management control, site and plant facilities, engineering and plant realization;
- transport and disposal of materials;

- communication and information;
- procedures and practices;
- working and local environment;
- engineering integrity of plant;
- operator performance, including recruitment, training, personnel capabilities and working discipline.

There are millions of root causes affecting plants, and system defences must be set up to guard against them and monitor key performance indicators.

RESIDUAL RISK

Absolute safety is an ideal state of a system where there are no threats to humans and their environment or to the hardware and software parts of the system. An objective of a safety programme is to bring any system as close to the ideal state of absolute safety as it is possible, within the system constraints and project/manufacturing objectives. In other words, it is accepted that there will be some residual risk of harm to people, property and the environment, but this has been reduced to what is considered to be a tolerable level as perceived by a responsible authority or other generally acceptable group.

This residual risk must be accepted as being inevitable even though in many cases it must appear remote. The analogy between the Earth and a process plant (Figure 1.2) is a useful one. The molten core of the Earth gives rise to volcanic eruptions through weaknesses in the surrounding crust, or earthquakes due to the movement of land masses. These mainly affect the local environment but can affect the wider environment.

Similarly a process plant contains hazardous materials. There is no way that such a system can be made absolutely safe. It is possible, however, to strengthen the materials of construction, reduce the inventory, improve the control systems, give warning that there is likely to be a release of material and implement an appropriate emergency plan.

The situation of both these systems is constantly changing. Sometimes the changes in the surrounding environment mean that an undesired event has a greater impact. Large cities have been built on fault lines in the Earth's crust. Populations have built up close to airports and hazardous chemical plants, such as at Bhopal. So there is a need for continuous vigilance and to effect further reduction in residual risk; for example, by strengthening buildings and fastening down

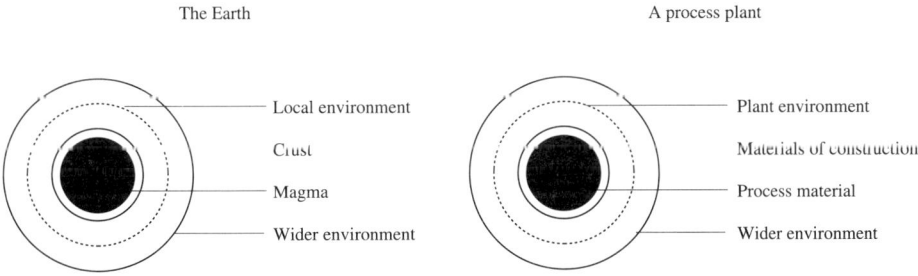

Figure 1.2   Analogy between the Earth and a process plant.

plant in an earthquake zone or improving control and mitigation and reducing inventories on a chemical plant. If the risk of harm becomes unacceptable, the plant must be closed down and made safe.

HAZARD ANALYSIS AND HAZARD REDUCTION
In order to reduce residual risk it is common to use *hazard analysis* to study a system to identify applicable hazards and their possible consequences. *Hazard reduction* can then be applied to eliminate some hazards or at least mitigate against their consequences to an extent that is comparable with project and manufacturing objectives incorporating the general safety objectives and policies of the company.

A hazard analysis involves the identification of undesired or adverse events that lead to the materialization of a hazard, the analysis of the mechanisms by which these undesired events could occur and usually the estimation of the extent, magnitude and likelihood of any harmful effects. In theory it is applied only to the identification of hazards and the consequences of the credible accident sequences of each hazard. Unfortunately non-credible accidents have a habit of occurring in practice.

During any study it is fairly easy to recognize the particular threat which is presented to plant or people. This is done by giving attention to a particular hazardous condition. This is not necessarily an undesirable event in itself, but has the potential to induce one or more undesirable or dangerous events. Hazardous characteristics embrace both hazards and hazardous conditions. Hence when reference is made to hazard identification, it is — more often than not — the identification of hazardous characteristics and the risk which is of concern.

The general stages in hazard reduction involve:
- the identification of hazards and hazardous conditions;
- the identification of the incident scenario;
- hazard review and determination of consequences;
- assessment of risk;
- reduction of the potential consequences of accidents;
- reduction of the frequency of occurrence of major accidents;
- control of external threats and unplanned change;
- attention to organization, management, training, procedures, information, etc;
- implementation, assessment and continued vigilance.

None of these steps are necessarily completed at any particular time, and hazard analysis is usually carried out as part of a formal safety review.

Process safety reviews interact with a plant from its inception as a project to its commissioning and throughout its period of beneficial production and eventual disposal. Most of these safety reviews are key milestones during project development and are virtually mandatory in order to demonstrate the successful process safety management of a project and plant. The project stages can be divided into conceptual engineering, basic engineering, detailed engineering, equipment procurement and construction and commissioning. There is an overlap between phases and on certain projects, where the process technology is well known, phases are combined. Safety reviews are carried out at various times during the life of the plant. Existing plants are often subjected to quite detailed studies, should the need be perceived or required under legislation —

TABLE 1.1
The relationship between hazard reviews and the incident scenario

Incident scenario		Concept studies	Preliminary studies	Detailed studies
Risk evaluation		Hazard indices	Short-Cut Risk Assessment	Quantified Risk Assessment (QRA)
Major events	Mitigation protection	Concept Hazard Analysis (CHA)	Preliminary Consequence Analysis	Consequence Analysis
Release of material			Preliminary Process Hazard Analysis (PPHA)	Safety Report
Dangerous disturbance	Recovery action			Operability Review
Emergency control				Hazop
Hazardous disturbance	Normal control			
Immediate causes				

as is enacted, for example, worldwide for major hazard installations. Clearly there is an increase in the knowledge available about a plant as its life continues. Consequently the type of review selected for a safety study of a mature plant is not so constrained by the availability of information as it is during the project stage.

The principal methods used for hazard and risk analysis are quasi-formal techniques, designed to facilitate the process of hazard identification and risk evaluation by directing the thought process of the analysts along certain lines. They use the incident scenario as their primary model, but no method aims to cover the whole scenario. A summary showing how some of the techniques to be described fit into the study of incident scenarios is given in Table 1.1.

In Table 1.1, various hazard analysis and risk assessment techniques have been identified and their use is linked both to the project stage and the incident scenario.

The safety reviews signify the progress of a capital project on its way to completion. The main hazard reviews which are commonly carried out by industry are not named consistently in different countries, and also vary according to need and information available. Consequently here there has been an attempt to standardize the terminology. In the following sections an introduction will be given to each of the techniques used, following the sequence in which they appear in the text.

CONCEPT HAZARD ANALYSIS

Concept Hazard Analysis (CHA) is used for the identification of hazards and hazardous conditions. The primary aim is to identify areas which are recognized as being particularly dangerous from incidents in the past, not necessarily involving the same process. It also identifies the need to explore any difficulties which might be experienced with unwanted reactions. Elimination or reduction of the hazards is a primary objective, because doing this at an early stage of the design is particularly cost-effective. The method can be used to help determine the suitability of the planned site and the acceptability of a particular process route, bearing in mind process safety and the green policies of the company. It may be used to guide selection of hazard control systems.

CHA is mainly used during 'conceptual engineering', which involves the technical and economic evaluation of a project's feasibility including process route, process chemistry, hazard identification, block flow and initial process flow diagrams, fundamental design basis for equipment, maximum intended inventory, control and possibly emergency control, plot plans and site selection. Variants of the method are also recommended for use during 'initial project review' and during 'basic engineering'.

CHA essentially involves a number of people sitting down at a meeting at which basic information is available about the process. Various keywords are then used to stimulate an appropriate discussion which is recorded together with actions and notes.

The keywords used all refer to the main hazards presented by the process. A sample of the principle keywords in common use is given in Table 1.2. New keywords can be added. In practice about ten basic selections usually suffice for a small study. It would not be expected that the same set of keywords would be used in the chemical, oil and nuclear industries.

This simple method is particularly useful when the process is only specified in broad terms and detailed flow sheets and layouts are not available. It certainly identifies all the major hazards on the plant and external threats at a time when the design is still fluid. Direct recording of the discussion is readily accomplished and a record of the meeting may be quickly circulated to members of the study team.

CHA is poor at identifying the immediate and root causes of incidents. It is not good at checking out the control system. It can be used for a general discussion on whether the company has the organizational and management support to carry out such a project and run the plant, given the level of management and manpower competence for this type of process. It can recognize the impact of the proposed plant on the status of the site with respect to national regulatory control and consents for effluents and so on. When it comes to details, however, more powerful methods must be used. Only crude estimates can be made of the risk of damage and would probably be carried out using hazard indices. The need can be prioritized for further studies to identify the suitability of a given site for a process novel to the company.

PRELIMINARY PROCESS HAZARD ANALYSIS
The technique of Preliminary Process Hazard Analysis (PPHA) was specifically developed for use during 'basic engineering', which involves process flow sheeting and material and energy balances, process flow diagrams, initial piping and instrumentation diagrams (P&IDs), process equipment performance specification and data sheets, block model of plot and site plans. It follows up the results of the Concept Hazard Analysis to provide further information on factors such as wanted and unwanted reactions, the deduction of hazards and hazardous characteristics on the plant, the identification of incident scenarios and the evaluation of emissions, effluents, wastes and off-specification products.

TABLE 1.2
Some keywords in common use for Concept Hazard Analysis

Dangerous substances and reactions:
- hazardous substances (specifically noting the prime hazard they represent)
- dangerous substances (as specified by national regulations)
- planned reactions
- unintended reactions

Fires and explosions:
- fires (name the type of fire)
- explosions (name the type of explosion)

Release and discharge:
- flammables
- thermal radiation/flame impingement
- toxics
- pollutants
- noise

Dangerous disturbances:
- exceeding mechanical limitations
- overtemperature and undertemperature
- overpressure or vacuum
- overload/stress/tension
- impact blow/drop
- critical defect in construction
- abnormal opening
- adverse change in product/discharge

Notable disturbances:
- specific problems associated with equipment
- reactions (planned and unplanned)
- material problems
- utility problems
- mode of operation

External threats

Management, organization and location

PPHA aims to duplicate the incident scenario down towards immediate cause (Table 1.1, page 7). The first part evaluates the incident scenario leading to a significant release of process material. The second part is a Preliminary Consequence Analysis that starts with the significant release of process material and evaluates the consequences of it.

Traditionally the evaluation leading to the release of material is carried out in an identical manner to that used in a Fault Tree Analysis (FTA), with the top event being a significant release of material by a discharge or rupture, either into the atmosphere or with products into downstream plants or into storage. The analysis proceeds downwards in the opposite direction to the path of the incident scenario, evaluating at each level the cause of the undesired events in the chain. This process continues until the analyst is satisfied that an estimate can be made of the frequency of a given incident scenario, or that the incident scenario is insignificant as a contributor to the overall frequency of the top event and further evaluation is not worthwhile.

This basic scheme has been adapted to start the search at a more convenient point: *the dangerous disturbance of plant* leading to rupture and discharge. The change is only minor but enables the analyst to start the search with a few, well-defined suggestions. It also contains elements of Hazop by starting the study at a deviation.

A PPHA starts by selecting a dangerous disturbance from one of the four general groupings of dangerous disturbances listed here:
- *disturbances which will result in rupture on exceeding mechanical limits*, such as explosions, over- and underpressure, over- and undertemperature, machine overload or stress, impact blow or drop;
- *a critical defect in construction*, such as might be left during manufacture or assembly or caused by a deterioration in construction;
- *flow through an abnormal opening to atmosphere*, either made or left in plant;
- *an adverse change in a planned product or other release*, either before leaving plant, on transfer or after leaving plant.

The analyst then identifies the way in which significant release of material can occur as a result of this disturbance. The hazardous deviation and the process deviation giving rise to this deviation are then identified. So in terms of the incident scenario, the analyst has identified the features denoted below in bold type.

SIGNIFICANT RELEASE OF MATERIAL
  FAILURE TO RECOVER SITUATION
DANGEROUS DISTURBANCE OF PLANT
  INADEQUATE EMERGENCY CONTROL
HAZARDOUS DEVIATION
  FAILURE TO CONTROL SITUATION (ON ALARM)
PROCESS DEVIATION
  INADEQUATE NORMAL CONTROL
IMMEDIATE CAUSES OF INCIDENT

The gaps related to control and recovery may then be filled in to gain a full understanding of the control measures taken either automatically or by the operators (production or maintenance). Appropriate *pro formas* have been developed for this purpose.

The immediate causes of selected scenarios may be identified but are further defined during the Hazop technique to be described shortly. A fault tree for such scenarios may be completed and the frequency of the significant release of interest can be computed.

It should be noted how the technique encourages an appreciation of the role of the operator in avoiding disaster. It also emphasizes that emergency control systems do break down and any hazard analysis must allow for this possibility.

A Consequence Analysis can then be used to identify the impact of the significant release. As various releases can have similar effects, this is best done following the identification of several releases. The aim is to prevent or minimize the consequences of catastrophic loss of toxic, reactive, flammable or explosive chemicals.

Evaluation of consequences is carried out using event trees. Figure 1.3 represents a simplified diagram evaluating the frequency of fatalities caused by the release of a specific quantity of flammable material which can produce toxic material on combustion.

The eventual outcome from an event tree can be a complex estimation of the likely impact of the event on the system's total environment. The analyst works in the same direction as

AN INTRODUCTION TO HAZARDS AND RISKS

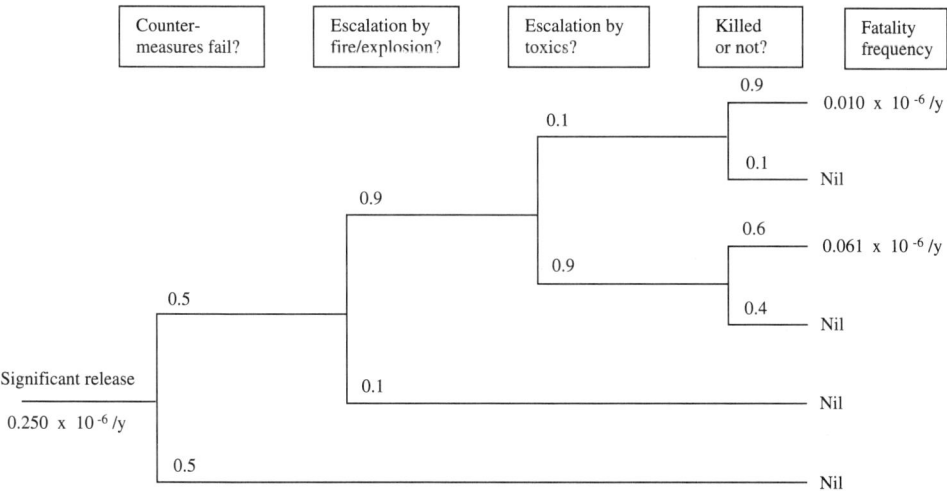

Figure 1.3 Simplified event tree estimating fatalities from a release.

the scenario unwinds, evaluating the effects of countermeasures to the release, its ignition, escalation by major events such as fire, explosion or toxic event and failure to mitigate against these events to reduce their occurrence or effects. The emergency response is a vital part of such a study.

Preliminary Process Hazard Analysis can be used to upgrade the process design manual and the equipment specifications for the detailed basic engineering of the process. If an upgraded process flow diagram is used for the study it is possible to effect an early improvement to the P&IDs by modification of the control and emergency control systems. The study should ensure adequate treatment of all effluents and emissions and consider the acceptability of all discharges.

The study aims to identify some, but not all, immediate causes giving rise to incidents. It does not aim to identify operability problems as this is the task of the subsequent Hazop. However, it should ensure that any major safety problem is identified and problem areas subsequently found can be eliminated at minor cost.

The Consequence Analysis can be used to improve the layout of the plant and the segregation of plant, transport, property, people and the environment. It can improve access and egress to the site and protection from external threats, ensure physical protection and segregation of key plant items and avoid or protect against adverse topographical, geological and meteorological factors.

A primary objective of Preliminary Process Hazard Analysis is to develop the study in such a form that it can be followed by risk assessment. In the first place the objective is to prioritize scenarios for further study using Quantified Risk Assessment. It is important at this time to carry out a sociotechnical system review and implement the beginning of a risk management programme.

The final report should provide an adequate and clear record of the safety studies carried out. It forms the basis for other reports to local authorities and regulatory bodies. The analysis serves as a good starting point for developing effective operating instructions, emergency plans and other procedures to be used in production and maintenance activities.

11

CRITICAL EXAMINATION OF SYSTEM SAFETY

A methodology has been developed for the Critical Examination of system safety. The plant is partitioned into convenient sections. The design intent of that section is then described, taking care to avoid too much detail. Each significant phrase or even word of the design intent of the design is queried using the questions:

- why do it? This suggests optional methods with the aim of improving process safety (and economics if required);
- what can go wrong affecting safety, production or quality? This can identify dangerous conditions associated with the design intent and their causes. Alternative systems can be generated, or modifications, or further controls.

Appropriate task sheets can help in the development of the study.

In generating alternatives it is helpful to use certain keywords amplified by a keyword dictionary. These are extremely powerful and normally applied as follows. The first aim of the designer should be to *eliminate or substitute* dangerous substances, processes or operations. In some cases it may be best to *combine* certain activities or plant items. To *avoid* is a less stringent action and means that it may be possible to evade certain conditions or actions deemed to be undesirable. *Modify*, *alter*, *rearrange* and *improve* are general suggestions. They suffer by not giving a specific direction of change as compared to *simplify*, *increase*, *reduce*, *prevent*, *segregate* and *isolate*.

The key benefits stem from examining how the process step is achieved, paying particular attention to process materials and considering:

- a change in the quantities or qualities of streams;
- use of extra or different process materials;
- changing the method, its operating conditions or activities or its sequence, frequency, absolute time or duration;
- using different equipment or materials of construction.

The impetus for change should be to carry out an operation better, make the frequency of a safety, quality or production lapse less likely, and lessen the consequences should this occur. The costs of effecting the change must be evaluated, and safety and production requirements will at times be at odds.

The aim of this study is to encourage the designer at some time to stop the current activity and reappraise the fundamentals of the process. Process safety studies always seem to be examined by considering what can go wrong. Sometimes analysing in this way should give way to thinking what can be done better to improve process safety and loss prevention. The aim is to try to think more positively and be less prone to consider add-on safety systems immediately when assessing safety problems.

HAZARD AND OPERABILITY STUDIES

THE PROCEDURE

A hazard and operability study (Hazop) is a formal systematic examination of a processing plant for identifying hazards, failures and operability problems, and assessing the consequences from such maloperation. This leads to fewer lapses in safety, quality and production provided that the plant is installed according to design and maintained in appropriate condition. A Hazop is carried out as a team activity. The P&IDs of a plant are examined one by one at the detailed design stage. The Hazop can also be used as a check on the operability of an existing plant. Each process line, vessel or operation on the plant can be examined as the team leader determines. In a typical study, each line selected usually runs from one main plant item to another, but frequently may include any heat exchanger in between.

Having selected a line, then a parameter is chosen. Typical parameters are flow rate, temperature, level, composition, flow quantity and a physical property such as viscosity. Then a number of guidewords are selected to apply with this parameter. Typical guidewords include:

- no, not, none — meaning the activity is not carried out or ceases;
- more of — referring to a quantitative increase in an activity;
- less of — referring to a quantitative decrease in an activity;
- part of — meaning the incomplete performance of an activity;
- reverse or the inversion of an activity;
- other (than);
- as well as — when another activity occurs as well as the original activity;
- sooner/later than — when an activity occurs at the wrong time relative to others.

The parameters and guide words are used to suggest deviations of process variables and their causes. Thus the term NO FLOW contains the guideword NO and the parameter FLOW. This deviation is normally treated as a process deviation (see Figure 1.1, page 2).

The team then identifies what are the meaningful causes of each deviation on this particular line. For example, it might be possible for a NO FLOW to be caused by isolation in error, wrong routing, blockage, incorrectly fitted non-return valve, large leak, equipment failure, incorrect pressure differential, delivery side overpressure, vapour lock and so on.

Then, knowing the cause and the deviation, it is possible to evaluate the consequences of this deviation.

Typically, but not necessarily, the analyst is working at the bottom end of the incident scenario:

SIGNIFICANT RELEASE OF MATERIAL
FAILURE TO RECOVER SITUATION
DANGEROUS DISTURBANCE OF PLANT
INADEQUATE EMERGENCY CONTROL
HAZARDOUS DEVIATION
FAILURE TO CONTROL SITUATION (ON ALARM)
PROCESS DEVIATION
INADEQUATE NORMAL CONTROL
IMMEDIATE CAUSES OF INCIDENT

It is possible, however, to work in this manner from any point in the scenario where a meaningful deviation occurs.

Hazop is the method of analysis most widely used in the process industries. It is recommended for use by legislators, regulators and the engineering institutions. This virtually mandatory use has stemmed from its proven capabilities for the task.

Its key feature is the ability to examine most process deviations, although all the causes and consequences of these deviations may not be diagnosed. It suggests necessary changes to a system or suggests procedures for eliminating or reducing the probability of operating deviations. It assists in the decision-making required to reduce all identified potential hazards to a tolerable level of risk as defined by the current company guidelines.

The study is not comprehensive and any individual or corporate effort will yield results directly proportional to the appropriate background experience of the participants. The system accommodates the use of standards and codes of practice, yet examines their relevance in specific circumstances of plant operation. A record is demonstrated indicating a comprehensive study has been made of the safety of the plant.

Detailed design involves complete design of all equipment with full specifications and engineering drawings. This includes information on materials of construction, electrical classification, relief system design and design basis, safety systems and ventilation system design. Also completed are the P&IDs and piping isometrics, final site and plot plans, civil and structural fabrication and erection drawings. So at the time that Hazop is carried out any detailed design changes are not normally expected. This emphasizes that the earlier safety reviews should be carried out prior to Hazop so that some changes can be identified at a more opportune time. Conversely it is essential to follow these safety reviews by carrying out a Hazop.

FAULT TREE AND EVENT TREE ANALYSIS

Fault and event trees are widely used as communication aids to demonstrate system failures and their development to managers, designers and operators. Their use in qualitative analysis demonstrates the effect of system failure modes and design changes. They can readily be used to quantify the probability or frequency of failure and this can be related to the risk of specific consequences occurring.

A fault tree is a method by which a particular undesired system failure mode can be expressed in terms of component failure modes and operator actions. The system failure mode to be considered is termed the 'top event' and the fault tree is developed in branches below this event showing its causes. The diagram contains two basic elements — 'gates' and 'events'. 'Gates' allow or inhibit the passage of fault logic up the tree and show the relationships between 'events' needed for the occurrence of a higher event. Gates must not be directly connected to other gates and a combination event, drawn as a rectangle, must be used at every logical gate.

Figure 1.1 (see page 2) is an example of a simple fault tree. Normally each immediate cause would be specified and each cause would be shown, entering the combination event labelled 'immediate cause' via an OR gate.

A fault tree is simply a snapshot of the system at a given moment. It can only be related to system states with difficulty. It is constructed by developing each level by deductive reasoning from the top event.

The method used for Preliminary Process Hazard Analysis serves as an excellent technique for the construction of fault trees. Quantification of the trees is aided by a knowledge of Boolean algebra. There are now a number of programs which quantify fault trees so a full understanding of Boolean algebra is not essential. However the numerical values obtained for the frequency of the top event should always be carefully reviewed, as mistakes on fault trees are fairly common during their initial development.

An event tree diverges as shown in Figure 1.3 (see page 11). It follows a cause through to the possible outcomes, branching at each point when there is more than one possible result from the precursor event, until the final outcomes of interest are reached. It is widely used in Consequence Analysis.

The top event from a fault tree is often used as the initiating event of an event tree, as shown in Figure 1.4. For example, the frequency of a significant release of material can be input to a Consequence Analysis.

TASK ANALYSIS

A Task Analysis involves first describing the overall goals of the task which can be any operation or stage activity. The task is then broken down into a set of subordinate tasks or operations which must be performed to achieve the goal at that level. For example, a process operation may be divided into appropriate stages. Each stage is expanded down to a level appropriate for the study being undertaken. This will often be the level of the individual steps in the task. A plan is then produced to show how this should be carried out. Such a plan might take the form of a hierarchical diagram or a tabular list.

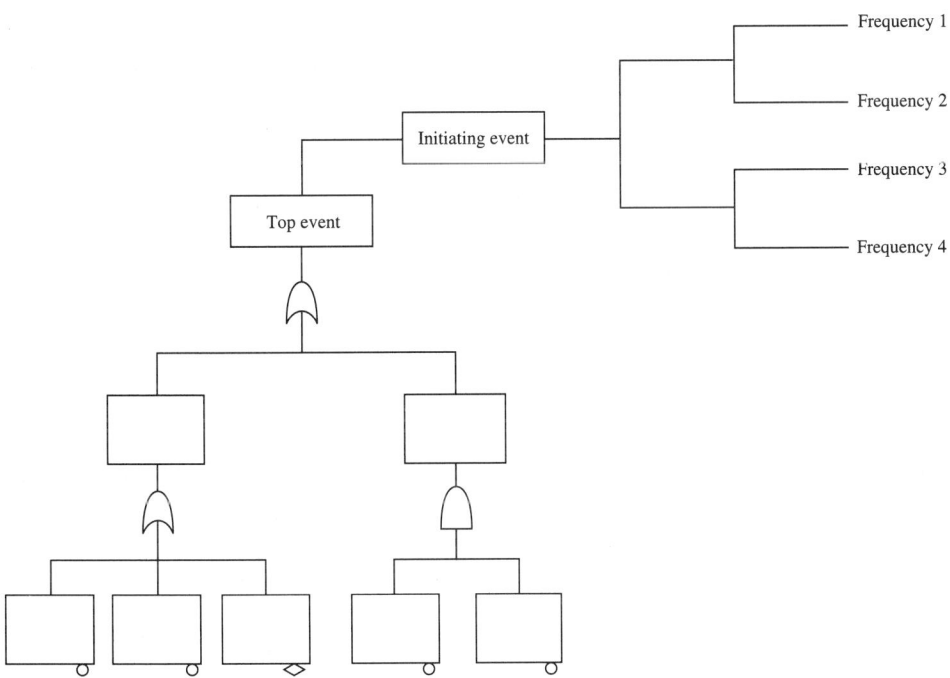

Figure 1.4   Fault trees can serve as the input to event trees.

An initial analysis is carried out on hazards from equipment and materials, difficulties created by the environment, the lack of appropriate controls and protections and so on. It can be subjected to Critical Examination. This analysis of the plan or system should examine possible deviations from the system, determine the likelihood of those deviations and determine any deviation at the start and end of the procedure.

Modifications are made to improve the method of working and to reduce the effect of deviations and appropriate controls, precautions and mitigation introduced. The analysis of the plan will lead to improvements in such features as:
- working methods, procedures and practices;
- work environment and exposure frequency;
- communication and information processing;
- skill and capabilities of people;
- management, supervision and control.

The task should be placed in context and factors influencing performance should be considered such as:
- task similarities, frequency and duration;
- level of training and experience of personnel carrying out the task;
- team structure, co-operation and supervision;
- communications and information processing;
- job aids and technical back-up;
- the working environment;
- pressure and time constraints.

Task Analysis should be extended to consider recovery of a situation and evaluate human reliability by various methods. Human error is claimed almost invariably to be the cause of process incidents. Obviously it contributes to such incidents in many ways including operator, designer and management errors. But it is not acceptable to blame the last people who touched the equipment, whether it be a process valve or the joystick of an aeroplane. The cause of such errors is generally a function of many root causes.

Task Analysis is extremely useful for assisting in the understanding of the general nature of an operating problem. It is invaluable in writing procedures and instructions. Unfortunately it is an extraordinarily time-consuming activity.

The use of Task Analysis to break down a system into steps improves the Hazop of a batch process and is strongly recommended.

RISK ASSESSMENT

The *risk* of an event is the likelihood of a specified undesired event occurring within a given period or in particular circumstances. The risk is usually considered to be a function of the frequency or probability of an event occurring and the consequences of its occurrence, particularly with respect to causing damage and injury. The *individual risk* is the frequency at which an individual may be expected to sustain a given level of harm from the realization of a specified hazard. The *societal risk* reflects the likelihood of accidents involving multiple casualties.

What is considered to be a tolerable risk by a company, its employees, the public exposed to risk from a given facility and the public at more remote locations from the facility differs. Values may be set within company guidelines which help employees in making decisions about whether the residual risk meets the requirements of the company. These standards will require review and there is always the possibility of confrontation with pressure groups about whether such values are acceptable. For this reason some companies avoid giving quantitative values in certain circumstances.

Risk evaluation is a complex process of determining the significance or value of the identified hazards and risks to those concerned with or affected by the decision. It is difficult to generate accurate absolute values of risk for process plant. Not only is appropriate data on equipment and human reliability difficult to obtain, but the model itself is subject to inaccuracies. Relative values are less prone to error and can be trusted to effect significant improvements in residual risk.

Risk evaluation involves the following steps:
- *hazard identification* to determine the incident scenarios, hazards and hazardous events, their causes and mechanisms;
- *frequency estimation* to determine the frequency of occurrence of identified hazardous events;
- *Consequence Analysis* to determine the extent and probability of the consequences of identified hazardous events;
- *risk evaluation* to determine the risk levels;
- *sensitivity analysis* to prioritize further studies of risk, evaluate the significance of risk levels and set a schedule for implementation.

The elements of the procedure are used both to generate information and as an aid to decision-making. The procedure should only be taken as far as is necessary to generate the information required or to make a decision. The extent of application of the various elements and degree of quantification employed therefore varies significantly from one situation to another.

A Short-Cut Risk Estimation is recommended for use with Preliminary Process Hazard Analysis where relative values are used to reduce residual risk. It is also useful to prioritize further studies. Full Quantified Risk Assessment can only be done with confidence following Hazop or a related exercise, and if personnel can affect the frequency of occurrence then further Task Analysis and reliability studies are necessary.

The understanding of risk is increasing at all levels of society and so is the debate about what is a tolerable level of risk in any given circumstances. Residual risk is the risk remaining after all proposed improvements to the facility under study have been made. Residual risk is always present and its study is continued throughout the life of the plant to at least maintain the initial project standards and revise them as company or regulations require.

FURTHER NOTES

The differences between the hazard analysis methods are small but quite important. Why is this the case? Consider conventional optimization techniques used in mathematics. For simple studies differentiation is sufficient or a simple search can be carried out, assuming the system has only one optimum value. Hill-climbing techniques can be used to speed up the search for an optimum. If the system is not unimodal and has several optima, quite complicated methods can be used with the search often starting from more than one point. Of course, if the resources permit, it is possible to analyse every possible case by such measures as the input of a selection of values for every variable.

Similar arguments apply in the search for a 'safe' plant. The search should be started at the best point appropriate to the information available and the objectives of the study. Subsequently it is desirable to check the system again, preferably using a different procedure and from a different starting point.

Here all the pre-Hazop techniques aim to identify problems quickly whilst the design is still fluid. Concept Hazard Analysis primarily identifies hazards and assembles available experience which is relevant to a particular plant. Preliminary Process Hazard Analysis concentrates on developing key scenarios and giving advice on hazard controls. A Critical Examination of the system aims to change the line of thinking away from what can go wrong by suggesting ways that the system can be carried out better, if possible by reducing the need for as many add-on safety systems.

Hazop studies should always be carried out on the detailed design or on any plant which has never been subjected to Hazop. Task Analysis and other detailed studies not yet considered, such as Failure Mode and Effect Analysis, should be used when such detailed studies are justified by the danger or by the savings made by carrying out such studies.

2. CONCEPT HAZARD ANALYSIS

CONCEPT HAZARD ANALYSIS METHODS

This chapter considers some hazard analysis methods appropriate to the following two activities:
- *Conceptual engineering*, which involves the technical and economic evaluation of a project's feasibility including process route, process chemistry, hazard identification, block flow and initial process flow diagrams, fundamental design basis for equipment, control and possibly emergency control, plot plans and site selection.
- *Basic engineering*, which involves process flow sheeting and material and energy balances, process flow diagrams, initial piping and instrumentation diagrams (P&IDs) process equipment performance specification and data sheets, block model of plot and site plans.

These hazard analysis methods are often termed pre-Hazop studies.

Concept Hazard Review follows the usual project review of the definition of the scope of the project and provides the means for an early assessment of safety, health and environmental hazards. It links in with other project work beginning at this time and contributes to key policy decisions such as siting and preferred route. The review determines the need for specific safety reviews and their timing.

Concept Hazard Analysis (CHA) is used for the identification of hazard characteristics in an attempt to identify areas which are recognized as being particularly dangerous from previous incidents in the past. It also identifies the need to explore any difficulties which might be experienced with unwanted reactions. As well as identifying environmental damage, the analysis may also consider whether the proposal fulfils the 'green' policies of the company. Elimination or reduction of the hazards is a primary objective, as doing this at an early stage of the design is particularly cost-effective.

Two methods have been adopted herein. Concept Hazard Analysis (Initial Review) or CHA(IR) is an early safety review in wide use in the process industries. It is often known as pre-Hazop or Hazop 1 — see Turney[1]. The developed method is slightly more refined; it is termed Concept Hazard Analysis or CHA and here it refers to a full study using the procedure.

Preliminary Process Hazard Analysis is described in Chapter 3. It is an improved version of CHA and gives a far better appreciation of the incident scenario. It can be used for some early quantified risk analyses.

A Critical Examination of system safety can be used either to eliminate or reduce the possible consequences of a hazardous event by an early study of the design intent of a particular processing section. This should be carried out at an early stage of the system design and well before the process design is completed. A brainstorming attitude should be employed to make radical changes, particularly in difficult areas. The method has been developed as a way of ensuring that inherent safety and related factors are considered at the design phase. This is described in Chapter 4 and is an essential aid to any concept engineering study.

CONCEPT SAFETY REVIEW

At the start of a project safety analysis, the project manager and others should carry out a Concept Safety Review (CSR). This should be carried out as early as possible, sometimes even during process development where one objective might be to select between feasible routes.

The group should preview and define the objectives and scope of the project. This includes general information about the development plan and the plant or object being analysed. It is particularly important to ascertain the perceived need against a range of options including process development, available processes and whether these will be licensed, the availability of alternative sites and modes of transport of raw materials and products, the availability of experience within the company and site and so on. It may be that a particular project does not require the study of all these items and it is as well to make such matters clear at the start.

Information should be obtained on the safety, health and environmental hazards of all chemicals and materials involved in the new process. This takes account of both individual and collective properties of materials. Helpful information is contained in regulations such as exist for COSHH and CIMAH in the UK and the OSHA Codes of Federal Regulations in the USA. It is essential to be aware of all relevant safety regulations applying to a particular project.

A general appreciation should be generated of the main hazards presented by the plant such as fire, explosion and release of harmful substances — toxic gases and liquids, effluent, radioactive and corrosive materials. These are listed as hazardous substance characteristics and amplified later by data sheets which include information on toxicity, permissible exposure limits, thermal and chemical stability, reactivity, corrosivity, fire fighting agents and so on.

The study reviews information on previous incidents for the plant being studied using both information available on incidents within the company and its affiliates and information available from global sources. For a project under development the latter information should be augmented by studies of the reaction route and incidents affecting plants using related reactions.

At each site under consideration it is necessary to consider on-site and off-site transport of raw materials, products and wastes including loading, off-loading, type of transport and route. The requirements for facilities and services, emergency planning, interaction with other plants and so on must be examined.

The study considers all organizational factors affecting the project, including the availability of experienced staff both within the company and at the site. This experience is reviewed in terms of general experience, experience of related plants and specific experience of the plant. The group discusses ways of overcoming any problems. It also identifies the impact of the plant on the general health and safety management policy of the site. Criteria should be established for all safety, health and environmental factors with which the plant must comply, together with relevant company standards, national legislation and other regulatory approvals and consents. Any effect on the position of the site with respect to effluents and emissions status under CIMAH or OSHA regulations must be reviewed. General project criteria should be defined including the codes of practice to be followed and the extent and timing of all safety reviews.

The CSR is a means by which improvements in design procedures are made known to the designers and by which it is ensured that current thinking on ways of improving design practices are implemented. The CSR is normally part of the project review but can be amalgamated with a CHAIR if appropriate for a particular project. Certainly all this data is essential for the next stage of project safety reviews which will now be described.

CONCEPT HAZARD ANALYSIS INITIAL REVIEW

The normal version of Concept Hazard Analysis used by industry is described here as the Concept Hazard Analysis Initial Review (CHAIR). It is more often known as a pre-Hazop study. In its simplest form this merely involves generating a simple record of one or more meetings between appropriate specialists at which various aspects of design and operation are discussed in a structured way. This record might merely consist of a simple system description identifying the plant, plant items and process description. Various keywords can then be used to stimulate an appropriate discussion topic which is recorded together with actions and notes.

A list of keywords is given in Table 2.1 (see pages 22–23). As occurs in all such methods there is a tendency for the number of keywords to be increased until eventually the method begins to lose its value. The principle ones in use are given in capital letters. The list of keywords may be extended to include operability problems such as loss of services, off-specification material and discharge. New keywords can be selected. In practice about ten basic selections usually suffice. It would not be expected that the same set of keywords would be used in the chemical, oil and nuclear industries.

The list in Table 2.1 has been developed in order to study the release of process materials. It involves identifying the substances and reactions, studying fires and explosions and identifying problems associated with releases and disturbances. Dangerous deviations are identified with the emphasis on their point of occurrence, which is immediately prior to rupture and discharge. Notable disturbances and external threats are often associated with the cause of difficulties and focus attention on a different area of the incident scenario. Equipment problems and services are a catch-all topic. The idea is that for any item of equipment there are many associated problems about which experience is often readily available.

A further list in Table 2.2 (see page 23) is based on information published by the Institution of Occupational Safety and Health. The similarities and differences indicate the spectrum of hazards considered by different groups of people. Such divergence of views is invaluable as it ensures the wider consideration of the issues affecting safety.

It is usual to select which keywords to use in a meeting at an early stage. Normally the number of keywords studied should be limited to about ten, but if subsequently someone present suggests a further keyword during the meeting then it should be used. The object is to promote a full exchange of ideas in a stimulating atmosphere. With more keywords the meeting drags on and becomes tedious.

Such simple methods are particularly useful when the process is only specified in broad terms and detailed flow sheets and layouts are not available. One advantage of this approach is that direct recording is readily accomplished and a record of the meeting may be quickly circulated to members of the study team. Another feature is that at this stage the design is still fluid and changes are inexpensive to make.

It is recommended that work starts prior to the meeting with an examination of the substances present in significant amounts and their classification. It is important to note the presence of any substance which might be deemed as a dangerous substance according to regulations (see CIMAH or OSHA Rule 29) or industry standards (API 750). Expert advice may be required on reactions known to possess particularly hazardous characteristics. Information should also be available on any previous history of major incidents associated with the process or the substances being used.

TABLE 2.1
Keywords in Concept Hazard Analysis Initial Review

Substances and reactions

SUBSTANCES	REACTIONS
• substances which are flammable	• planned reactions
• substances hazardous to health/toxic	• unintended reactions
• substances hazardous to environment	• reaction with common contaminants
• dangerous substances (CIMAH or OSHA)	• combustion products
SEPARATIONS	• reaction with additives, MSAs, etc
• accumulations	

Fires and explosions

FIRES	EXPLOSIONS
• flash fire	• BLEVE
• torch fire	• vapour cloud explosion
• pool fire	• physical explosion
	• chemical explosion

Release and discharge

FLAMMABLES	NOISE
THERMAL RADIATION	ADVERSE DISCHARGE
TOXICS	• product
BACTERIA	• discharge
RADIOACTIVITY	• accumulation/spill
POLLUTANTS	HANDLING
• emissions	ELECTRICAL/RADIATION
• fugitive emissions	• electrical discharge
• periodic or emergency emissions	• radiation
• effluents	• laser
• waste	

Dangerous disturbances

EXCEEDING MECHANICAL LIMITATIONS	IMPACT BLOW/DROP
PHYSICAL EXPLOSIONS	CRITICAL DEFECT IN CONSTRUCTION
CHEMICAL EXPLOSIONS	• defect in construction
REACTION	• deterioration of construction
OVERTEMPERATURE	• vibration/loosening
OVERPRESSURE	• collapse
UNDERTEMPERATURE	ABNORMAL OPENING
UNDERPRESSURE	• made in plant
Overheating	• left in plant
Overcooling	• sneak path
OVERLOAD/STRESS/TENSION	ADVERSE CHANGE IN PRODUCT/DISCHARGE
	• off-specification product/discharge
	• change after leaving plant
	• incorrect transfer
	• abnormal vent/spill

TABLE 2.1 (continued)
Keywords in Concept Hazard Analysis Initial Review

Notable disturbances

EQUIPMENT PROBLEMS	UTILITY PROBLEMS
• specific disturbances	• off-specification services
REACTIONS	• loss of supply
MATERIAL PROBLEMS	MODE OF OPERATION
• off-specification feeds	• entry of vessels
• loss of supply	• start-up
	• shutdown
	• maintenance/inspection
	• abnormal operation
	• emergency operation

External threats

Accidental impact	Theft, hooliganism
Act of God, natural causes	Force majeure, sabotage
Extreme weather	Fire
Abnormal environmental extreme	Explosion, electrical explosion
External interference, loosening	Toxic event
Contamination	Corrosion, erosion
Vibration	Source of inadequate visibility
Drop, fall	Excess contrast, glare

TABLE 2.2
Further keywords for Concept Hazard Analysis Initial Review

Fall of person from height	Drowning
Fall of object/material from height	Excavation work
Fall of person on same level	Stored energy
Manual handling	Explosions due to chemicals or dust
Mechanical lifting operations	Contact with cold/hot surfaces
Compressed air	Chemicals/substances
Use of machines	Biological agents
Operation of vehicles	Noise
Stacking	Ionizing radiation
Housekeeping	Non-ionizing radiation
Lighting	Vibration
Fire, including static electricity	Hand tools
Electricity	Confined spaces
Adverse weather	Cleaning

METHODOLOGY OF A MEETING

The methodology of the CHAIR meeting is described in Table 2.3. The information available on 'Substances and reactions' is first considered. Normally this then directs attention to 'Release and discharge' considering toxic materials and in the case of flammables possible 'Fires and explosions'. The study examines selected 'Dangerous disturbances'. It continues to consider 'Notable disturbances' and 'External threats'. The aim is to highlight major problems or even decide that a particular site might be possible. Another option might be to decide firmly against a specific route or use of a given chemical. The worst credible accident is usually identified, but protections and safeguards are only developed in detail appropriate to the information available. For CHAIR this might only go as far as proffering advice on emergency control systems.

AN EXAMPLE — A LIFTING PROBLEM

This example is based on an actual case study and produced a process which is greatly improved. In the original proposal a flask (containing radioactive material) arrives at a plant on a flatroll (Figure 2.1) and is lifted over a wall into the plant. The wall is there for ventilation purposes. Under the keyword 'Drop/impact' the team identified that there was a hazard if the flask is dropped from a height of greater than 5 m. The flask had been shown to remain intact if dropped from less than 5 m. But the proposed design required a lift of 6 m to get over the wall. By questioning whether this wall was the only way of achieving the ventilation requirements — that is, by challenging the design — the team arrived at the solution shown in Figure 2.2 (page 26). This achieved the ventilation requirements by means of a roller shutter, and gave a lift height of less than 4 m, so rendering the operation deterministically safe.

The study demonstrates how when carrying out any safety review it is important to agree the objectives in advance. One of these objectives should always be that when a potential hazard is identified then try to design it out if at all practical.

TABLE 2.3
Methodology of a Concept Hazard Analysis

(1) Assemble a study team.

(2) Define the objectives and scope of the study.

(3) Agree a set of keywords.

(4) Partition each process flow diagram or block diagram into reasonably sized sections.

(5) Use each keyword to identify dangerous events or disturbances and discuss their consequences.

(6) Determine if the hazard can be designed out or the hazard characteristics reduced.

(7) Determine any protections and safeguards.

(8) Determine whether further objectives can be resolved at this time.

(9) Determine comments, actions and recommendations.

(10) Report using *pro forma*.

CONCEPT HAZARD ANALYSIS

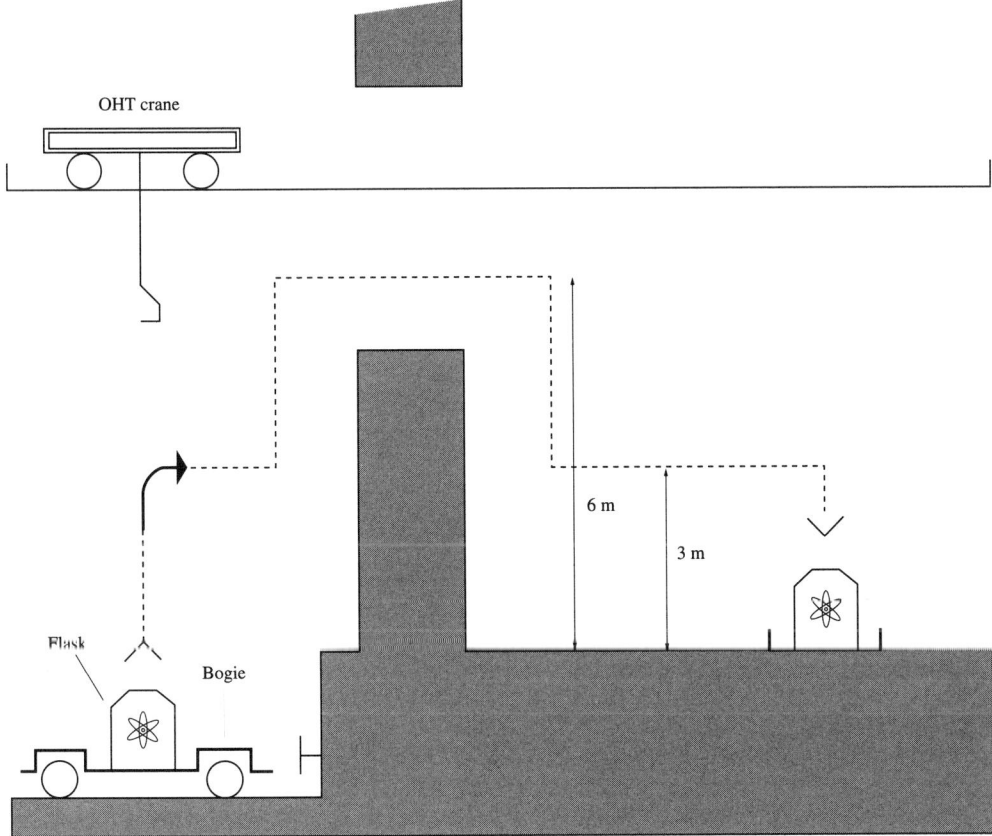

Figure 2.1   The initial handling problem.

All safety studies prior to Hazop give the opportunity to ask the 'obvious' or 'silly' questions about a project which may not otherwise be asked at an early stage in the plant design. Through such questions the plant can be driven in the direction of being safe by design, rather than being made safe at a later stage by the 'bells and whistles' of add-on safety. Because CHAIR is applied at a very early design stage, any changes recommended can be achieved with little immediate costs or increase in the project programme, and with potentially large long-term savings in both areas.

THE IDENTIFICATION OF SERIOUS HAZARD CHARACTERISTICS
Certain hazardous characteristics of processes have been noted as causing accidents over the years. The list is merely illustrative of such factors.

HAZARDOUS CHEMICALS
There are 5000 chemicals in technical use. Occupational Exposure Standards and Maximum Exposure Limits are promulgated in the UK by the Health and Safety Executive under the require-

25

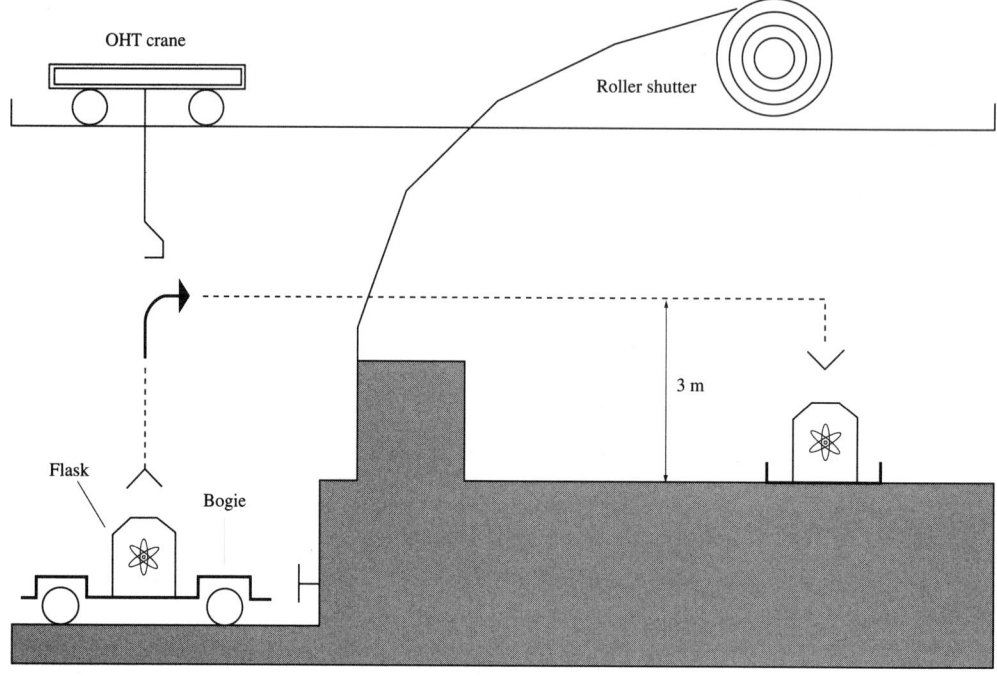

Figure 2.2   Modified system for handling the flask.

ments of the COSHH Regulations and in the USA by the Occupational Safety and Health Administration (OSHA) under various Codes of Federal Regulations. Information on all such chemicals is readily available, although the actual effect of them on humans is subject to considerable doubt.

Lists of dangerous chemicals are produced regularly by appropriate bodies — for example, within the European Union. Such lists cover environmental threats on discharge into the aquatic environment and substances or groups of substances deemed to have the potential to cause a major accident. When the Environmental Protection Agency (EPA) in the USA established its list of highly dangerous substances under the Emergency Planning and Community Right to Know Act there were 402 chemicals on the initial list. Similar lists with triggering inventories are given under the CIMAH Regulations in the UK and in OSHA Process Safety Management of Highly Hazardous Chemicals, 29 CFR 1910.119. The list of the American Petroleum Institute (API 750) is based on any plant having the potential to release five tons of gas or vapour in a period of four minutes.

Clearly the size of any inventory and the process conditions have a major impact on the possible consequences of the release of materials and energy from these processes. It is in general best to be wary of processes operating in or near the explosive range of materials or operation at extremes of temperature and pressure and close to the limits of the materials of construction. Particularly hazardous characteristics are inherent whenever there is an increase in the dispersion of flammable or toxic liquids and solids, or the separation of such material from inerts.

CONCEPT HAZARD ANALYSIS

HAZARDOUS REACTIONS
Extreme operating conditions as a requirement of any reaction is always of concern. Bretherick[2] gives an extensive list of particularly hazardous reactions and this should be the first source of information.
The following general comments give an indication of the problems generated by some reactions.

Oxidation reactions
Oxidation reactions are highly exothermic. Oxidants extract electrons from the substance being oxidized. Beware particularly of nitric acid which functions as an oxidant even when cold and diluted, and hydrogen peroxide which is catalytically decomposed by traces of heavy metals.

Reduction reactions
Reduction reactions involve the transfer of electrons from a reducing agent to the substance being reduced. Powerful agents must be used under inert gas blanketing to prevent the reduction of the oxygen content of the air.

Redox reactions
Redox reactions cover the well-known set of extremely exothermic reactions between active oxidants and reducers.

Exothermic reactions
In general all exothermic reactions pose a danger as they can either generate an increase in pressure through vaporization of liquids or raise the temperature of their containers so that they rupture. A runaway reaction can reach higher temperatures because it may go faster with temperature and the rate of heat removal may be limited. More material is available to convert or is converted so that more heat can be generated. Sometimes polymerization reactions are complicated by agitation and heat transfer problems which arise from the considerable increase in viscosity caused by the increasing molecular size as polymerization proceeds.

Endothermic reactions
Endothermic reactions cannot run away but the products may be potentially hazardous. Such products are thermodynamically unstable and may decompose rapidly or unexpectedly — for example, cyanogen, acetylene, allene, diazomethane.

Unstable structures
Some unstable chemical structures are noted in Table 2.4 (see page 28). In general, advice should be obtained from an appropriate specialist about any chemical which is outside the experience of those involved in any study. Compounds with a zero oxygen balance or compounds with a high positive oxygen balance when mixed with fuels should be noted.

Unwanted chemical reactions
Unwanted chemical reactions can be classified as follows:
- intended reaction with sudden or delayed start;

27

TABLE 2.4
Unstable chemical structures (Source: Bretherick, L., 1990, *Bretherick's Handbook of Reactive Chemical Hazards* (Butterworths, London, UK))

Alkenes	Tetrazoles	Hydrazine
Dienes	Alkene oxides	Hydroxylamine
Alkynes	Alkyl hydroperoxides	Nitrous oxide
Alkenynes	Diacyl peroxides	N-halogen compounds
Azo compounds	Acyl azides	Hypochlorites
Diazo compounds	Nitrous compounds	Chlorine oxides
Diazonium salts	C-nitro compounds	Chloric acid
Aziridines	N-nitro compounds	Redox compounds
Nitriles	Nitrate esters	Aminometals
Triazoles	Oximes	Oxosalts

Complex compounds containing a metal with co-ordinated nitrogenous ligands such as nitrite, nitrate, chlorate or perchlorate

- intended reaction which may cause unwanted reactions on runaway;
- side and secondary reactions accompanying intended reaction;
- unintended reaction when correct reactants do not give intended reaction;
- unintended reaction due to mixing of wrong substances;
- unintended reaction due to impurities, contaminants or catalytic effects.

The last category is particularly important when reaction can occur with common contaminants such as air and water. Major accidents regularly occur due to these impurities. Air when mixed with flammables can result in fires or explosions. Water was one of the ingredients in the runaway reaction at Bhopal. Clearly the frequency of occurrence of a substance, its inventory level and the probability of it coming into contact with people are important considerations in addition to the hazard potential. It should be noted that unwanted reactions also occur outside reactors. They occur, for example, in reactor inlet lines, following reactor outlet, in heaters and separators and in sewers.

Exposure to chemicals
The causes of exposure to chemicals include the following undesired events:
- unplanned release through an available opening;
- discharge on rupture either within or outside of design conditions;
- a change in a planned release;
- an emergency discharge, even when planned;
- improper waste disposal;
- change or accumulation of release after discharge;
- incorrect or improper use of product;

- product is off-specification or otherwise unsuitable for use;
- entry into a vessel with hazardous material present or subsequently added.

Planned emissions from the plant include fugitive emissions which are the vapours emitted continuously or non-continuously from a non-point source. This excludes readily identified leaks from point sources such as stacks, chimneys and vents. Rates of emission vary widely. A major source is valve-stem leakage which can account for approximately 70% of all losses. For example, some 10% of the total number of valves might be leaking to give a flammable organic compound concentration greater than 10,000 ppm. Typical uncontrolled emission rates on chemical plant from valves vary from about 5 g/h for gas duties to 0.5 mg/h for heavy liquids. Pressure relief valves might leak some 100 g/h, drains and sewer vents about 30 g/h.

Periodic emissions are discontinuous fugitive emissions that may occur at a constant or intermittent frequency. The mass emission rates are not necessarily identical for each episode. Typical sources include manual samplers, continuous analysis by on-site instrumentation, tanker loading, purging prior to maintenance, opening up equipment and the emptying and draining of equipment.

The study of such emissions should always be carried out and expert advice sought where appropriate from an occupational hygienist. Materials are usually most hazardous when they are airborne and achieve effective entry into the body by inhalation. They may damage air passages and lungs during this entry. Skin penetration and ingestion are other routes by which certain chemicals enter the body.

ROOT CAUSES

Any undesired event can represent a hazard and this includes the latent errors which can affect any part of the incident scenario and involve consideration of the following subsystems:
- external systems such as industrial and government bodies, contractors, emergency responders, public;
- system climate including legislation and regulations, political climate and pressure groups, economic climate and business factors, business focus, corporate and safety culture, technical knowledge;
- organization and management including decision-making hierarchy, commitment to safety, interaction with external and internal systems, resource provision, emergency provision;
- management control including resource control, responsibility and accountability, management of change, supervision and control, planning, monitoring and quality control, appraisal;
- site and plant facilities including the site and its layout, engineering and process design, commissioning and realization of plant, transport and storage, use and disposal of materials;
- communication and information including information quality, safety information, channels and media interface, incident reporting, emergency response;
- procedures and practices, working practices and procedures, safety studies, quality control, emergency procedures;
- working and local environment considering welfare, safety culture, supervision and support;
- engineering integrity considering quality of plant, availability and maintenance, safety and operating margins, plant upgrading and modifications, standards and codes;
- operator performance including recruitment, training, personnel capabilities, working discipline.

There are millions of root causes affecting plants, and system defences must be set up to guard against them and monitor key performance indicators. At the time of a Concept Hazard Analysis it may be appropriate to consider all the above categories and note where deficiencies appear in the current organization, management, technical know-how and capabilities of personnel. In this way the engineering and administrative controls can be reviewed.

THE METHANATOR — A CASE STUDY FROM A HYDROGEN PLANT

A hydrogen plant is required to produce a hydrogen feed stock for the hydrocracking of aromatics to benzene and the conversion of benzene to cyclohexane. The plants are to be built as an integrated complex. Hydrogen is to be produced by the steam reforming of methane. The basic reaction is carried out by allowing methane and excess steam to react at a pressure of 25 bar and temperature of 800°C over a catalyst in the reformer. The reaction takes place in 40 tubes, each some 30 m long and 10 cm wide, located in a furnace. This reaction produces oxides of carbon and hydrogen together with excess water and a small amount of unconverted methane. Figure 2.3 shows a block diagram of the hydrogen plant.

From the reformer the outlet gases are to be cooled (by generating superheated steam) before conversion of much of the carbon monoxide by reaction with water to produce carbon dioxide and more hydrogen. This is to be carried out in a high temperature shift reactor operating at about 400°C followed by a low temperature shift reactor operating at about 180°C. This stream is then cooled to 30°C and water separated in a gravity separator to sewer. The main alternative process considered was pressure swing adsorption of hydrogen, but this has now been rejected.

The bulk of the carbon dioxide is to be removed by absorption using an aqueous solution of sodium carbonate and bicarbonate at a stream pressure of 22 bar. The main process stream is then passed through a gravity separator to remove water (including small amounts of dissolved sodium salts, hydrogen and methane).

The final gas contains a small percentage of oxides of carbon that must be reduced to some 20 ppm as one of the downstream plants using this hydrogen feed stock contains catalyst that reacts with oxides of carbon. It is proposed that the oxides of carbon are removed in a methanator — that is, a fixed-bed catalytic reactor operating at an inlet temperature of 400°C and 20 bar.

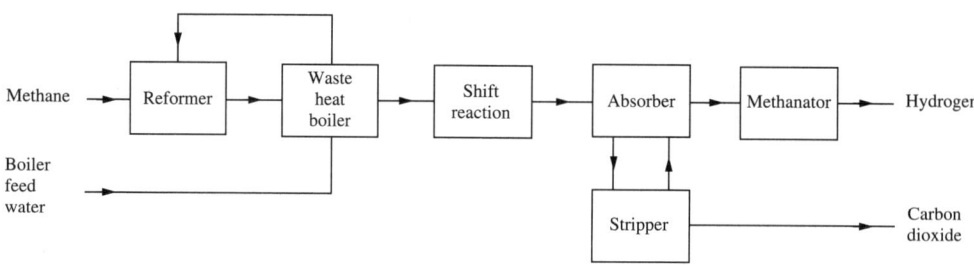

Figure 2.3   Block diagram of hydrogen plant.

The purpose of the methanator — to remove oxides of carbon down to parts per million to protect catalyst in downstream plants from damage — should be noted and given consideration when off-specification product is discussed during the review.

A preliminary process flow diagram (Figure 2.4, pages 32–33) is available for a section of this hydrogen plant in which methanation is carried out. The methanation section involves the treatment of a gaseous mixture of hydrogen 98%, methane 1%, oxides of carbon 1% which is saturated with water and at 40°C and 21 bar. The stream is heated to 400°C before passing through a fixed-bed catalytic reactor, the methanator R101. The reactor outlet stream is then cooled in the preheater/cooler E101 before cooling to 30°C in E102. Any condensed water is removed in a knockout pot C102 before compression to 40 bar.

The main reaction in the methanator involves the conversion of oxides of carbon to methane:

$$CO + 3H_2 \rightarrow CH_4 + H_2O$$

$$CO_2 + 3H_2 \rightarrow CH_4 + 2H_2O$$

The reaction goes virtually to completion — that is, down to under 20 ppm of oxides of carbon in the final product which is hydrogen.

The methanator is started up from cold by preheating it with a stream of nitrogen heated in the main reformer when that unit is started up. It is then left hot until the reactor is ready to come on stream, usually about two hours. This stream is obtained off the main process stream just below a steam superheater. Note that this stream is only available as nitrogen during the start-up of the reformer. At other times it is the untreated product direct from the reformer; it is very hot and has a very high level of oxides of nitrogen.

A list of streams and substance characteristics should be prepared beforehand by the process engineering group or equivalent. A brief review of the main streams is generally helpful and describes the process. The report should be updated as actions are taken or resolved with respect to safeguards and the assembly of further information. As fresh hazardous conditions are identified these can be incorporated within the record for appropriate action.

Hydrogen is flammable and tends to self-ignite on release as a leak through flanges and joints. It readily dissipates to atmosphere even when diluted with 20% carbon dioxide. If hydrogen is ignited it will burn as a jet. This long flame is likely to impact on other vessels and items of plant.

The reaction is exothermic and this causes the main problem on the methanator which arises if oxides of carbon are not removed from a level of 20% in the absorber down to at most 10%. The reaction continues to go to completion and consequently very high temperatures can be generated. Failure due to overtemperature occurs either in the reactor, in the outlet pipeline or in the preheater/cooler.

Other problems arise from gas blowby from the knockout pot, D102, which sends a considerable amount of gas to sewer. Note that there is always a small amount of gas present in the sewer as gas dissolved in the water separates out. Liquid blowby from D102 sends liquid in the gas to the compressor, C102. This can generate excessive stress in this machine, possibly causing rupture with explosive violence. The results of the analysis are given in Tables 2.5 and 2.6 (pages 34–36 and 38–41). Certain points are not included here in order to assist in the writing of this text.

HAZARD IDENTIFICATION AND RISK ASSESSMENT

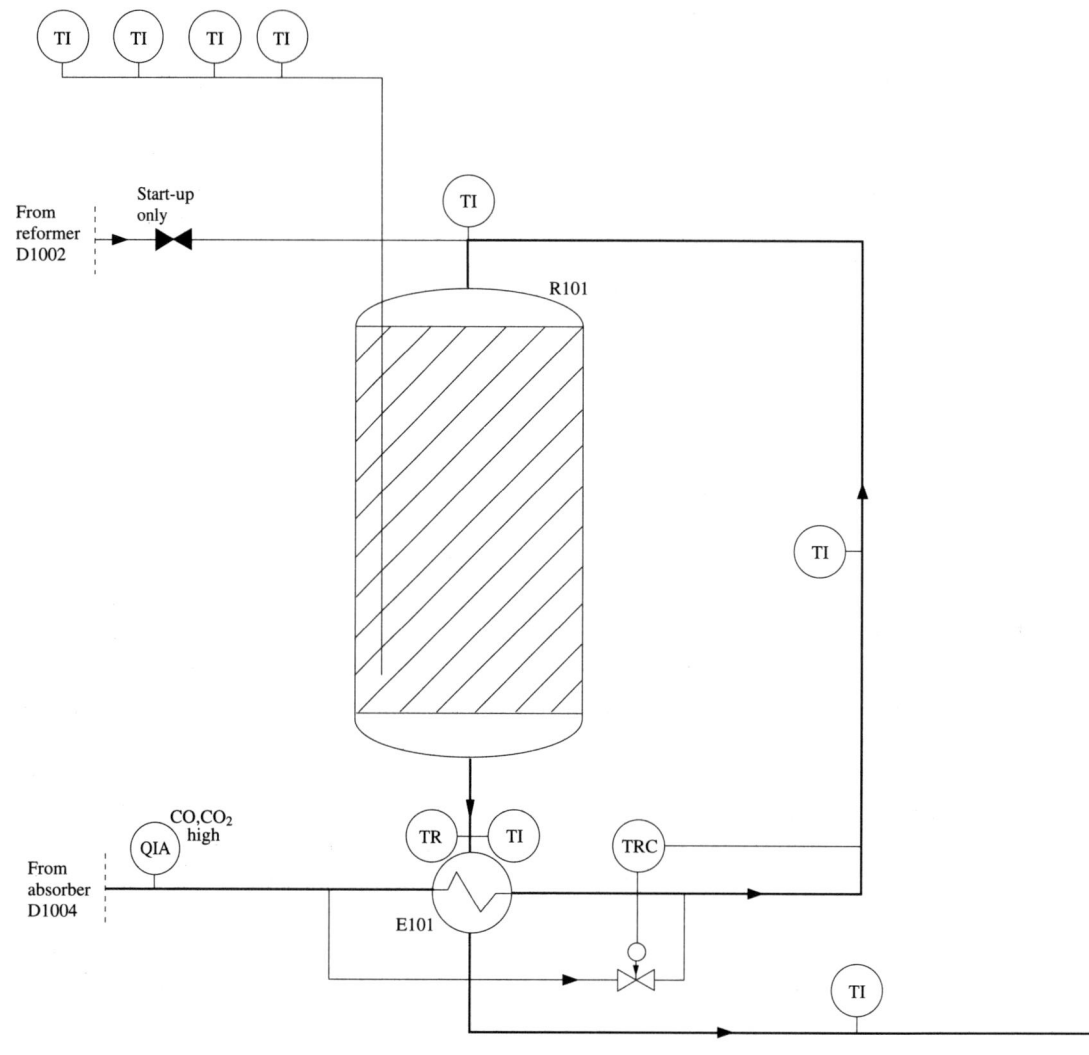

Figure 2.4 Initial P&I diagram of hydrogen plant. (See page 243 for instrumentation definitions.)
10 = Nitrogen, 10 bar; 11 = Cooling water supply; 12 = Cooling water return.

Item number	R101	D102	E101	E102	C102
Title	Methanator	KO pot	Methanator preheat	Methanator cooler	H_2 compressor
Op. temp, °C	450	30	400	100	50
Op. pres., bar	20	18.7	21	19	40

CONCEPT HAZARD ANALYSIS

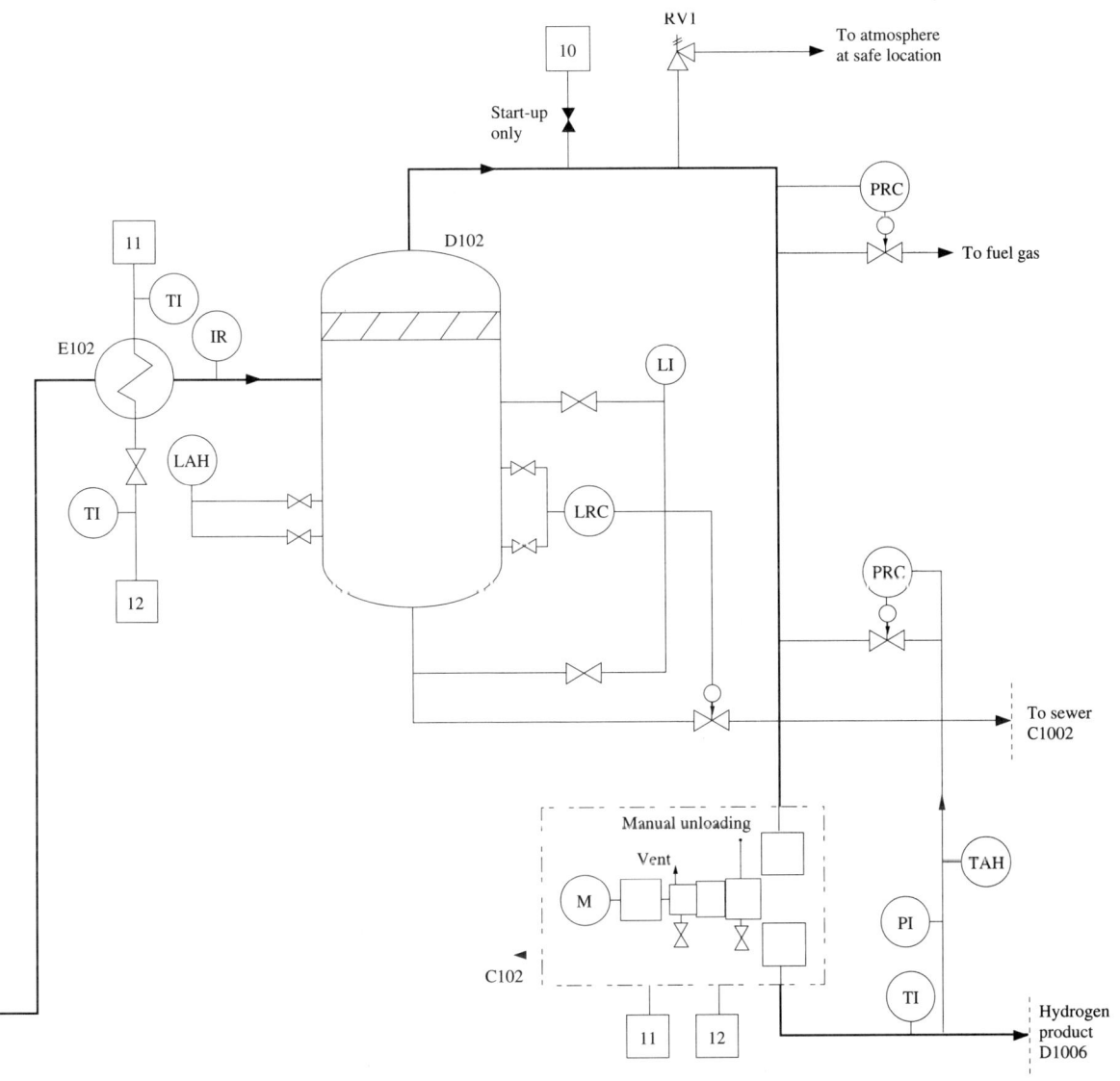

33

TABLE 2.5
Concept Hazard Analysis Initial Review of the methanator

Keyword	Discussion	Action/recommendation
FLAMMABLES	Hydrogen is flammable and tends to self-ignite on release as a leak through flanges and joints. Otherwise it readily dissipates to atmosphere. If diluted with 20% CO_2 it will still disperse satisfactorily.	Consider fire-fighting requirements
FIRE	Loss of integrity on methanator circuit likely to give rise to torch fire. Escalation probable if torch impacts on other equipment or piperack. Catalyst can cause fires if exposed after use. Keep it blanketed by nitrogen if vessel opened.	Evaluate length of flame from 6 inch pipe at various locations. Action to depressure needs evaluation. Maintain oxygen-free atmosphere at shutdown.
EXPLOSION	An external explosion is unlikely as hydrogen disperses upwards and plant is open. An internal explosion could occur if plant is not thoroughly purged at start-up.	Nitrogen purge required at start-up and shutdown
DISCHARGE	The emission of material on overpressure can be to atmosphere or to flare	Check if H_2 is environmentally friendly
DISCHARGE	Excess production to fuel gas or flare or vent	Check effect of hydrogen in diluting the fuel. Check any radiation if vent ignites.
CORROSION	Sodium salts might be present in feed due to carry-over from absorber. A ceramic bed would aid dispersion and remove some particulate matter.	Consider ceramic bed and note impurities when determining materials of construction
EFFLUENTS	Water containing sodium salts in effluent. Small quantities of flammables as dissolved gases which separate out in sewer.	Effluent to waste treatment. Sewer vented by high standpipes fitted with flame arrestors.
POLLUTANTS	Emergency relief and venting of excess hydrogen to atmosphere is preferred. The outlet might ignite. Liquid effluent from knockout (KO) pot is water which contains sodium salts and some dissolved flammables.	Check radiation from flare on ignition. Discharge to vented sewer and send to effluent treatment.

TABLE 2.5 (continued)

Keyword	Discussion	Action/recommendation
REACTION AND OVERTEMPERATURE	Methanation is highly exothermic. As the reaction goes to completion any additional oxides of carbon in the feed will result in increased outlet temperature. If the absorber fails to remove any CO_2 the oxides of carbon increase to 15% by volume.	Attention to eliminate this hazard can be taken at the absorber with multi-train design and excess capacity. A trip system is required to bypass the reactor if high outlet temperature or high oxides of carbon. Check mechanical design to withstand maximum feasible temperature.
OVERPRESSURE	Overpressure can occur due to lack of demand for product or downstream blockage. Upstream plant has slow response to reduce capacity. Hence need for venting as well as process relief.	Check relief requirements if product off-spec or lack of demand. Consider venting to atmosphere or fuel gas. Report back on relief valve decisions.
OVERLOAD OF COMPRESSOR	Liquid blowby from the KO pot could cause an explosion in the compressor	Install trip system on KO pot
ADVERSE DISCHARGE FROM KO POT	Gas blowby on the KO pot would overload the sewer system with flammables	Consider trip system and ensure sewer vents can cope with any overload. LAL essential.
ADVERSE CHANGE IN PRODUCT	High oxides of carbon affect downstream plant. Trip compressor and vent upstream if methanator trips out.	Develop an integrated system of alarms and trip system. Check tolerance of downstream plants for high COx.
START-UP OF METHANATOR REACTOR, R101	At present it is necessary to provide an additional source of heat to warm up the methanator bed. The safest way would be to use the heat exchanger circuit but demands to preheat boiler feed water and generate steam can fight this choice. Actual activation of the catalyst needs evaluation.	Consider a start-up line from the reformer to preheat the methanator, R101. Check other features regarding the activation of the catalyst.
UTILITY PROBLEMS • Loss of air • Loss of cooling • Loss of electricity	Do not depressurize or allow flow through methanator. See KO pot, D102. All flows in stop so absorber fails. Hence the methanator must trip out.	Open the bypass on air failure Open the bypass on electricity failure.

TABLE 2.5 (continued)

Keyword	Discussion	Action/recommendation
EXTERNAL THREATS	A local fire would necessitate the immediate depressure of the plant. Segregate plant items by distance not barriers.	Check plot plans of rest of complex.
	Space required around methanator for packing the reactor and bringing in drums.	Check packing operation to ensure no impact danger.

The review has generated a considerable amount of information about the plant. Obviously when the plant is built or previously fully designed even further information is available. Consequently it may be decided to carry out the initial review in more detail as is illustrated in the following section in which the analysis is termed Concept Hazard Analysis (CHA).

CONCEPT HAZARD ANALYSIS

A detailed Concept Hazard Analysis (CHA) may be commenced at a stage when the preliminary process flow diagram is available. This version of CHA aims to identify the main hazards which the proposed plant will generate or face, and recommend safeguards and protections. The approach can vary considerably from a general identification of hazards giving further advice to the process design team, to carrying out a thorough look at each section of plant.

It is recommended that this study should usually be carried out on a process flow diagram on which the main add-on safety systems appear. An early P&ID is an alternative. The development of P&I diagrams is described in Appendix 1.

As before, the plant sections are evaluated. The perceived dangers are noted together with suggestions for safeguards (as a general aim rather than a reality). Appropriate comments are added for action. As well as identifying general hazards the opportunity is taken to include any specific hazards which have been associated with the equipment in the past.

It is particularly recommended to consider equipment problems at this time. This can be accomplished using the available experience present at the meeting. It is helpful if an equipment knowledge base is available which can suggest dangerous disturbances. The meeting can then perform a crude what-if analysis on the system. Such an analysis asks a prepared set of questions beginning with the phrase 'what if ... ?'. For example, 'What if the feed to the reactor has a higher concentration of component A?'. The subsequent answer provokes appropriate concern. In general the aim is to avoid considering Hazop style deviations or immediate causes; if this is warranted, however, it should be done. The technique is there to be used according to the general experience available in the group. Anything is acceptable. For example, the loss of services may be considered even though this fits with determining the immediate cause of incidents rather than of consequential events.

The study will vary considerably according to the knowledge which is available and which the participants have about the process at this stage. Many projects considered by industry are modifications to process plant, costing up to £1M. For these considerable information will be available. In other projects the study can be used to transfer information from, for instance, process licensors. In the case of a development project, the study can highlight key safety areas requiring further study.

The methanator example given below is typical for a plant for which a lot of knowledge is available, not all of which is included herein. On many studies less extensive documentation would be produced. Note that the essence of the scheme is to carry out a study which is relevant to the stage which a specific project has reached in its development.

The record of the study can also be used to commence a list of unresolved hazards as part of the risk management programme. Any hazard or hazardous condition can be noted on the record as it is identified and 'removed' from the list as it is resolved. This record can be maintained throughout the project, or the life of the plant if the company is so minded. One approach is to develop a safety schedule for the plant and upgrade information as reviews are carried out.

CONCEPT HAZARD ANALYSIS OF THE METHANATOR

The methanator is an existing process — that is, it exists as an available process under licence which has been built elsewhere. For this particular process it would be possible to carry out this study at an early stage of process design. It might also be decided to cut out this stage and go for a better structured Preliminary Process Hazard Analysis.

This study was carried out with the CHAIR results available. The study was carried out on Figure 2.5 (pages 42–43) and the results appear in Table 2.7 (pages 44–45). The results have been edited for presentation purposes.

The analysis was carried out by first examining general factors such as flammables, pollutants, reactions and dangerous disturbances before using available experience to identify equipment problems. Early specification of the need for start-up of the methanator ensures this is considered when evaluating the heat exchanger network.

The major effects on this particular plant stem from the presence of flammables, the exothermic reaction in the methanator, and the problems associated with having a compressor. Hydrogen escapes upwards into the atmosphere and this property makes it arguably one of the less dangerous flammable materials on release. The historical record shows few cases where an actual release caused considerable problems. Internal explosions are more serious and one weakness in the study as presented is insufficient attention being given to this during start-up and shutdown. The reaction is of interest as it is readily driven to completion. This means that it is an increase in the material available for reaction rather than an increase in feed temperature which is a primary concern. The separation of flammables in the sewer was not considered at the time of the original design many years ago. However, it was noted when the project was studied at the University of Sheffield in more detail ten years ago. Since that time greater awareness has arisen since the disaster in the sewers of Guatemala in 1992.

Note how useful it is to discuss typical equipment problems at this stage. Familiar problems associated with gravity separators and compressors come to light. Fair enough, most of the protective systems suggested would be incorporated by standard practice and are recommended in the equipment knowledge base. However, at least this experience has been shared and thereby advice is given to the process designers. This is invaluable to those people starting their careers or coming into contact with technology with which they are not familiar.

TABLE 2.6
Concept Hazard Analysis of the methanator (hazardous process characteristics)

	Keyword	Dangerous disturbance	Cause/consequences
(1)	FLAMMABLES	Release on rupture	Release may self-ignite and TORCH FIRE. Escalation to piperack likely with missiles.
(2)	FLAMMABLES	Release on emergency discharge	Release at safe height: possible ignition and FIRE
(3)	FLAMMABLES	Normal discharge to sewer	CHEMICAL EXPLOSION in sewer
(4)	REACTION	Exothermic runaway reaction in methanator	High level of oxides of carbon cause runaway with rupture and possible PHYSICAL EXPLOSION
(5)	REACTION	Air in combustion vessels	Combustion in vessels. Causes CHEMICAL EXPLOSION.
(6)	REACTION	Inadequate reaction due to catalyst failure, low temperature feed or methanator bypassed	Off-specification H, to downstream plant. This can cause runaway with CHEMICAL EXPLOSION.
(7)	POLLUTANTS	Effluent to sewer	Water with high sodium salts
(8)	POLLUTANTS	Effluent caused by firewater	Firewater will flood. River receives minor contamination.
(9)	POLLUTANTS	Noise	Noise in compressor area
(10)	OVERPRESSURE	Overpressure in hydrogen plant	High pressure causes PHYSICAL EXPLOSION
(11)	OVERTEMPERATURE	Overtemperature in methanator	Runaway reaction — see above
(12)	OVERTEMPERATURE	Overtemperature in compressor	Excess recycle of hydrogen around compressor can result in PHYSICAL EXPLOSION
(13)	MECHANICAL HAZARD DUE TO OVERLOAD	Overload of compressor	Stress in compressor caused by two-phase feed due to liquid blowby can result in PHYSICAL EXPLOSION
(14)	ABNORMAL OPENING	Vibration at compressor	Loosening of flange gives release. TORCH FIRE.
(15)	ABNORMAL OPENING	Spurious relief	Loss of material to safe point. Could ignite as minor TORCH FIRE.

Suggested safeguards	Comments/action
Segregation by distance. Depressure or steam purge.	Study best way of reducing damage. Project to advise on fire-fighting.
Segregation by distance	Check possible radiation levels on model
Vent sewer with standpipes	Check other plants for incidents and advise
More robust design of absorber. Trip methanator on high temperature. Alarms on temperature and CO_2 high.	Check action if trip fails. Check sneak path on start-up. Project to advise.
Purge plant before start-up. Ensure catalyst covered by N_2 as replaced.	Get more information on catalyst. Project to advise.
Alarm on high CO_2 outlet. Vent if off-spec and shut down compressor. Ensure methanator start-up. Connect methanator trip to compressor trip.	Design heat exchanger circuit to preheat methanator
Sewer to effluent treatment	Check effect on current treatment
	Check other sewers in area for contamination. Project to advise.
Building would cause explosion hazard	Operators to wear protection in danger area
Two relief valves in circuit. High pressure alarm.	Flare may be needed on fuel gas if demand low. Project to advise.
High temperature alarm on loop	Evaluate as no safeguard provided. Project to advise.
Trip on high level in KO pot. Level alarms in KO pot.	Explosion unlikely but note compressor may be damaged
Vibration probe	Project to note
Consider need for lock open valve after RVs or bursting disc before RVs	Project to advise

TABLE 2.6 (continued)
Concept Hazard Analysis of the methanator (hazardous process characteristics)

	Keyword	Dangerous disturbance	Consequences
(16)	EQUIPMENT: METHANATOR	Excess heat of reaction	See item (4)
(17)	EQUIPMENT: METHANATOR	Blockage	Blockage of sodium salts at top of reactor (causes channelling)
(18)	EQUIPMENT: METHANATOR	Low bed temperature on start-up	Failure to preheat bed causes off-spec product
(19)	EQUIPMENT: METHANATOR	Activation of catalyst	Failure to activate
(20)	EQUIPMENT: HEAT EXCHANGER	Internal leak causing off-spec product	Off-spec product affects downstream plants
(21)	EQUIPMENT: HEAT EXCHANGER	Less heat causing no reaction in methanator	Off-spec product affects downstream plants
(22)	EQUIPMENT: COOLER	Loss of cooling water	High temperature to KO pot
(23)	EQUIPMENT: KO POT	Liquid blowby	See overload (item (12))
(24)	EQUIPMENT: KO POT	Gas blowby	Increased flammables in sewer (item (3))
(25)	EQUIPMENT: COMPRESSOR	Overload	Stress due to loss of lube oil flow or two-phase flow from KO pot
(26)	EQUIPMENT: COMPRESSOR	Overtemperature at inlet	High recycle flow under control (item (12))
(27)	EQUIPMENT: COMPRESSOR	Overtemperature at outlet	Failure of cooling water to after-coolers
(28)	EQUIPMENT: COMPRESSOR	Overpressure at inlet	High recycle flow not under control (item (10))
(29)	EQUIPMENT: COMPRESSOR	Overpressure at outlet	Failure of control system
(30)	EQUIPMENT: COMPRESSOR	Maintenance of compressor	Loosening by maintenance requires absolute isolation of a high pressure low molecular weight gas

Suggested safeguards	Comments
Bed of ceramics on top of reactor	Check requirements at base
Improved heat exchanger network or start-up line	Establish requirement in design intent. Project to advise.
Separate hydrogen stream may be needed	Obtain information from manufacturer. Project to advise.
Analyse outlet stream for CO_2 and alarm	Determine policy for off-spec gas (see items (6) and (20))
Temperature alarm. Analyse for CO_2 in outlet.	Plant requires complete shutdown if the methanator cannot be preheated (see item (17))
Not critical	Cooling water temperature must be monitored. Install TI.
	Check requirements for dephlegmator. Project to advise.
Low level alarm. Trip system on low level.	Project to advise
Alarm on low lube oil pressure. Shutdown by trip system.	Re-advise when compressor selected
High temperature alarm on outlet	After-coolers should be on diagram. Re-advise when compressor selected.
Relief valve on outlet. High pressure alarm.	Project to advise after checking downstream plants
Maintenance policy must be agreed together with standby provision.	2×60% compressors preferred. Need double block and bleed systems plus nitrogen purge.

CONCEPT HAZARD ANALYSIS

41

HAZARD IDENTIFICATION AND RISK ASSESSMENT

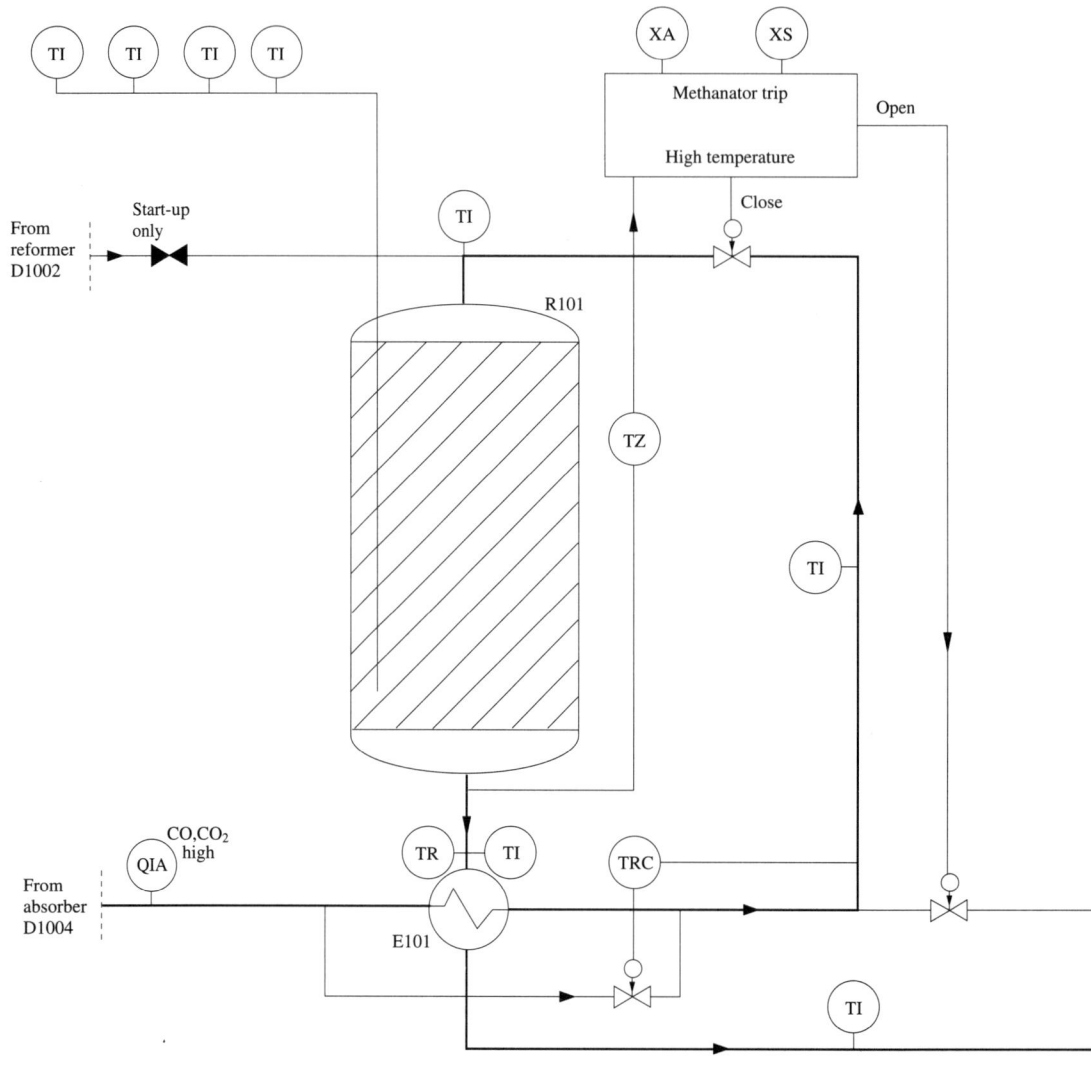

Figure 2.5 P&I diagram of hydrogen plant, Revision A. (See page 243 for instrumentation definitions.) 10 = Nitrogen, 10 bar; 11 = Cooling water supply; 12 = Cooling water return.

Item number	R101	D102	E101	E102	C102
Title	Methanator	KO pot	Methanator preheat	Methanator cooler	H_2 compressor
Op. temp, °C	450	30	400	100	50
Op. pres., bar	20	18.7	21	19	40

CONCEPT HAZARD ANALYSIS

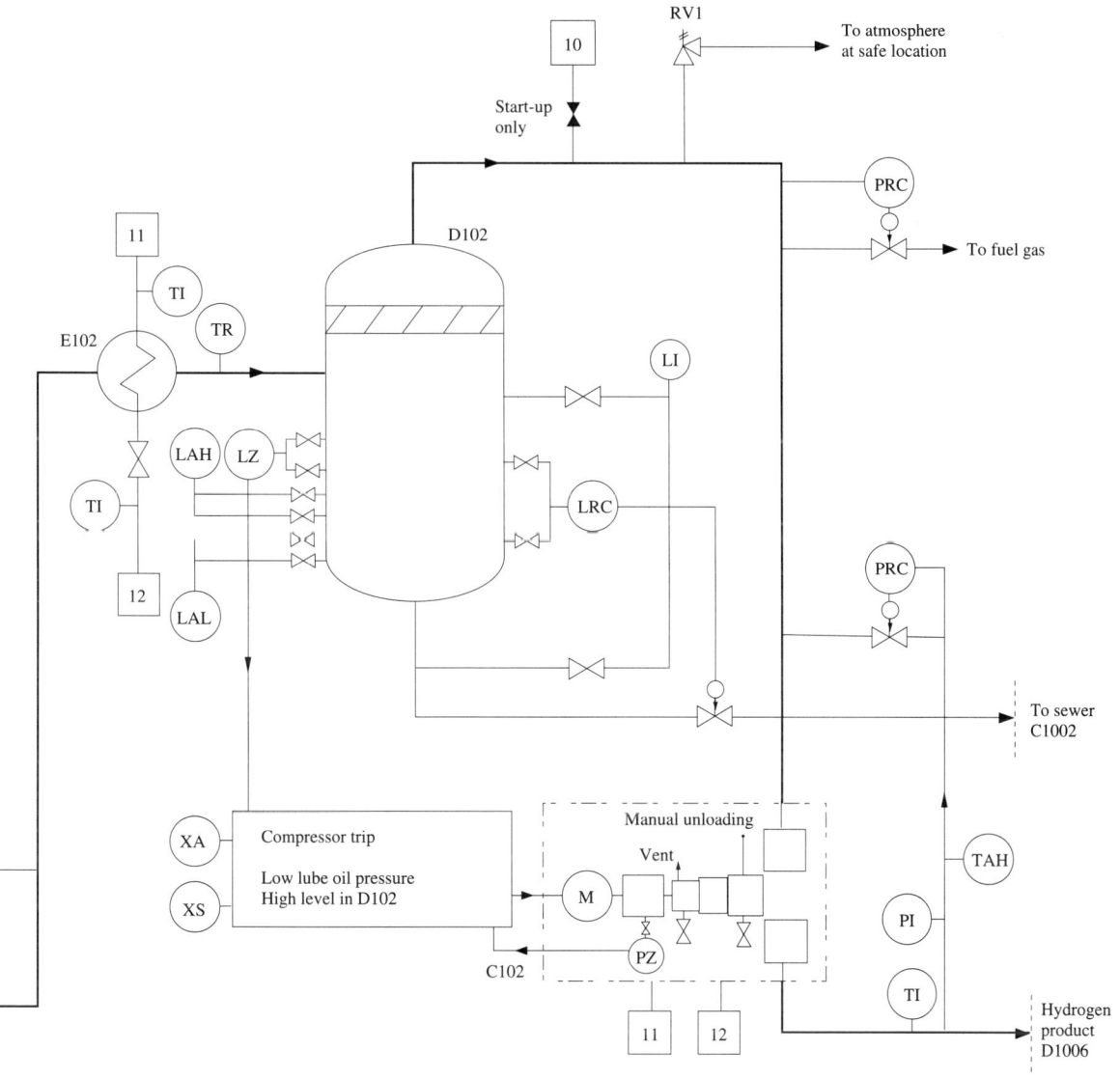

TABLE 2.7
Concept Hazard Analysis of the methanator (substance characteristics)

Stream number	Description	Substances
(1)	Feed to methanator preheater	Hydrogen containing oxides of carbon, methane, water and some sodium salts
(2)	Emergency discharge	As above
(3)	Feed to methanator	As above
(4)	Methanator product	Hydrogen, methane, water and some sodium salts. Oxides of carbon, 20 ppm.
(5)	Preheater outlet	As above
(6)	Methanator product cooler	As above
(7)	Feed to compressor	Hydrogen methane, water, oxides of carbon
(8)	Excess H_2 to fuel gas	As above
(9)	Compressor outlet	As above
(10)	Liquid from KO pot	Water, some sodium salts, dissolved H_2, CH_4
(11)	Methanator	Nickel-based proprietary catalyst
(12)	Start-up stream	Start-up stream of hot nitrogen
(13)	Nitrogen	Nitrogen purge line

The only item not thoroughly explored in this study was the provision of a standby compressor and its maintenance. This was excluded solely for presentation purposes as it effectively halves the size of drawings. A number of omissions have also been made in order to continue this story.

The CHA concentrates on hazards and the use of general experience on operating standard equipment. Of course there is not the same assurance that most of the important incident scenarios have been identified and no effort is made to determine the immediate causes of the incidents. If an earlier meeting using CHAIR has been held it is suggested that the method can be

CONCEPT HAZARD ANALYSIS

Inventory throughput	Operating conditions	Hazard notes	Action
Gas	25°C, 22 bar	FLAMMABLE with tendency to self-ignite on release. Highly buoyant.	
Gas	25°C, atmos.	FLAMMABLE and at some stage it will ignite	
Gas	400°C, 21.5 bar	OFF-SPEC feed with high oxides of carbon can cause runaway reaction	
Gas	450°C, 21 bar	OFF-SPEC. Oxides of carbon above 50 ppm can damage downstream plant.	
Gas	75°C, 20.5 bar		
Gas/liquid	25°C, 20 bar		
Gas	25°C, 20 bar		
Gas	25°C, 1.2 bar		
Gas	30°C, 40 bar		
Liquid	25°C, 20 bar	FLAMMABLE. Dissolved gases separate out in sewer.	Check venting arrangements
Solid		REACTIVE. This nickel-based catalyst may cause combustion on contact with air if flammables present.	Obtain more details
Gas	500°C, 25 bar	OFF-SPEC. After start-up this stream is the same as the reformer product with 15% oxides of carbon.	
Gas	15°C, 10 bar	Start-up only	

used by the process engineer without a further meeting to carry out a CHA. As the method builds on the data from the review it is easier to develop than other procedures. The results are readily understood and can be circulated for comment. But the method lacks a rigid structure, and in effect starts with a number of what-if questions prompted by the keywords — that is, what-if overtemperature in methanator? This is followed by consideration of whether the deviation is meaningful, the consequences and appropriate protection and safeguards. Clearly if the key what-if phrase is not asked then the danger may not be identified.

3. PRELIMINARY PROCESS HAZARD ANALYSIS

THE PARTS OF A PRELIMINARY PROCESS HAZARD ANALYSIS
A Preliminary Process Hazard Analysis (PPHA) follows up the results of the Concept Hazard Analysis to provide further information on factors such as wanted and unwanted reactions, the reduction of hazards and hazardous characteristics on the plant, the identification of incident scenarios and the evaluation of emissions, effluents, wastes and off-specification products. The incident progresses through time from the immediate disturbed state of the plant via a sequence of steadily worsening states. Between each plant state opportunities exist to eliminate or mitigate the progression of the incident. Passive mitigation/control systems are constantly in place whilst active mitigation/control systems respond as events progress. PPHA aims to duplicate the incident scenario down to immediate cause, as shown in Table 3.1. The component parts of this table will be referred to in detail as the description of the method progresses. It is best completed in two parts: the first part evaluates the incident scenario leading to a significant release of process material; the second part is a Consequence Analysis that starts with the significant release of process material and evaluates the consequences of such a release.

Traditionally the evaluation leading to the release of material is carried out in an identical manner to that used in a Fault Tree Analysis, with the top event being a significant release of material by a discharge or rupture, either into the atmosphere or with products into downstream plants or into storage. The analysis proceeds downwards in the opposite direction to the path of the incident scenario, at each level evaluating the cause of the undesired events in the chain.

This process continues until the analyst is happy that a satisfactory estimate can be made of the frequency of a given incident scenario, or that the incident scenario is insignificant as a contributor to the overall frequency of the top event and further evaluation is not worthwhile.

This basic scheme has been adapted to start the search at a more convenient point: the dangerous disturbance of plant leading to rupture and discharge. The change is only minor but enables the analyst to start the search with a few well-defined suggestions, described shortly.

Evaluation of consequences is carried out using event trees. The eventual outcome is an estimation of the likely impact of the event on the system's total environment. The analyst works in the same direction as the scenario unwinds, evaluating the effects of countermeasures on the release, its ignition, escalation by major events such as fire, explosion or toxic event and failure to mitigate against these events to reduce their occurrence or effects. The emergency response is a vital part of such a study.

In order to assist the analyst in this task, a structured approach has been developed together with a prepared format for documentation. Both a generic fault tree and a generic event tree are used. These are amplified using two knowledge bases which assist in the task. The Hazcheck knowledge base (Appendix 2) follows the basic structure of the analysis and the development of an incident scenario. It can suggest the general causes and consequences of undesired events. The equipment knowledge base provides more specific information on equipment and

TABLE 3.1
General incident scenario

IMPACT: HARM AND DAMAGE	• consequences categories (Appreciable to Catastrophic) • minor consequences/near miss
MITIGATION MEASURES FOR A RELEASE FAIL	• inadequate post-incident emergency response • emergency response inadequate
ESCALATION BY TOXIC RELEASE	• accumulation after release • secondary loss of toxic material • emergency discharge treatment system fails • initial loss of toxic material
ESCALATION BY FIRE OR EXPLOSION	• escalation by fire or explosion • delayed ignition of flammable mixture • immediate ignition of flammable mixture
COUNTERMEASURES FOR A RELEASE FAIL	• release fails to disperse • response by mitigating countermeasures inadequate • immediate response to release inadequate
SIGNIFICANT RELEASE OF MATERIAL	• release of material by rupture or discharge • release creates hazard or hazardous condition
FAILURE TO RECOVER SITUATION	• immediate attempts to isolate loss fail • inadequate further action to effect control or process abort
DANGEROUS DISTURBANCE OF PLANT	• dangerous disturbance resulting in rupture on exceeding mechanical design limits • critical defect or deterioration in construction • flow through abnormal opening to atmosphere • adverse change in a planned product or other release
INADEQUATE EMERGENCY CONTROL	• emergency control systems fail to correct the situation • further control action fails
HAZARDOUS DEVIATION	• hazardous trend in process conditions • construction defective or deteriorated in service • abnormal opening in equipment • change in planned discharge or vent
FAILURE TO CONTROL SITUATION (ON ALARM)	• operators fail to correct the situation • maintenance fail to correct the situation • normal control systems fail to correct the situation
PROCESS DEVIATION	• deviation in condition of process materials • deviation in materials of construction or plant integrity
INADEQUATE NORMAL CONTROL	• normal control systems fail to correct the situation • monitoring fails to detect undesired trend
IMMEDIATE CAUSES OF INCIDENT	• action by plant personnel inadequate • defects directly cause loss of plant integrity • plant or equipment inadequate or inoperable • control system or emergency control inadequate • change from design intent during plant life-cycle • environmental and external causes of disturbance

TABLE 3.2
Methodology of Preliminary Process Hazard Analysis

Develop incident scenario to significant event:
- assemble the plant information
- partition the plant and select a section
- select a dangerous disturbance
- identify and note the significant event resulting from the dangerous disturbance
- identify and note each hazardous disturbance giving rise to this dangerous disturbance
- complete the inadequate emergency control actions and failure to recover the situation
- identify process deviations and their immediate causes. Fill in appropriate inadequate control actions
- select another dangerous disturbance to study
- complete the studies for the section identifying each significant event
- complete the studies for the plant identifying each scenario to significant cause

Continue with the studies to evaluate risk:
- determine the likelihood of each basic event and each significant event
- evaluate the likely outcomes of significant events using Consequence Analysis and considering the emergency response
- carry out a Short-Cut Risk Assessment
- prioritize incident scenarios for further study and where necessary repeat the study of immediate causes in Hazop detail

facilities. It contains information on types of equipment, the dangerous disturbances leading to major releases, and typical control and emergency control systems.

STRUCTURE OF A PRELIMINARY PROCESS HAZARD ANALYSIS

Table 3.2 shows the general structure of a PPHA. The main sections will now be described.

ASSEMBLE PLANT INFORMATION
Plant information should include process information on fundamental process chemistry as follows:
- notes on dangerous reactions and side-reactions;
- data on hazardous materials;
- a process flow diagram showing control measures and safeguards or an initial P&ID;
- equipment specification sheets;
- notes on inventory levels;
- any available operating information.

A functional model of the plant may be helpful for batch and loading/off-loading activities. It is important that early studies should be carried out as a precursor to PPHA and a formal Concept Hazard Analysis (usually the initial review) is essential to identify the following:
- inherent hazards of the process chemicals;
- probable pollutants;
- possible difficulties associated with reactions and equipment;
- site suitability and probable environmental impact;

TABLE 3.3
Dangerous disturbances

Disturbances resulting in rupture on exceeding mechanical limits:

- physical explosion
- overpressure
- overtemperature
- machine overload or stress
- chemical explosion
- underpressure
- undertemperature
- impact blow/drop

Critical defect in construction:

- critical defect left in construction
- item loose or critical/vibration
- critical deterioration in construction

Flow through abnormal opening to atmosphere:

- flow through abnormal opening left in plant
- flow through abnormal opening made in plant

Adverse change in a planned product or other release:

- adverse change before leaving plant
- incorrect transfer
- adverse change after leaving plant
- abnormal vent/spill

- external events;
- a clear specification of the objectives;
- a full process specification of feeds, products and wastes;
- constraints on emissions and effluents;
- specification of utilities.

At some stage such studies should show the effect of various losses of inventory to identify the main hazards and gauge the extent of the consequences of such losses. Appropriate site and surrounding information should also be provided. It is a mistake to provide excessive data for any meeting at which the PPHA is carried out. The material should be provided merely as part of the technical briefing of selected personnel.

PARTITION PLANT INTO CRITICAL SECTIONS
The plant is partitioned according to the main plant items and their associated ancillary equipment. In general a section of plant can be selected which includes the relevant heat exchangers and input/output process pipelines. Each section should be looked at as a whole.

SELECT A DANGEROUS DISTURBANCE TO STUDY
The optimum starting point for this analysis is at a point on the incident scenario termed 'Dangerous disturbance of plant'. If the analyst is more comfortable starting elsewhere in the scenario then this is acceptable. A list of main dangerous disturbances is given in Table 3.3. Where appropriate, more details on these disturbances can be obtained from an equipment knowledge base as described in Appendix 1, pages 249–251. The changes in operating parameters which occur on the incident scenario are distinguished as the danger increases by the following terms:

- *dangerous disturbance*, as when the temperature has passed the point of overtemperature and a trip system has failed to correct the situation;

TABLE 3.4
Pro forma for Preliminary Process Hazard Analysis showing initial mark-up

IMMEDIATE CAUSES	INADEQUATE NORMAL CONTROL	PROCESS DEVIATION	FAILURE TO CONTROL (ON ALARM)	
		Stage 1		
Stage 3	Stage 3		Stage 3	
Comments: Stage 1 — identify dangerous disturbances, significant event, hazardous disturbances, process deviations. Stage 2 — identify inadequate emergency control and failure to recover the situation. Stage 3 — identify immediate causes, inadequate normal control and failure to control on alarm.				

- *hazardous disturbance*, as when the temperature is reported as high-high temperature and emergency control might be effected;
- *process deviation*, as when the temperature is reported as high and an alarm might signify the need for intervention by the operator.

The latter deviation corresponds to a Hazop deviation which will be identified in Chapter 5.

It should be noted that the identification of specific points where alarm and emergency control are activated enables a record to be made of the safe upper and lower limits for such parameters as temperature, pressure, level, inventory, composition and flow.

IDENTIFY THE SIGNIFICANT EVENT RESULTING FROM THE DANGEROUS DISTURBANCE
The significant event resulting from the dangerous disturbance will usually be the loss of process material to atmosphere, either upon the rupture of a vessel or pipework or other release of material through an available opening. The release may be planned or unplanned. It may take place with considerable release of energy, as when following an explosion.

It is emphasized that other significant events can be defined. It has already been noted that a product or intermediate stream may change so that it creates danger. In some circumstances it may be important to analyse the possible dangers of human entry into a vessel. The rupture of equipment can lead to leaks between streams and on occasion this may warrant treatment as already identified.

IDENTIFY EACH EVENT ON A *PRO FORMA* FOR PPHA
A *pro forma* is provided to note events (Table 3.4) and it follows the general incident scenario (page 2). The stages generally used in completing the form up to a significant event are shown in

HAZARDOUS DISTURBANCE	INADEQUATE EMERGENCY CONTROL	DANGEROUS DISTURBANCE	FAILURE TO RECOVER SITUATION	SIGNIFICANT EVENT
Stage 1		Stage 1		Stage 1
	Stage 2		Stage 2	

Table 3.4. They are:

Stage 1 — identify dangerous disturbances, significant event, hazardous disturbances and process deviations;

Stage 2 — identify inadequate emergency control and failure to recover the situation;

Stage 3 — identify immediate causes, inadequate normal control and failure to control on alarm.

In many ways this procedure follows that adopted in Hazop but from a higher starting point in the incident scenario. The gaps in the *pro forma* are filled in, working from deviation to cause and consequences. It is important that the study does not attempt to resolve immediate causes to the same extent as in a Hazop. This is too time-consuming and furthermore the less critical problems will be identified later. So it would be enough to note a high inlet temperature to the reactor without identifying the control problem or design restriction responsible, *if* the meeting is satisfied that this will cause few difficulties during subsequent Hazop.

PRELIMINARY PROCESS HAZARD ANALYSIS OF THE METHANATOR

The methanator is the one used for Concept Hazard Analysis. It is essentially a finishing operation to remove oxides of carbon from a stream of hydrogen containing low percentages of methane and water. For convenience Figure 3.1 (pages 52–53) is upgraded from Figure 2.5.

In order to carry out a PPHA, the plant was divided into three sections: Node A — the reactor circuit of methanator and associated heat exchanger system; Node B — knockout pot; Node C — compressor. The analysis is reported on the working sheets Tables 3.5 to 3.8 (pages 54–57), which expand from the initial headings as illustrated. These results are selected from further documentation and are not intended to be comprehensive.

HAZARD IDENTIFICATION AND RISK ASSESSMENT

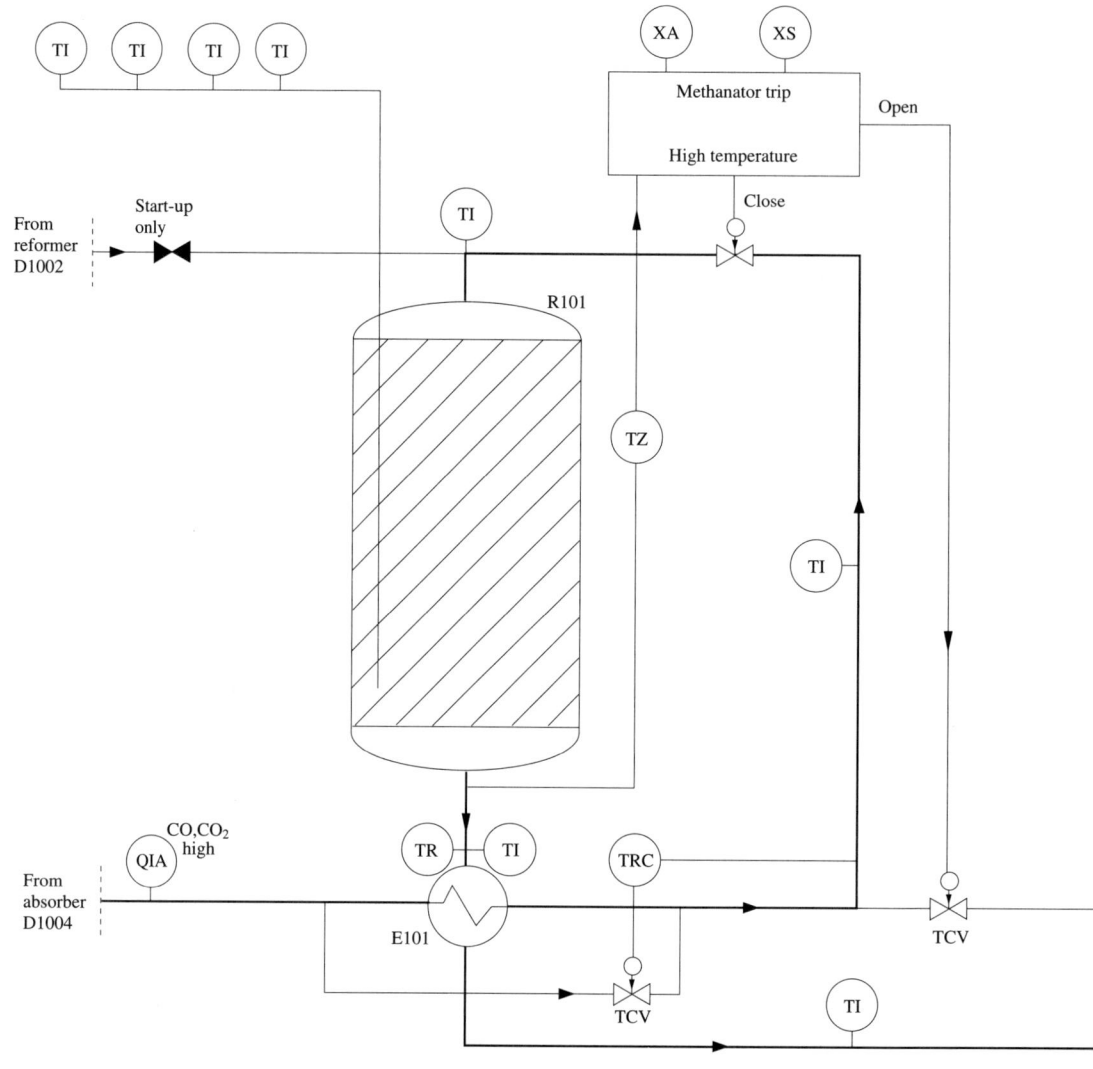

Figure 3.1   P&I diagram of hydrogen plant, Revision A. (See page 243 for instrumentation definitions.) 10 = Nitrogen, 10 bar; 11 = Cooling water supply; 12 = Cooling water return.

Item number	R101	D102	E101	E102	C102
Title	Methanator	KO pot	Methanator preheat	Methanator cooler	H_2 compressor
Op. temp, °C	450	30	400	100	50
Op. pres., bar	20	18.7	21	19	40

PRELIMINARY PROCESS HAZARD ANALYSIS

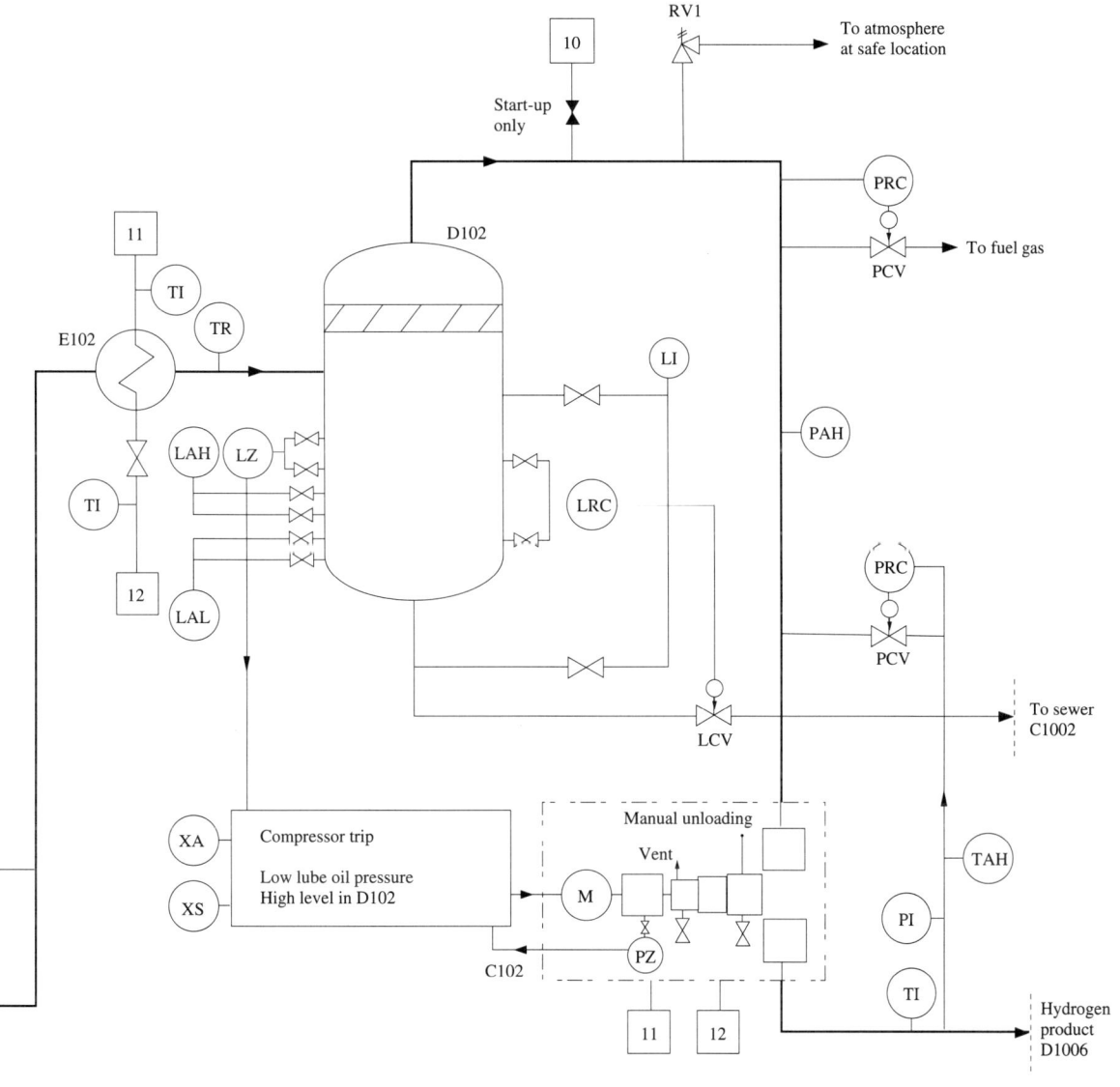

TABLE 3.5
Preliminary Process Hazard Analysis sheet. Plant — methanator; reactor circuit.

IMMEDIATE CAUSES	INADEQUATE NORMAL CONTROL	PROCESS DEVIATION	FAILURE TO CONTROL (ON ALARM)	
		High temperature in reactor		
Recommendations, comments, actions: • The increase in oxides of carbon results in more input material to the exothermic reaction.				

TABLE 3.6
Preliminary Process Hazard Analysis sheet. Plant — methanator; reactor circuit.

IMMEDIATE CAUSES	INADEQUATE NORMAL CONTROL	PROCESS DEVIATION	FAILURE TO CONTROL (ON ALARM)	
High inlet temperature to reactor	Control system inadequate or mis-set		Operator fails to stop trend on TAH by adjusting control (slower acting and runaway unlikely)	
High CO_2 in stream from absorber	Fault on absorber	**High temperature in reactor**	Operator fails to stop trend on COx or TAH by correcting absorber	
Impurities in feed down start-up line (sneak path)	Inadequate isolation after start-up		Operator fails to stop trend on TAH by isolation	
Recommendations, comments, actions: • Do not depressurize on high temperature unless no flow through methanator. • Operator alerted by several alarms. New TAH in and out. • Check if start-up line needed if heat exchanger circuit modified.				

HAZARDOUS DISTURBANCE	INADEQUATE EMERGENCY CONTROL	DANGEROUS DISTURBANCE	FAILURE TO RECOVER SITUATION	SIGNIFICANT EVENT
High-high temperature in reactor		Overtemperature in reactor		Release by rupture on (overtemperature)
Recommendations, comments, actions (cont): • An increase in inlet temperature is less severe as it does not increase the heat output from the reaction.				

HAZARDOUS DISTURBANCE	INADEQUATE EMERGENCY CONTROL	DANGEROUS DISTURBANCE	FAILURE TO RECOVER SITUATION	SIGNIFICANT EVENT
High-high temperature in reactor	Operator fails to stop inflow. High temperature shutdown system fails.	Overtemperature in reactor	Operator fails to stop all flows	Release by rupture on (overtemperature)
Recommendations, comments, actions (cont): • Alter outlet location of start-up line. Add PAH and TR. Double block and bleed. • Check catalytic activation. • Improve absorber reliability.				

TABLE 3.7
Preliminary Process Hazard Analysis sheet. Plant — methanator; knock-out pot.

IMMEDIATE CAUSES	INADEQUATE NORMAL CONTROL	PROCESS DEVIATION	FAILURE TO CONTROL (ON ALARM)	
(1) LCV fails open. (2) Controller fails.	Normal control fails	Low level in KO pot	Inadequate action to increase level on LAL	
(1) LCV fails shut. (2) Controller fails. (3) Liquid discharge is inadequate.	Normal control fails	High level in KO pot	Inadequate action to reduce level on LAH on increase from correct level in KO pot	
(1) Vents blocked. (2) Excess flow from KO pot.	Failure of vents in sewer to cope with flow	Accumulation in sewer of flammables	Operator unaware of problem unless prewarned by KO pot failure	
Recommendations, comments, actions: • Explosion in sewer can occur if inadequate venting of hydrogen by standpipes fitted with flame arrestor. • Consideration should be given as to whether a shutdown system is needed in the event of loss of liquid level in the KO pot.				

TABLE 3.8
Preliminary Process Hazard Analysis sheet. Plant — methanator; compressor.

IMMEDIATE CAUSES	INADEQUATE NORMAL CONTROL	PROCESS DEVIATION	FAILURE TO CONTROL (ON ALARM)	
(1) Failure of PRC. (2) PCV failed open. (3) Lack of demand.	PCV fails open or design inadequate	High temperature in recycle loop	Inadequate action to reduce recycle rate on TAH in recycle loop	
(1) Inadequate support. (2) Vibration in compressor.	No instrumentation, only occasional inspection by operator	Vibration of system	Failure to note vibration	
(1) PCV failed open on plant outlet. (2) PC failed on discharge to fuel. (3) Inadequate venting to fuel.	Pressure control system fails to cope	High pressure	Increase in pressure and PAH	
Recommendations, comments, actions: • The high temperature problem can be resolved by a simple change. • Move the TAH nearer to the compressor.				

HAZARDOUS DISTURBANCE	INADEQUATE EMERGENCY CONTROL	DANGEROUS DISTURBANCE	FAILURE TO RECOVER SITUATION	SIGNIFICANT EVENT
Low low level in KO pot	Failure of operator to close drain	Gas blowby to sewer	Continued failure to maintain liquid level in KO pot	Liquid entering compressor generating high stress levels
High high level in KO pot	Failure of shutdown system to stop compressor. XA acting as LAHH should make operator aware that system is activated.	Liquid blowby to compressor	Failure to reduce level in KO pot	Significantly high release of flammables to sewer
Accumulation in sewer of flammables	Operator unable to reduce flow to sewer or unaware of problem	Flammables spread through sewer	Recovery efforts fail or operator unaware of problem	Unplanned release of flammables near source of ignition

Recommendations, comments, actions (cont):
- Check venting capacity in the sewer. A flammable vapour detector in the sewer is of no value.

HAZARDOUS DISTURBANCE	INADEQUATE EMERGENCY CONTROL	DANGEROUS DISTURBANCE	FAILURE TO RECOVER SITUATION	SIGNIFICANT EVENT
High high temperature in recycle loop	No further warning: operator to stop compressor	Overtemperature in recycle loop and compressor	Operator fails to rectify fault or stop compressor	Rupture of compressor or pipe
System partially loosened	Operator fails to identify vibration and stop compressor	Loosening of pipe and fittings	Failure of operator to stop compressor	Release of material
High high pressure	Relief valve fails or cannot cope with demand	Overpressure of upstream plant	Operator fails to correct or unload compressor	Release of process material on rupture

Recommendations, comments, actions (cont):
- Vibration probes should be installed, project to identify location.
- Install non-return valve (NRV) on supply to downstream plant.

The main hazards on the methanator circuit were readily seen to be high pressure or high temperature in the reactor. Clearly it is important to check the latter thoroughly as there is the threat of a highly exothermic reaction in the reactor. The initial study exposed the dangers of high concentrations of oxides of carbon entering the reactor, either due to problems in absorption or in the operation of the low temperature shift reactors or passage of oxides of carbon down a sneak path from the reformer. Incidentally the start-up line was added as a panic measure when it was found during commissioning that the methanator could not be started up using the original design. No safety studies were undertaken at the time as it was not then the practice.

This study helps to identify the need to balance demand between the hydrogen plant and its customers downstream. It has been noted that this is often missed when this process is used as a Hazop exercise. Also the need for a downstream carbon monoxide/carbon dioxide analyser has been noted.

The study of the knockout pot seems at first glance to be largely routine, indeed almost unnecessary. However it did show up the dangers of dissolved gas entering the sewer. An analyser in the sewer is pointless as there is always a flammable atmosphere. However, vented standpipes are essential. If there is an explosion in the sewer, it is unlikely that the damage would amount to more than the manhole covers lifting along its path. Some discussion has in the past centred on whether to install an atmospheric degassing pot and majority opinion is generally against this idea.

The study of the compressor was fairly standard and would not usually be expected to generate many ideas. What did not emerge in this study was the need to maintain production of the whole complex if at all possible, and the recommendation to install — instead of one compressor — two 60% duty machines or similar. Availability of the plant is often not considered to be part of plant safety (apart from when considering emergency equipment). However, many incidents have occurred during start-up and shutdown, and in some cases deterioration of plant can be increased during these activities. It is often sensible to note any loss of production during safety studies as this is generally a way of justifying change as well as improving critical tasks.

It will be seen later that the answers obtained for frequency of major events supported existing safety practice for a plant of this nature. All risk assessment studies were carried out following the safety review meeting. A number of faults will emerge from further analysis including for this plant one which may be quite serious. This does not invalidate the study but merely underlines the need for further studies at the Hazop stage.

PRELIMINARY CONSEQUENCE ANALYSIS OF METHANATOR SECTION

The major failure of equipment on the methanator section can occur due to a combination of operating pressure and high temperature, or the guillotine failure of a six-inch pipe. Frequency and probability information can be attached to each basic event on the PPHA sheet and a fault tree generated (Figure 3.2). The release of process material through overtemperature of the methanator is likely to occur with a frequency of 0.0005 per year. This will result in a continuous release of hydrogen containing small amounts of methane. Subsequently this stream would contain larger quantities of carbon monoxide and carbon dioxide as the system depressurized.

The maximum continuous rate of release through one end of a six-inch pipe would stem from any point of release. The planned emergency action would initially involve stopping the flow of methane into the reformer and the shutdown of all burners. Initially steam would continue to enter the reformer in order to purge the system. Action to depressurize the system would vary

PRELIMINARY PROCESS HAZARD ANALYSIS

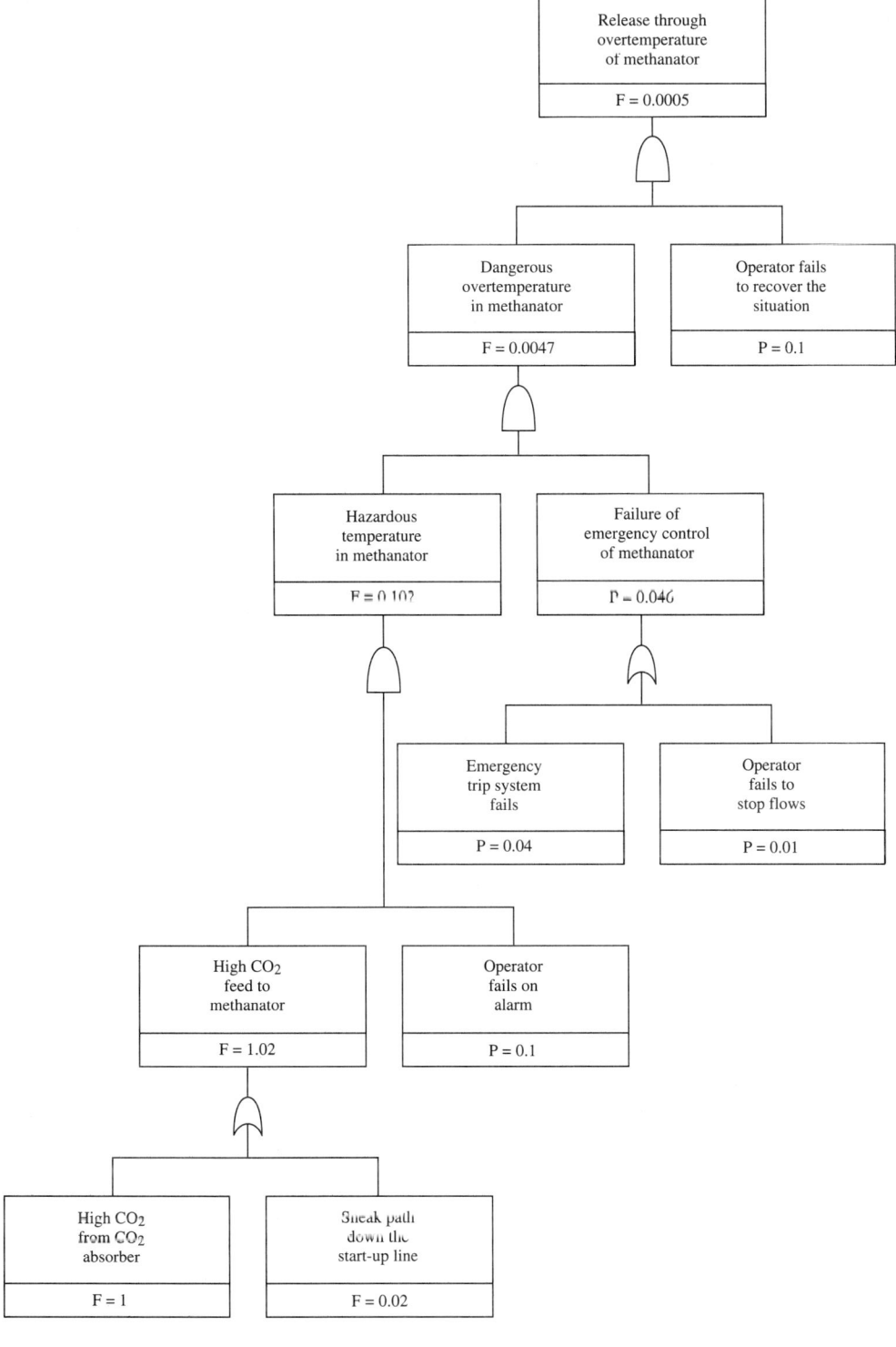

Figure 3.2 Fault tree for runaway reaction.

TABLE 3.9
Preliminary Consequence Analysis for the methanator circuit

Plant: Hydrogen
System: Methanator circuit

WHAT IF UNDESIRED EVENT?	FR/PR	GATE	FAILURE TO MITIGATE	
What if significant release of material on catastrophic failure of the methanator circuit?	E–3	AND	Countermeasures for a release fail: insufficient time for response	
What if ignition of flammable mixture giving torch fire?	0.5	AND	Countermeasures fail to control fire: fire too great to be put out immediately	
What if escalation by fire with further spread of fire to pipe rack?	0.2	AND	Countermeasures fail to control fire: fire brigade fails to put out fire (no barrier)	
What if ignition of flammable mixture and pool fire?	1.0	AND	Countermeasures fail to control fire: fire brigade fails to put out fire	
What if escalation by explosion and missiles land on cyclohexane plant?	0.2	AND	Countermeasures for a release fail	
FR = failure rate; PR = probability rate; L = likelihood; S = severity; P = priority.				

according to the immediate damage and the location of the leak. This determines which manual valves should be opened to vent the process.

Hydrogen has a tendency to ignite immediately on release. This is a complex phenomenon based on its dielectric constant and the friction generated by its release. In the absence of reliable data, it is assumed that the probability of ignition is 0.5. This value is considered to be high. The reformer furnace is the only other ignition source, situated 20 m (normally) upwind of the methanator. Also hydrogen is extremely buoyant and disperses rapidly. So a hazardous flash fire is extremely remote, say for 1 in a 100 releases. If the release of hydrogen occurred at a frequency of 10^{-3} times per year, then it is considered — given a probability of someone working in the area at the time of the event of 0.05 — that the chance of a fatality due to this cause would be 5×10^{-7}. This hazard is considered to be insignificant for process workers.

| \multicolumn{7}{l}{**Release:** A major release of hydrogen from the methanator circuit. Release will continue for thirty minutes before initial release depressurized or changed to a discharge of steam.} |

PR	L	S	P	CONSEQUENCES	IMPACT	COMMENTS/DATE
1	3	1		Release causes hazardous condition: cloud of flammable material	Minor plant damage. No damage to the environment.	Results correspond to near-miss
1	4	3	B	Escalation by torch fire	Major plant damage. Injuries if personnel in direct contact. Low chance of fatality.	Assumes input of feed to hydrogen plant stopped
0.5	5			Escalation by further release of material		Depressuring initiated but slow process. Consider locating pipe rack further from methanator circuit.
0.2	6	3	C	Escalation by pool fire, generating possible explosion with missiles	Major plant damage. Further injuries possible. Greater chance of fatality.	Check on action to be taken with respect to pipeline flows on emergency
1	6	4	C	Escalation by further release of flammable material. Aromatics washed into sewer.	Severe plant damage. Probably fatality on works. Injuries on works. Minor environmental pollution of waterway.	Not studied for cyclohexane plant as part of this exercise. Develop full emergency plan.

A torch fire would result from any ignition of the escaping gas. This could burn through other equipment containing gas — for example, the shift converters — and increase marginally the extent of any release. Less likely, but possible, is that the flame might impinge on the main pipe rack passing through the plant. This would then break, giving rise to a pool fire of aromatics. Missiles are also possible, but fairly unlikely in this eventuality. The analyst must check the plant layout against such possibilities and evaluate the fire-fighting procedures.

Any fire would largely be confined to the hydrogen plant. This is assessed as a major category incident, largely because of the loss of the plant for at least a year. Some injuries to personnel are possible but a fatality is unlikely. The fire could spread to the adjacent aromatics plant and gave rise to a severe category incident. There would be little possible danger to the public or office staff, however, as the site offices and the works boundary are some 500 m away.

There is no toxic hazard from the release of hydrogen should it fail to ignite, although passing birds might find the experience both unpleasant and buoyant.

All these assessments are reported in Table 3.9, pages 60–61. A what-if approach has been used in this instance. The relevant event tree (Figure 3.3) assumes a maximum frequency of release of 10^{-3}/year. The risks as evaluated indicate that the design of the process is acceptable. Further information on the evaluation of short-cut risk is given later. The special terms used in Table 3.9 are as follows. FR and PR are the frequency and probability of an event. L is the likelihood as measured by the negative power of the frequency — that is, a value of −1 to −5 corresponds to 10^{-1} to 10^{-5}. C is the consequence or severity in a range from 5 catastrophic to 1 minor. P is a prioritization index evaluated as a function of likelihood and consequence, with A requiring immediate attention, B probably requiring further study and C probably not needing further study. For further information, see Chapter 10.

SUMMARY OF MAIN DANGEROUS DISTURBANCES AND THEIR IMMEDIATE CAUSES
Table 3.10 (pages 63–64) serves as a summary of the main findings from the PPHA of the methanator which led to the development of Figure 3.4 (pages 66–67) from Figure 3.1. As a result of these studies the process was modified as shown in Figure 3.4. This design must still be validated by further risk assessment and by Hazop studies. All problem areas have not been evaluated to allow for further presentation aims in Chapter 5. It is also necessary to evaluate further passive protection measures and check the plant integrity as to the most likely points of failure.

RISK ASSESSMENT USING RISK EVALUATION SHEETS
Subsequent to the PPHA it may be necessary to carry out a risk assessment using short-cut risk evaluation and reported when appropriate on risk evaluation sheets.

The first objective in the use of the risk evaluation sheets is to highlight areas where it is desirable to constrain either or both the occurrence and severity of the incident. Appropriate changes can usually be readily identified. However, in some cases it is necessary to prioritize a number of scenarios for further study and investigate such scenarios using hazard analysis and risk assessment methods as appropriate.

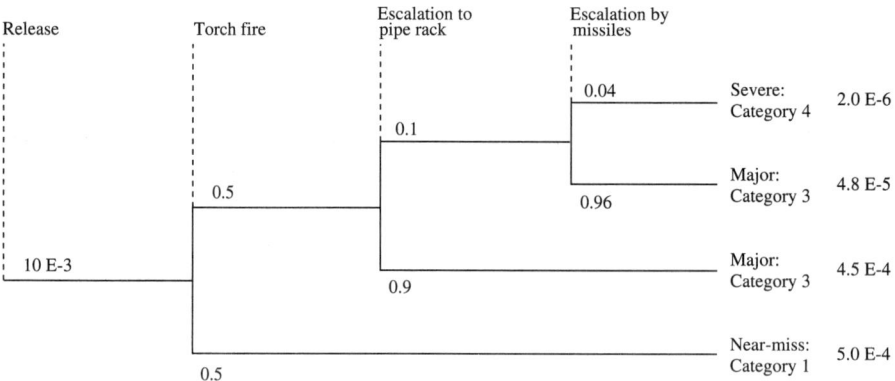

Figure 3.3 Event tree for major release from methanator system.

TABLE 3.10
Summary of main dangerous disturbances and their causes

Dangerous disturbance	Causes	Actions
Overtemperature in reactor and runaway reaction	High COx from absorber.	Improve operability by incorporating two-stage absorption unit. Project to advise.
	Sneak path down start-up line.	Redesign heat exchanger circuit, or alter location of start-up line outlet. Add PIA, TIA, double block and feed.
	High inlet temperature to reactor on failure of TRC or TCV.	TAH on inlet to reactor. Note that runaway reaction is unlikely due to completion of reaction.
	Operator unaware of problems.	Oxides of carbon now monitored with alarm (see below). TAH on reactor outlet.
	Failure of emergency trip on loop or valve failure.	Location of trip valves might be changed with main isolation prior to E101. Excess material then passes to flare or other safe route. This creates less load on E101 and PRC to fuel gas.
Overtemperature in heat exchanger E101 under abnormal operating condition	Reactor runaway, see above	Check design for maximum conditions, particularly at the hot inlet. Project to advise.
Overpressure in reactor and inadequate pressure relief	Downstream blockage.	Very unlikely as hydrogen in clean duty, but see below.
	Lack of demand for product and PRC to fuel gas cannot cope.	Emergency depressurizing valve required.
	Off-specification product.	See above and compressor study.
	Trip valves isolate relief valve.	Action depends on downstream design but location of relief valve moved to before E101.
	Operator unaware of problem.	PAH adjacent to PRC to fuel gas.
Release of material followed by torch fire	Overtemperature and overpressure. Ignition probable due to static generation.	Main problem escalation to pipe rack. Detailed study required although initial layout is satisfactory.
Deterioration of materials of construction	Impurities in feed	Check effects of sodium salts and impurities in water

TABLE 3.10 (continued)
Summary of main dangerous disturbances and their causes

Dangerous disturbance	Causes	Actions
External cause	Possible missiles from adjoining aromatics complex; impact from crane, etc	Ensure layout allows for good access for replacing catalyst and appropriate lifting davits are provided. Consider emergency isolation of whole of hydrogen plant into different sections. Project to advise.
Overtemperature in exchanger E102 under emergency conditions	Spurious failure of trip valve or its activation	Design exchanger to cope with this maximum duty. Check effect of loss of cooling water supply under this condition. Decision placed on hold. Process to advise.
Planned release to sewer followed by separation of flammables		Sewer must be vented with high standpipes. Alternatively send to flare system with water separation provision. See also below.
Gas blowby to sewer	LCV fails open	Additional block valve would help. Alternatively a trip system on low level. Detector inappropriate as there are always flammables in sewer. Decision on hold regarding use of sewer. LAL is essential (not shown on Figure 3.4).
Discharge into N_2 line	Process pressure exceeds nitrogen pressure	Provide removable spool and isolation. This may not be practicable if hot so consider option of double block and bleed with pressure alarm. Process to advise.
Liquid blowby giving liquid in compressor and high stress	Failure of LHA and trip system	Reasonable protection. Also check out manufacturer's recommendations for compressor trip, etc, when details available.
Overpressure at inlet to compressor	PRC fails open	Reverse flow from aromatics plant. Install non-return valve (NRV) and check downstream protection.
Overtemperature at inlet to compressor	Excess circulation around compressor	Check aftercoolers or other design change not revealed until Chapter 5
Emergency discharge via relief valve	Discharge on overpressure	Discharge may ignite. Ensure this presents no radiation hazard. Project to advise.

A risk evaluation sheet (Table 3.11, page 70) has been completed for the methanator/preheat section. This sheet relates to Figure 3.4. For this particular study the risk was considered to be tolerable to the company and its workforce. The results are considered to be if anything pessimistic.

PRELIMINARY PROCESS HAZARD ANALYSIS OF WASTE HEAT BOILER

This study is presented as an exercise for the reader to attempt. It requires the analysis of a waste heat boiler system.

A simplified storage/waste heat boiler system is shown in Figure 3.5 (pages 68–69). Boiler feed water (deionized and deaerated) is pumped from storage under level control as determined in the steam drum. It can be assumed that the preheater in the circuit does not affect any hazards being considered in this study.

The level control system on the steam drum incorporates cascade control where one controller provides the command signal to another controller. In a less complex system, if the level in the drum fluctuates, then slugs of cold water might enter the steam drum causing the steam rate to deteriorate. In the system shown a flow ratio controller, FfC, endeavours to keep the input flow of feed water equal to the output flow of steam. The level thus remains fairly constant. The principal advantage of such a system is that feed water corrections are made before level changes, as a flow system has an inherently faster rate of response. The level controller is tuned to correct for any long-term drift in level but to have little effect on feed water flow during short-term excursions in level.

The drum is fitted with a high level alarm, a low level alarm and a low-low level alarm. The waste heat boiler is heated by a process stream on a large continuous plant which cannot be instantly shut down without some residual heat remaining in this stream. On a low-low level alarm the whole plant must be shut down fairly quickly or the tubes of the waste heat boiler will not be covered by water, with possibly serious results. The exact impact of this effect has not been studied.

Part of a PPHA is carried out to identify the current incident scenario for the occurrence of 'Waste heat boiler tubes exposed'. A separate analysis is needed to examine problems in the boiler feed water storage.

The results of the study are given in Table 3.12 (pages 72–73). In this case the process deviation was taken as a low level in the accumulator. Some analysts may consider a better selection to be the low flow from the pump. There is no objection to this or to extending the number of deviations considered. After all, in Hazop there is a lot of repetition. Obviously all depends on the objective of the analysis as some hazardous disturbances always lead to other disturbances.

Other possible events not considered in Table 3.12 include inadequate design of the boiler feed water supply relative to the maximum load on the boiler and failure of the PRC or relief valve. Problems may also arise from the capacity of the downcomers/upcomers of the boiler preventing boiler feed water reaching the boiler. This possible design fault should be thoroughly checked.

A simplified fault tree is given in Figure 3.6 (page 71). The prime aim of PPHA is hazard identification, not quantification. A detailed study would have to consider reasons for the failure of procedures or operator inactivity. It would also be necessary to investigate the factors affecting

HAZARD IDENTIFICATION AND RISK ASSESSMENT

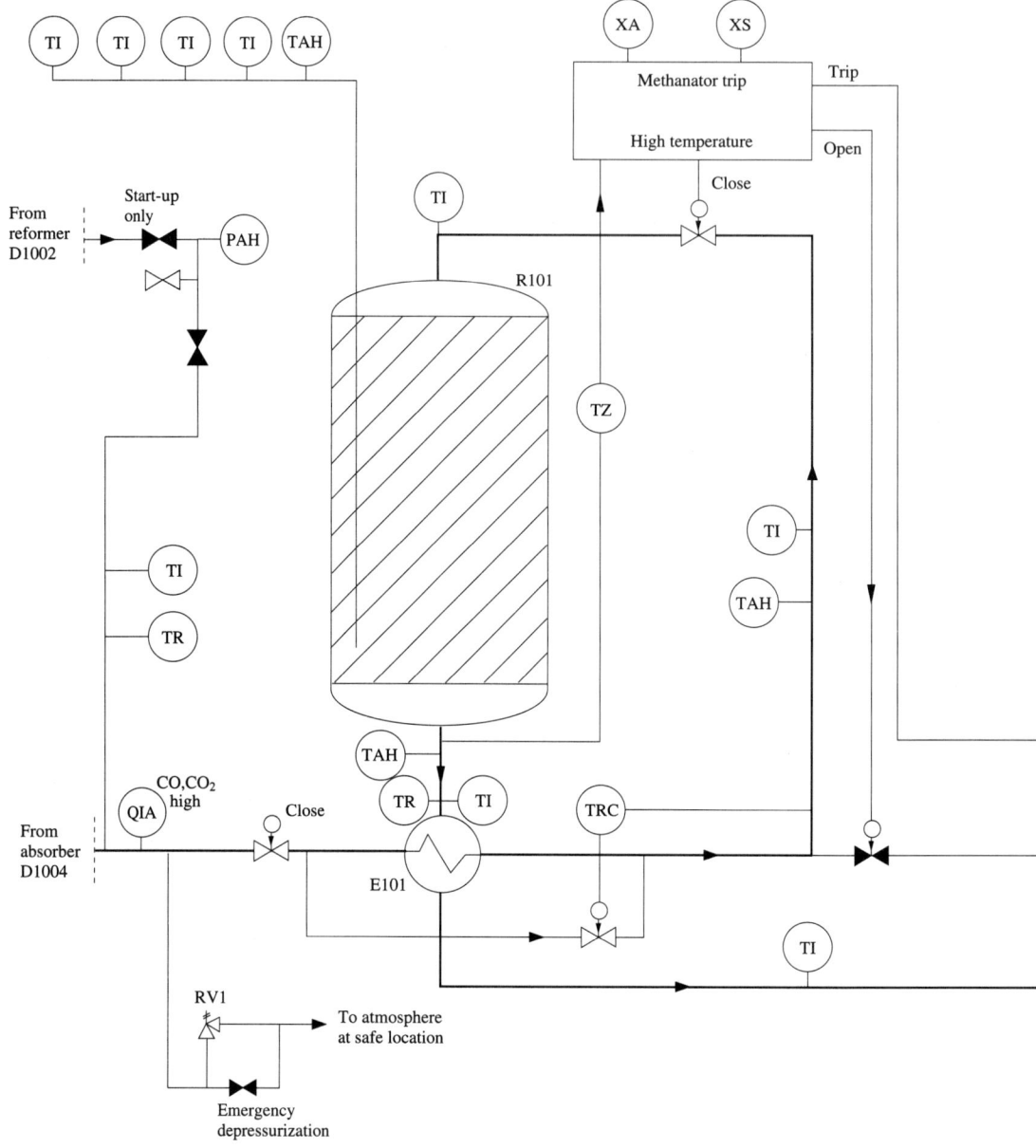

Figure 3.4 P&I diagram for methanator, Revision B. (See page 243 for instrumentation definitions.)
10 = Nitrogen, 10 bar; 11 = Cooling water supply; 12 = Cooling water return.

Item number	R101	D102	E101	E102	C102
Title	Methanator	KO pot	Methanator preheat	Methanator cooler	H_2 compressor
Op. temp, °C	450	30	400	100	50
Op. pres., bar	20	18.7	21	19	40

PRELIMINARY PROCESS HAZARD ANALYSIS

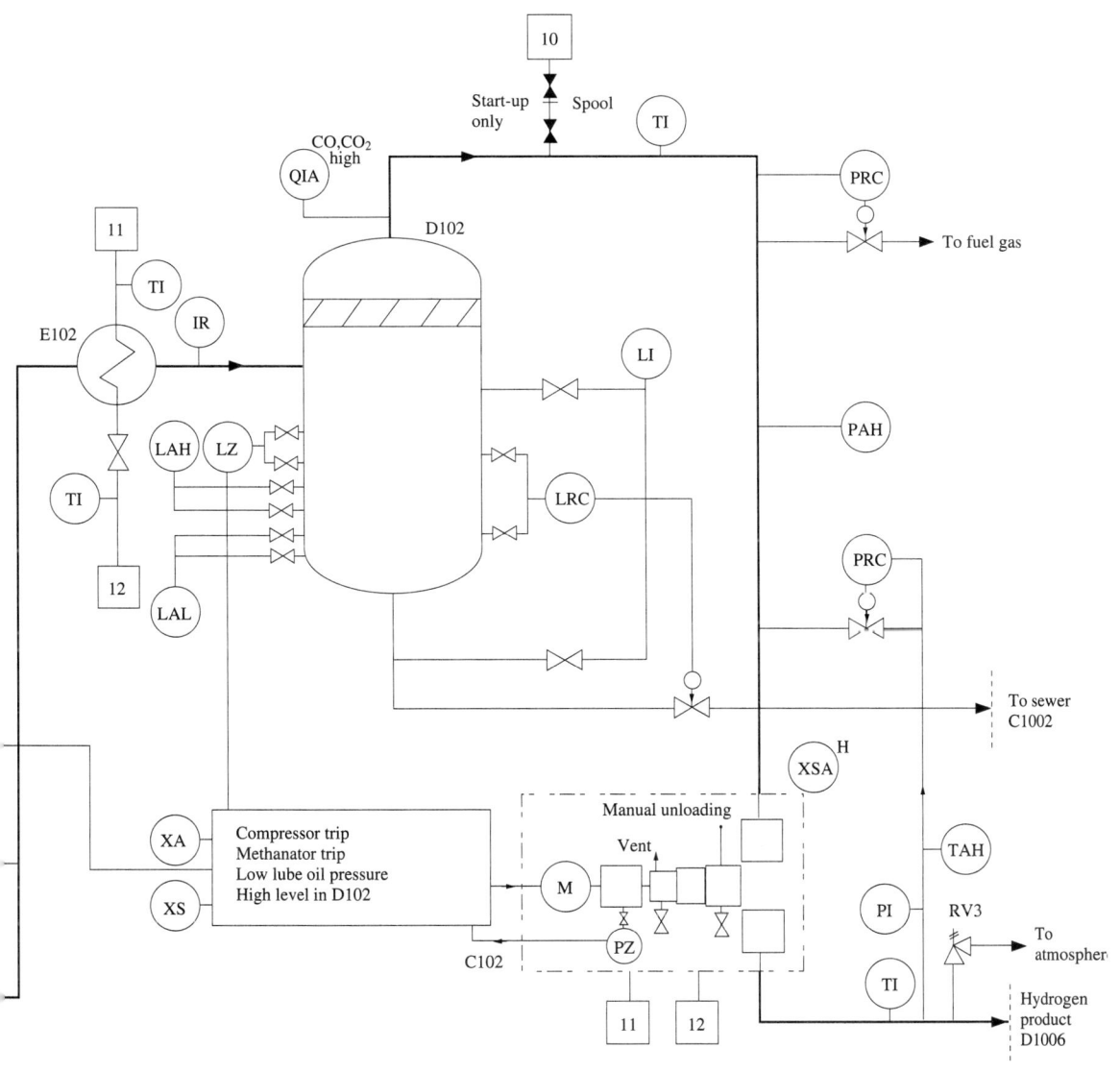

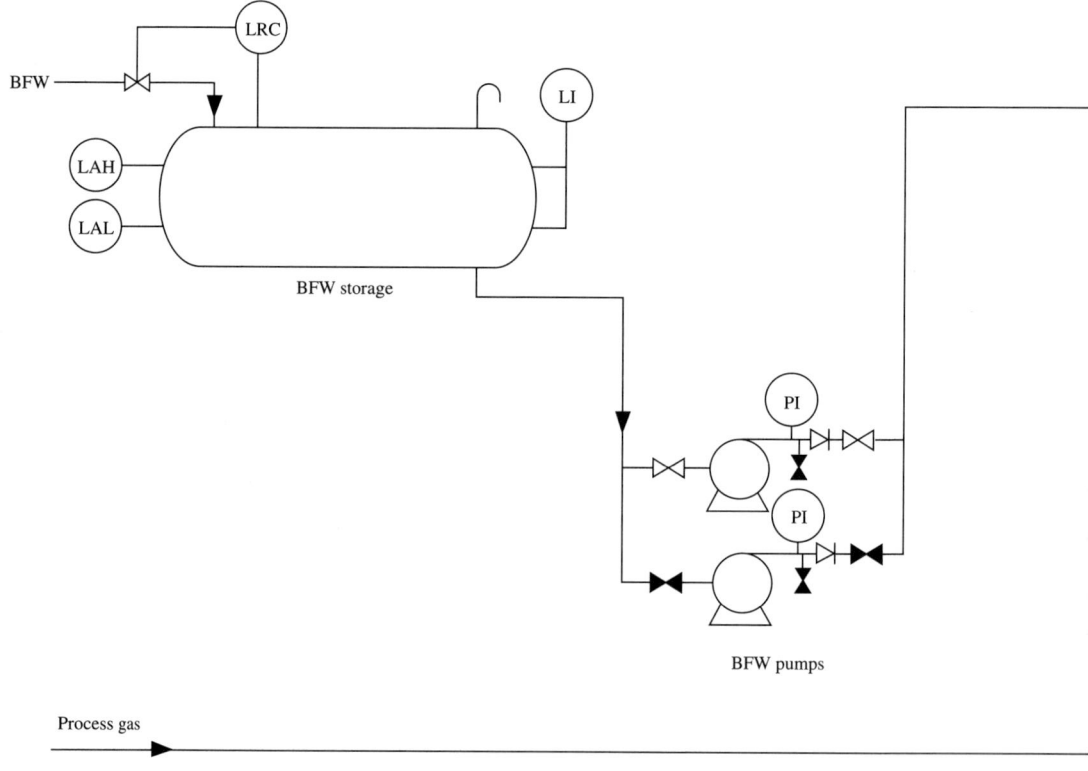

Figure 3.5   P&I diagram for waste heat boiler. (See page 243 for instrumentation definitions.)

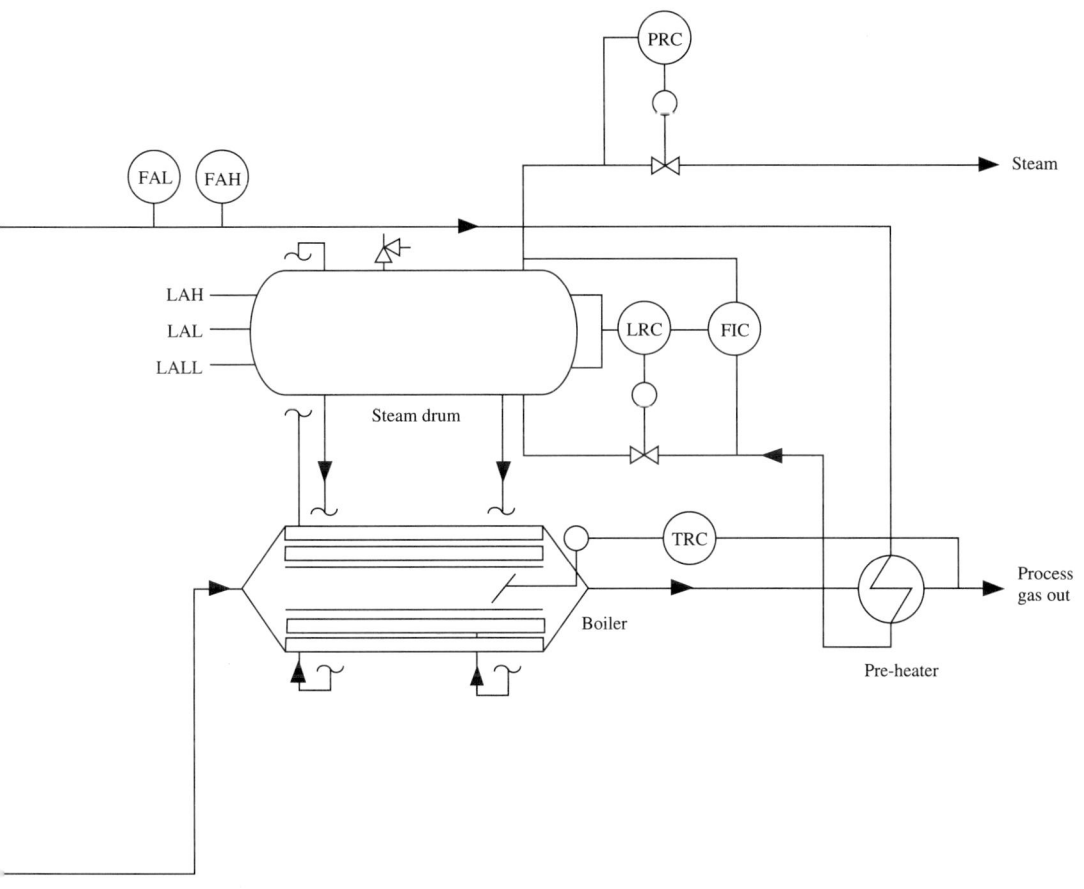

TABLE 3.11
Preliminary Process Hazard Analysis risk evaluation sheet

Project: Book	Reference: GLW	Date: 1.4.94	
Plant: Hydrogen	Location: Sheffield on Sea	Page: 1 of 4	
Unit: Methanator	Equipment: Reactor and preheat	Priority	
Function: Fixed bed reactor converting oxides of carbon to hydrogen		S	L
CONSEQUENCES OF ESCALATION	Fire escalates to pipe rack and C plant	4	−6
FAILURE TO PREVENT FURTHER ESCALATION	Failure to avoid domino due to lack of time and ineffective fire-fighting		0.01
CONSEQUENCES OF SIGNIFICANT EVENT	Torch fire on section of plant	3	−4
FAILURE TO MITIGATE OR AVOID ESCALATION	Failure to avoid ignition: self-ignites as hot AND release not attenuated in 15 minutes		
SIGNIFICANT EVENT	Release through overtemperature		E−4
FAILURE TO RECOVER THE SITUATION	Operator fails to stop all plant flows		0.1
DANGEROUS DISTURBANCE	Overtemperature in reactor		0.0047
INADEQUATE EMERGENCY CONTROL	Failure of operator to stop flow to methanator Failure of shutdown system		0.01 0.04
HAZARDOUS DISTURBANCE	High-high temperature in reactor		0.102
FAILURE TO CONTROL ON ALARM	Operator fails to reduce trend on TAH or COxH		0.1
PROCESS DEVIATION	High temperature in reactor		0.12
INADEQUATE NORMAL CONTROL	Operator fails to resolve fault on absorber Operator cannot isolate line		1 1
IMMEDIATE CAUSES	High CO_2 in stream from absorber Impurities in feed: sneak path down start-up line		1 E−2

Recommendations/comments/actions
(1) Public not affected by domino escalation.
(2) Business damage would be extensive if spread to complex.
(3) The operator can increase the probability of a release by wrong action and special supervision is required on any methanator problem.
(4) Do not depressurize on high temperature unless sure of no flow through methanator.
(5) Operator now alerted by several alarms, eg QAH, TAH.
(6) Check if start-up line needed if heat exchange circuit modified.
(7) Note new start-up instructions and procedures.
(8) Check catalyst activation.
(9) Improve absorber design to enhance reliability by building in additional lean solution hold-up capacity.
(10) The Short-Cut Risk Evaluation justified the changes from the PPHA. *The system must be re-evaluated once the procedures for start-up and catalyst activation are known.*

PRELIMINARY PROCESS HAZARD ANALYSIS

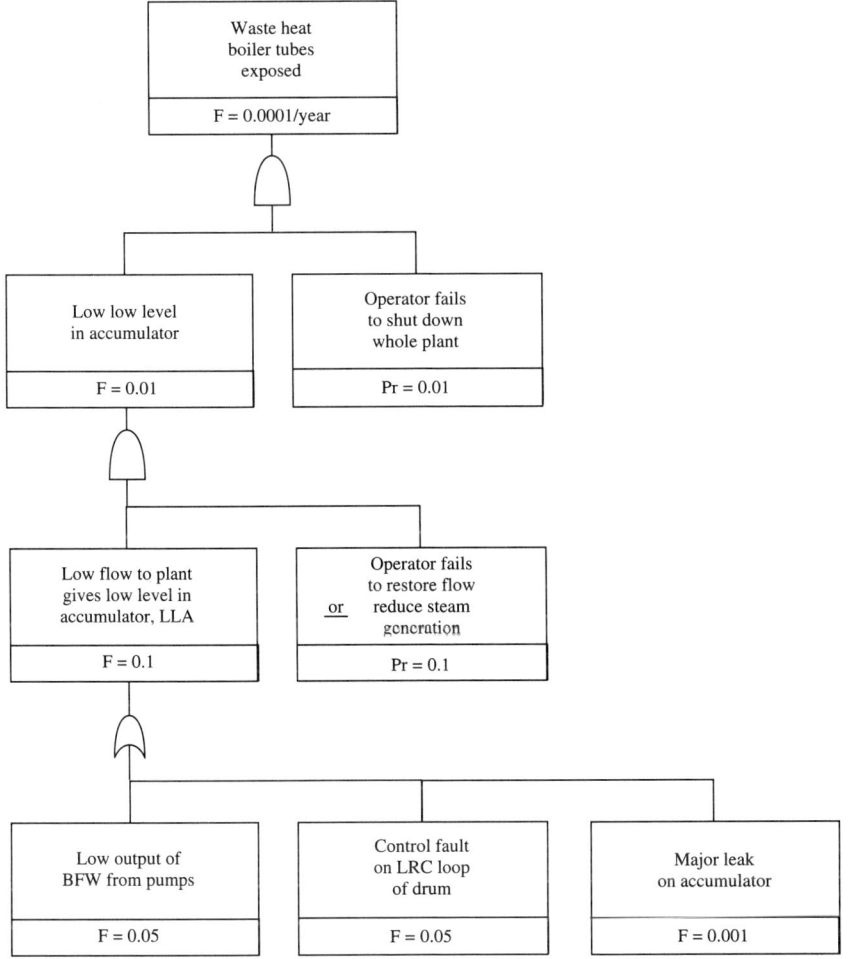

Figure 3.6 Fault tree leading to exposure of waste heat boiler tubes.

the setting of the low-low level alarm. Should the shutdown be automatically initiated at this point? It is believed that the results presented for the occurrence frequency of the top event are pessimistic.

It is emphasized that this exercise has not been subjected to rigorous study and the PPHA has not been completed. No design alterations have been considered. The study has been left for completion as a tutorial exercise. There is no doubt that further improvements could be made by a more rigorous study. It is also a useful small plant on which to carry out a Hazop. However, the P&ID should first be developed to an appropriate standard.

TABLE 3.12
Preliminary Process Hazard Analysis: main disturbance

IMMEDIATE CAUSES	INADEQUATE NORMAL CONTROL	PROCESS DEVIATION	FAILURE TO CONTROL (ON ALARM)	
Low output of BFW from pumps				
Low level in storage	Failure to recover situation on storage LRC or supply	Low flow with alarm followed by low level		
Leak on inlet to pump	Failure to identify spill	Low flow with alarm followed by low level	Operator fails to restore flow of BFW	
Incorrect valve status	Inadequate use of procedure	Low flow with alarm followed by low level	OR	
Pump failure		Low flow with alarm followed by low level	Operator fails to restore steam generation	
Leak on outlet		Low level		
Control fault on LRC loop of drum	Inadequate LRC on drum	Low flow with alarm followed by low level		
Major leak on accumulator	Inadequate maintenance or external cause	High flow with alarm followed by low level		

HAZARDOUS DISTURBANCE	INADEQUATE EMERGENCY CONTROL	DANGEROUS DISTURBANCE	FAILURE TO RECOVER SITUATION	SIGNIFICANT EVENT
Low-low level in accumulator with LLA	Operator fails to shut down whole plant in time	Waste heat boiler tubes exposed	Shutdown action too late	Rupture of boiler tubes on overtemperature

4. CRITICAL EXAMINATION OF SYSTEM SAFETY

INTRODUCTION
Process safety is dominated by consideration of what might go wrong. If a problem is identified then it is reduced in its effect by adding appropriate instrumentation and control. It is seldom that a more positive approach is taken which involves major change. This is partly due to failure to consider radical alteration at an early stage of design when it is appreciably less expensive to make modifications and there may be less pressure on project time. However, should an accident happen because such a change was not made then the costs can be very great. This not only includes any damage, replacement and compensation but also the costs of the investigation, loss of production, increased insurance and so on.

Suggestions abound in the literature about making plants inherently safe and user-friendly. For example, deterministic preventative systems make the plant move in a safe direction in the event of a specific fault sequence. Plant can be selected which is easier to control with responses which are faster when needed and slower or less steep to change for dangerous trends. A less hazardous chemical can be substituted for a chemical considered more dangerous. A simple plant has fewer units, less points of leakage and its control is readily understood by the operators.

The Critical Examination of system safety gives a structure to making such suggestions with the declared belief that there must be a better and less dangerous way to carry out the process.

CRITICAL EXAMINATION
Method Study grew out of time and motion studies in the 1950s. The process was examined in sections to see where improvements could be made and questions were asked about the efficiency of the method. A principal component of this new method was the technique known as Critical Examination. It used a questioning approach asking What-When-Where-Who-How-Why questions as summarized in Table 4.1.

The technique began to be applied to automatic operations which did not directly involve activities by operators, and continuous chemical plants were examined to see if by alteration it was possible to reduce capital and operating costs. Occasionally attempts were made to consider safety but the technique did not really suit identifying what could go wrong. It was oriented more towards what could be done better. It did, however, lead to the major breakthrough known as 'Hazop'.

During the 1970s Hazop was developed further. At the same time so was job safety analysis. Undoubtedly such methods led to work being carried out more safely. They also led to considerable extra design and study effort with an increase in the use of safety procedures and quality control. They increased the requirements during production activities. Sometimes all the additional protective systems served as a hindrance to carrying out the task. For example, some

TABLE 4.1
Critical Examination record sheet

Proposal	Alternatives	Conclusion
What is to be achieved? What is its purpose? Why is it necessary?	What else could be achieved? Why not ... ?	What should be done? Why that way?
When is it to be achieved? Why then?	When else could it be achieved? Why not ... ?	When should it be done? Why that way?
How is it to be achieved? How can it be done better? How might it be done worse?	How else could it be achieved? Why not ... ?	How should it be done? Why that way?
Where is it to be achieved? Why there?	Where else could it be achieved? Why not ... ?	Where should it be done? Why that way?
Who achieves it? Why them?	Who else could achieve it?	Who should do it?

protective clothing is bulky and awkward to move around in; spark-proof tools are less effective at gripping. Also when the pressure is on to keep up production and complete a task then people carrying out the duties may ignore those features of a task which are not perceived as essential.

The current approach in design is to analyse all features related to the process step or task step. It must build in safety features during the design phase and ensure that appropriate working practices can be followed. But it may not stop and study some of the basic concepts and ask if the operations could be done better.

Critical Examination is very different from any other method used in subsequent safety studies. The impetus is to recommend change rather than maintain the *status quo*. Despite this there remains the tendency with any design to keep along the same process path as first envisaged by the designer. Thus the reaction conditions might first be selected and the rest of the system developed with these in mind. An incorrect optimization of one section dominates optimization of the whole, despite separation costs greatly exceeding other costs on most chemical plants. The other difficulty is that there remains a tendency to use add-on safety measures as the first solution.

Some features suggested by Critical Examination are suggested by other methods of hazard and task analysis. This is excellent news as any search procedure should overlap with others. However, for a complex system having found one optimum, it is vital to recommence the search from a different starting point and where appropriate use a radically different search procedure.

It does not make sense to carry out a Critical Examination solely considering safety criteria. Risk is a complex criteria involving:
• the financial risk in terms of loss of production and production capability;
• the safety risk in terms of harm to people and damage to property and the environment;
• the damage to the business for either reason.

Many problems affecting safety occur when the plant is not functioning normally and both automatic control systems and operators find themselves working in areas in which they have little previous experience. Also there is considerable pressure on the people to get the plant back on-line.

This chapter looks at large chunks of the process or considers special studies of high risk operations. The selection of tasks is discussed in more detail when Task Analysis is considered in Chapters 7 and 8.

THE GOAL OF ANALYSIS
The overall goal of the operation to be carried out should be stated. A first activity in Critical Examination is to write down a statement of the design intent describing the function and the method, and stating clearly what is to be done or achieved and how it is to be accomplished. This should be noted in a sentence or so which identifies in minimum detail the change of state achieved by an operation, reaction or activity. This would normally indicate the operating conditions and the equipment involved but not further details.

For some processes or task activities, additional statements about each step may be necessary covering the What, Where, When, How and Who of the proposal. If the plant is not in normal operation for the purpose of the study, then this must be stated. A statement may be frequently added indicating any dangerous condition, defined here as one leading to a dangerous disturbance of plant.

Each significant phrase (or even word) of the achievement is probed by asking 'Why?' (sometimes repeatedly) to query the proposal or existing facts and the purpose of the proposal. The aim is to expose the strengths and weaknesses of the present situation.

It has been found in a study of an overall process that the key benefits usually stem from examining how the process step is achieved, paying particular attention to the following.
- Materials:
— consider a change in the quantities or qualities of streams;
— use extra or different process materials.
- Method:
— change the operating conditions or activities;
— change the sequence, frequency, absolute time or duration.
- Equipment:
— use different equipment.

The impetus for change should be to carry out an operation better, to make the frequency of a safety, quality or production lapse less likely, and to lessen the consequences should this occur. The presence of features which make the plant/process inherently safe in the case of a fault sequence is to be encouraged. The costs of effecting the change must be evaluated, and safety and production requirements will at times be at odds.

The analysis of the plan should lead to improvements in such features as:
- working methods, procedures, practices;
- work environment, exposure frequency;
- communication and information processing;
- skill and capabilities of people;
- management and control.

A Critical Examination can be used to reduce capital and operating costs, improve production efficiency and output, boost quality and increase process and occupational safety. The changes can affect people, materials, equipment and both working environment and environs. Unfortunately it is almost certain that every change will not move each variable in the same direction, so before every decision all the factors must be considered. Experience and common sense stops this becoming a daunting prospect.

IDEAS FOR CHANGE

BRAINSTORMING

The classical method for generating ideas is termed 'brainstorming'. For example, if the objective is to design a plane to 'cut wood' a meeting would be held in which all the alternatives would be put forward which can carry out this function. They would vary from the obvious — knife, razorblade, saw, axe, chisel — via the fairly reasonable — rasp, sandpaper, grinder — to the extraordinary — beaver's teeth, lasers, woodpeckers. Each of these is then costed and studied for development. Note how some lead to similar products — for example, sanders and surform planes.

SUCCESSIVE QUESTIONS

Ideas emerge from successive questioning of a design intent; Table 4.2 presents an example. The table was developed in this form by Elliott and Owen[3]; their paper is still recommended reading

TABLE 4.2
Generating alternatives by successive questioning

Successive questions	Alternative ideas generated with emphasis on elimination
Design intent: A distillation column is required to separate compounds A and B	
Why distil? To separate A from B	(i) Separate them some other way — for example, by fractional crystallization (ii) Don't separate them at all
Why separate? Because the recycle reactor won't crack A mixed with B	(i) Find an alternative market which will take A and B (ii) Change the process so that it does not make B
Why not crack? Because the furnace temperature isn't high enough	(i) Change the reactor conditions so that A and B can be cracked
Why this temperature? Because tube materials won't stand a higher temperature	(i) Find another tube material to withstand higher temperature (ii) Find a catalyst to permit cracking at a lower temperature

for all who use Critical Examination. Successive questioning is easy to apply and can be readily followed as a group activity. Further examples will be presented later in this chapter. In the example in Table 4.2, it would be easy to incorporate safety considerations by asking whether the change introduced is a safer as well as a more economic solution.

CHECK-LISTS AND KEYWORDS

It is also possible to generate options when examining the design intent by using keywords such as those given in Table 4.3, pages 79–81. This directs the study into a more judgmental format.

The list was developed with an attempt to avoid excessive repetition. The keywords are extremely powerful and normally applied in the sequence presented. Thus the first aim of the design should be to *eliminate* or *substitute* dangerous substances, processes or operations. In some cases it may be best to *combine* certain activities or plant items. To *avoid* is a less stringent action, and means that it may be possible to evade certain conditions or actions deemed to be undesirable. *Modify, alter and rearrange* and *improve* are clearly very general suggestions. They suffer by not giving a specific direction of change as compared to *simplify, increase, reduce, prevent, segregate* and *isolate*.

The applications given here are process oriented. Once the user gets familiar with the way of thinking, the use of the keywords is all that is needed to generate alternatives. Certainly it would be a great mistake to extend them or make the approach more formal; this would return to the procedure used in the 1960s which became tedious and too time-consuming for all but critical tasks.

GOALS

The method is usually applied to studies of design intent. However it can be applied to many situations such as arise in the following management goals:
- hazard reduction;
- inventory reduction;
- inherent safety;
- reduced exposure limits;
- fire safety;
- material handling;
- total environment;
- risk reduction;
- emergency action.

Some of these areas may be tackled when subject to review.

Two case studies are considered in this chapter. The case of the methanator tackles the problem early in the design. The study on the rig was part of a much later review and probes suggestions for risk reduction in greater depth.

CRITICAL EXAMINATION OF SYSTEM SAFETY

A methodology has been developed for the Critical Examination of system safety (CESS). This consists of the steps given in Table 4.4, page 81.

TABLE 4.3
Critical Examination: keyword dictionary

Keyword	Examples of use
ELIMINATE/ SUBSTITUTE/ COMBINE	• eliminate unnecessary materials or those posing a major hazard • eliminate all unnecessary details • substitute chemicals, change all or part of a route, use a lean technology • eliminate or substitute additives, solvents, heat exchange mediums • change by elimination all or part of the equipment or processing method • eliminate leakage points: use a weld not a bolted fitting, etc • eliminate a prime mover or heat exchanger or agitator • eliminate a separation stage or combine steps • eliminate a step by changing the sequence or combining tasks • eliminate intermediate storage, an installed spare • eliminate manual handling • eliminate sneak paths, openings to atmosphere • eliminate waste, emissions, effluents or products that are harmful in use • eliminate an ignition source, particularly a permanent flame • eliminate ingress of impurities into the system • eliminate common connections between process and utilities • eliminate hazardous events that can result from turning a single valve
AVOID	• avoid extremes of operating conditions • avoid operating in flammable atmospheres • avoid possible layering of materials, inadequate mixing • avoid flashing liquids, particularly in extensive heat exchanger networks • avoid production of large quantities of dangerous intermediates • avoid unwanted reactions in and outside reactors • avoid operating near extremes of materials of construction • avoid operating conditions leading to rapid deterioration of plant • avoid maintenance on demand and in short time periods • avoid critical transport of materials • avoid items of plant readily toppled by explosions • avoid stage, step or activity by doing something as well as or instead • avoid storage of material that spontaneously combusts • avoid static accumulation of hazardous materials • avoid adding materials during start-up/shutdown not normal to the process
ALTER/ REARRANGE	• alter the composition of waste, emissions and effluents • alter or rearrange the sequence, method of working • alter the aims of activities • alter process conditions, materials, phase, composition • alter the time or duration of an activity (faster/slower, earlier, later?) • alter the frequency of an activity (more/less, why then?) • alter quality, quantity, rate, ratio, speed of any part of an operation or activity • alter who does an activity (why them?, more/less people?) • alter additives, solvents, heat exchange mediums

TABLE 4.3 (continued)
Critical Examination: keyword dictionary

Keyword	Examples of use
ALTER/ REARRANGE (continued)	• alter equipment or processing method • alter process units elevation • alter response of process to change • alter working environment • alter system to eliminate deadspots • alter operations that may involve unwanted side reactions • alter products that are harmful in use
PREVENT	• prevent emissions and exposure by totally enclosed processes and handling systems • prevent waste and damage of raw materials and people • prevent exposure by use of remote control, personal protection
INCREASE	• increase separation efficiency • increase process toleration and flexibility • increase separation/segregation distances • increase safety margins, propagation time • increase use of fail-safe equipment • increase use of interlocks to ensure procedural sequences are followed • increase conversion in reactions
REDUCE	• reduce opportunities for operator error • reduce possible damage and downtime, knock-on and domino effects • reduce inventory; less storage, hold-up, smaller size of equipment, less piping, particularly for chemicals posing a major hazard • reduce amount of energy in system • reduce pressure and temperature above ambient • reduce exposure to hot surfaces • reduce fugitive emissions • reduce emissions and exposure by improved containment, piped vapour return, use of covers, condensation of return, use of reactive liquids, wetting dust • reduce noise and vibration levels • reduce frequency of opening, improve ventilation, change dilution or mixing • combine activities at shutdown • reduce size of possible openings to atmosphere • reduce quantity of information to operator • reduce levels of toxics • reduce generation of static electricity • reduce effects of utility failure and those on associated plant • reduce use of unconventional equipment
SEGREGATE	• segregate by distance, barriers, sheaths, duration and time of day • segregate plant items to avoid certain common-mode failures • segregate fragile items from roads, etc • segregate electrical cables from pipes containing hot or hazardous materials

TABLE 4.3 (continued)
Critical Examination: keyword dictionary

Keyword	Examples of use
ISOLATE	• isolate plant by shutdown systems, emergency isolation valves
MODIFY/ SIMPLIFY	• modify any topics above • simplify all necessary details • modify batch operation to continuous operation or vice versa
IMPROVE	• improve total quality control • improve plant integrity, reliability and availability • improve ventilation, lighting and clean lines • improve control or computer control. Select user-friendly controls • improve protection by guards, protective equipment, safety devices • improve transport and handling of materials • improve response • improve efficiency of use of equipment • improve quality of engineering, construction, manufacture and assembly • improve quality of working environment • improve training, better utilization and motivation of people • improve knowledge and skills of people • improve procedures and working practices • improve site facilities for normal and emergency use • improve quality of information about process operability to operator

TABLE 4.4
The methodology for Critical Examination of system safety

Partition the plant or procedure and select a section or task step
State the design intent: avoid too much detail
Partition significant phrases or words of the design intent
Why do it? • question each significant phrase or word in turn, justifying the proposed design • question the responses • generate alternatives with the aim of improving process safety (and economics if required) • make comments and recommendations
What can go wrong affecting safety, production or quality? • identify dangerous conditions associated with the design intent and their causes • generate alternatives, modifications and controls • make comments and recommendations
Report and recommend actions

A study team should be assembled including representatives from both design and operations who can cover the main disciplines involved. Team members should be senior enough to make on-the-spot decisions and be able and willing to attend a number of meetings. A recognized safety specialist may prove advantageous and, of course, individuals with expert knowledge of specific areas can be called in as and when required. It may not be practicable to involve those who do the work — that is, operators. The value of their possible input should be recognized, however, and at some stage reference made to their opinion.

Such studies require careful chairing to stop the meeting getting bogged down or rambling too widely. Although some brainstorming is acceptable, it must be constrained by some judicial thoughts. Also it is all too easy for strong conflicts to arise. All changes proposed to improve safety, production and quality will certainly not pull together; either economics or safety may suffer. Given good guidance and common sense it is possible to obtain sensible and useful results.

CESS is split essentially into two sections. First the proposal is questioned to generate alternatives. Then dangerous conditions are identified and further options, modifications and controls are suggested. The study should make use of appropriate task sheets to carry out the work. It is vital that a summary be prepared giving clear information on all actions.

TABLE 4.5
Analysis of a reactor system

Plant: Hydrogen plant	**Project:** Case study	**Drawing no:** Figure 2.4, page 32
System: Methanator	**Location:** Sheffield	**Date:** 1.6.94

Process step: Methanator R101, a fixed bed heat exchanger E101

Design intent: What is to be done, and how. A process having a fixed bed catalytic reactor operating at 20 bar and 400°C inlet, 450°C outlet is to convert small amounts of oxides of carbon (max 2%) in a stream of hydrogen into water to give a hydrogen product stream containing 10 ppm max of COx.

Why do it?	**Generated alternatives**
(1) Why remove oxides of carbon? Oxides of carbon react with the downstream catalyst on the aromatics plant. The methanator is provided as part of the plant specification for the hydrogen plant.	Eliminate the unit here and install it on the aromatics plant. This could provide material at reaction temperature for the aromatics plant.
(2) Why this aromatics plant catalyst? Specified by licensers.	Development work could be carried out to produce a catalyst less readily poisoned
(3) Why this process? It involves no further addition of materials and is applied as a finishing process.	Remove all oxides using absorption by changing the absorbent or using an extra stage. Use a less hazardous process such as pressure swing adsorption or membrane filtration.
(4) Why this catalyst and operating temperature? The reaction goes to completion at this temperature and the use of the size of bed has been optimized.	A different catalyst could be developed which methanates at a lower temperature

CRITICAL EXAMINATION OF SYSTEM SAFETY

CASE STUDY — CRITICAL EXAMINATION OF A METHANATOR

A study has been carried out of the methanator considered in Chapters 2 and 3. This study does not challenge the proposed location or layout. Only the reaction section is considered. This reaction is a finishing process designed to remove oxides of carbon. The reaction is exothermic going to completion. Near-miss incidents have arisen at other locations when excess oxides of carbon enter in the feed.

The information in Table 4.5 suggests how the process might be done better. This is a simplified sheet leaving out comments and recommendations. Some suggestions might prove worthwhile on a new design. It is doubtful whether any would have been generated by a conventional safety study.

The study continues in Table 4.6 to look at what might go wrong, and in this case the study is directed at plant safety. Note that all these features should be generated by a Concept

TABLE 4.6
Further analysis of the reactor system. What can go wrong, affecting safety, production or quality?

Dangerous deviation and does it matter?	Can it be avoided or modified and what must be done?
Very high temperatures are generated by excess oxides in feed to reactor	Consider an increase in the capacity of the absorber system. An extra unit is uneconomic. Improve the reliability of the absorber.
Reactor failure could arise given a very high temperature in the reactor	Consider whether to improve the metallurgy of the methanator to withstand the maximum temperature generated. Consider emergency quench on the methanator.
Reactor requires shutdown on high temperature	Consider interbed sensors to activate trip. Spurious failure of this system would be very annoying so appropriate reliability must be built into the system.
The methanation installation should be shut down on high temperatures which means that downstream plants must be shut down	Keep the upstream plant operating whilst methanator difficulties are resolved. Evaluate carefully the sensitivity of downstream plant to oxides of carbon to assess its shutdown needs.
No reaction if methanator bed is below 350°C. Off-specification material affects downstream plant so product supply and compressor should be stopped.	Ensure either the feed can at all times be preheated to 350°C or the bed can be prewarmed safely. Plant should be able to run at start-up by venting impure hydrogen with the methanator off-line. On placing on-line, product should be tested prior to use downstream.
Rupture releases flammables which self-ignite and torch	Check for impingement on other vessels and improve segregation between plants if practicable. Note that the flame may be invisible and train operators accordingly.

Hazard Analysis. Nevertheless the emphasis on change generates some different solutions, although here the emphasis is more on add-on safety. Just one useful idea at this stage can be very beneficial, as the process and plant design are still fluid.

The methanator is not an ideal subject for study as it is an add-on to the main hydrogen plant. Nevertheless the ideas in Table 4.6 for changing its location and strengthening the vessel seem useful.

CASE STUDY — MODIFICATION OF A GAS EXPORT RISER ON AN OFFSHORE RIG

It is vital that, at some point in any design, the opportunity is taken to generate change in order to improve process safety, production or quality. This applies regardless of whether the project is a new plant or a modification. In this case study a Fire Risk Analysis on an existing North Sea oil rig had identified that there was a serious chance of very large jet fires as a result of a release from the gas export riser. Such fires may cause structural failures of the double pipe (DP) steel jacket in which it is located. The riser can be isolated by emergency shutdown valves (ESDV) on board and a subsea isolation valve (SSIV) at the pipeline end manifold (PLEM). Shutdown is instigated manually on the detection of a release, either on hearing or smelling escaping gas or by the activation of the on-board gas detectors. Following detection there is still the possibility of a jet fire as the contents depressurize from 180 bar over a period of, say, ten minutes for a 50 mm hole.

The initial recommendation was to install jacket passive fire protection on the DP jacket so that the temperature of the structural jacket members in a 300 kW/m^2 jet fire would be kept below 300°C for a minimum of 30 minutes. Such protection serves to mitigate the effects of an incident so that personnel can be evacuated. But it does little to reduce the likelihood of an incident.

A Critical Examination of the design intent established the validity of the current safety systems (Table 4.7). The study was continued to generate ideas for methods of reducing the frequency of the undesired event and of mitigating the consequences (Table 4.8, page 86). The use of an open-ended sheath (Table 4.9, page 87) tackles some of the original safety issues but does not properly address the issue of ducting a gas release to deck level and the difficulty of inspecting the pipeline.

Finally the logical progression of enclosing the gas export riser by a concentric envelope or sleeve was examined (Table 4.10, page 88). The sleeve might extend from the topside at 23 m to below the water level at 10 m. It should be sealed top and bottom and filled with pressurized nitrogen. The pressure should be monitored with high and low pressure alarms set to indicate a gas leak from the riser and from the envelope. This gas leak should be contained by the envelope pipe. By surrounding the gas riser with a heavy steel sleeve the external surface of the riser is protected from the sea water environment and the internal surface of the envelope is protected from the export gas environment. The presence of nitrogen eliminated the problem of outer wall corrosion of the riser. The failure frequency of the riser has also been reduced by a reduction in the number of welds. Whilst welds on the end seals are added, the welds to guides on the supporting caisson have been removed. The failure of the system may have been reduced by two orders of magnitude. This work was carried out by Steve Kirby as part of an MSc dissertation at the University of Sheffield.

TABLE 4.7
CESS report of current design

Installation: STERIG			Ref: CE\1.1	
Subsystem: Gas export riser			Study team:	
Design intent: A gas export riser (250 mm diameter) delivers process gas (181 bar) via the subsea pipeline end manifold (PLEM), passing through the DP jacket.			SPK	
No.	Why do it?		Generated alternatives	Ref.
(1)	Question	Why does it pass through the jacket?	Move the position of the gas riser further from/outside of DP jacket	A.1/B.1
	Response	In order to provide protection from impact and collision		
(2)	Question	Why at 181 bar?	Reduce pressure by altering process optimization	A.1/B.1
	Response	Set in order to achieve desired delivery		
(3)	Question	Why that pipe size?	Reduce pipe size/use two pipes	A.1/B.1
	Response	Economics: in order to achieve desired throughput		
No.	Comments and recommendations			Action
(1)	Probable increase in risk of release due to collision/impact; therefore, no overall safety benefit is considered likely. Reject.			
(2)	No real advantage considered, plus process considerations restrict use of option. Reject. Neither alternative actually reduces the hazard. Reject.			
No.	What can go wrong affecting safety, production or quality?		Modification and control	Ref.
(1)	Dangerous condition	Pipe leak/pipe rupture leading to jet fire impinging on DP jacket	Provide passive fire protection	C.1/D.1
	Cause	Corrosion/collision/impact		
No.	Comments and recommendations			Action
(1)	Cost and weight implications prohibitive. Does nothing to reduce the frequency of the dangerous condition. Safety implications in making move. Reject.			

TABLE 4.8
CESS report on mitigation by current design

Installation: STERIG			Ref: CE\1.2	
Subsystem: Gas export riser			Study team:	
Design intent: A gas export riser is isolated via an emergency shutdown valve (ESDV) on board and a subsea isolation valve (SSIV) at the pipeline end manifold (PLEM). (Separation distance 400 m.)			SPK	
No.	Why do it?		Generated alternatives	Ref.
(1)	Question	Why is ESDV on board?	Move ESDV down the gas export riser	A.2/B.2
	Response	In order to enable good accessibility		
(2)	Question	Why is the SSIV at the PLEM?	Move the SSIV up the riser	A.2/B.2
	Response	Regarded as the optimum position		
No.	Comments and recommendations			Action
(1)	In order to comply with regulations the ESDV must be on board in a position of 'good access'. As such the ESDV cannot be moved. Reject.			
(2)	Moving the SSIV up the riser increases the risk of damage to the valve from dropped objects. The platform is also at an increased risk from releases downstream of the SSEV. These increases in risk outweigh any advantages from moving the SSIV. Reject.			
No.	What can go wrong affecting safety, production or quality?		Modification and control	Ref.
(1)	Dangerous condition	Pipe leak/pipe rupture leading to jet fire impinging on DP jacket	(a) Increase thickness of pipe wall (b) Connect to blowdown system (c) Surround riser in open-ended sheath	C.2/D.2
	Cause	Corrosion/collision/impact		
No.	Comments and recommendations			Action
(1)	(a) Simple approach. Reduction in failure frequency due to impact/corrosion is likely. However, an increase in failure frequency is possible due to material defect. Benefits must be assessed. (b) Reduces blowdown time of the section, but only serves as an emergency response to an actual release. It would also require a connection to the gas riser itself. This is not a favoured approach. (c) Prevents direct impact of the jet fire on the jacket, along with providing a possible decrease in the ignition probability. Further investigation required.			A.N.O. A.N.O.

TABLE 4.9
CESS report on use of open-ended sheath

Installation: STERIG		Ref: CE\1.3	
Subsystem: Gas export riser		Study team:	
Design intent: The gas export riser should be encased in an open-ended sheath.		SPK	
No.	Why do it?		Generated alternatives	Ref.
(1)	Question	Why a sheath?	Increase the thickness of pipe wall	A.3 B.3
	Response	In order to prevent jet fire impact on the jacket		
No.	Comments and recommendations			Action
(1)	Increasing the thickness of the riser wall is considered unlikely to reduce the failure frequency by any considerable amount (less than an order of magnitude). Reject. The sheath offers advantages but it should be noted that it only acts as a 'last line of defence' — that is, in mitigating consequences of a release, it does not prevent a release to atmosphere.			
No.	What can go wrong affecting safety, production or quality?		Modification and control	Ref.
(1)	Dangerous condition	Release with delayed ignition	Enclose ends of sheath	C.3 D.3
	Cause	Pipe leak/rupture due to corrosion/impact/collision		
(2)	Dangerous condition	Corrosion	Enclose ends of sheath	C.3 D.3
	Cause	Severe working environment		
No.	Comments and recommendations			Action
(1)	The disadvantages of the open-ended sheath — that is, that of ducting gas to deck level — can be overcome by enveloping the riser completely in a sheath. Further investigation warranted.			A.N.O.
(2)	As previously, with the added advantage of protecting the riser from the severe working environment, thus eliminating releases due to corrosion.			A.N.O.

TABLE 4.10
CESS report on complete enclosure

Installation: STERIG		Ref: CE\1.4	
Subsystem: Gas export riser		Study team: SPK	
Design intent: The gas export riser will be enclosed by a concentric envelope, sealed top and bottom. The space between the riser and envelope will be filled with pressurized nitrogen.			
No.	Why do it?		Generated alternatives	Ref.
(1)	Question	Why enclose the riser completely?	Refer to CE\1.1, CE\1.2 and CE\1.3 for alternatives and reasons for rejection	A.4
	Response	Reduces the failure frequency and consequences of a release		
(2)	Question	Why filled with pressurized nitrogen?	Use alternative inert	A.4
	Response	Eliminates corrosion and allows monitoring for leaks		
No.	Comments and recommendations			Action
(1)	Failure frequency is reduced as the outer wall of the riser is no longer exposed to sea water, hence corrosion is eliminated. The containment also means the consequences of a release are mitigated, as released gas cannot pass to atmosphere, and thus, controlled shutdown can be achieved.			
(2)	No advantage is seen in changing the inert. The inert atmosphere aids the elimination of corrosion. Monitoring of leaks is possible as the annulus pressure would alter. See below.			
No.	What can go wrong affecting safety, production or quality?		Modification and control	Ref.
(1)	Dangerous condition	Undiscovered leak into the space between the riser and the sleeve	Improve the detection system by installing pressure monitors on the envelope	C.4/D.4
	Cause	Pipe rupture		
(2)	Dangerous condition	Uncontrolled leak into the space between the riser and the sleeve	Modify the sleeve system to include a venting system	C.4/D.4
	Cause	Pipe rupture		
(3)	Dangerous condition	Large temperature difference between the riser and the envelope	Reduce the temperature drop so that it does not pass below $-10°C$ during riser blowdown	C.4/D.4
	Cause	Low sea temperature and high gas temperature or riser blowdown		
No.	Comments and recommendations			Action
(1)	Pressure monitors linked to an alarm would allow leaks in either the riser or envelope to be detected via fluctuations in pressure.			A.N.O.
(2)	Venting system would allow safe discharge of gas.			A.N.O.
(3)	Design of forged end seals should be such that a substantial safety margin exists to account for forces due to differential expansion.			

This case study has centred around the need for a plant modification to a section of an offshore installation. It is arguably the case that ill-considered modifications to existing plant (along with maloperation and inadequate maintenance) are a far more common cause of accidents than modifications attributable to the design. Where the design does often fail is in presenting the opportunity for accidents to occur, as well as discounting the fact that the consequences of such incidents are often made more severe by the original design specifications.

This updated technique has advantages in both style and impetus over the more traditional safety reviews. The need for no more than limited process information is the key to the approach; it enables examination to begin at an early stage. Numerous options in relation to an activity can be considered at a reduced cost, and at a time when implementation is relatively easy. The impetus of the study is set by the questioning approach adopted, the essence of which is to discover problems and their fundamental causes in terms of the design philosophy. It has been shown to work both for conceptual design and for a modification.

5. HAZARD AND OPERABILITY STUDIES (HAZOP)

WHAT IS A HAZOP STUDY?

A hazard is something which may cause harm and is an undesired event. An operability problem is a nuisance, or a break in production, with 'no' or minimal safety significance. A Hazop study is a formal, systematic examination of a processing plant in order to identify hazards, failures and operability problems, and assess the consequences from such maloperation. Usually this leads to fewer lapses in safety, quality and production provided that the plant is installed according to the design and maintained in appropriate condition.

Hazop is the most widely used method of analysis used in the process industries. It is recommended for use by legislators, regulators and the engineering institutions. This virtually mandatory use has stemmed from its proven capabilities for this task.

The basic philosophy behind the Hazop technique is that if a process operates within its intended design philosophy then undesired hazardous events should not occur. It is a primary assumption that the original process design and the equipment standards applied are correct. Such matters can be challenged during the Hazop should the team feel this is necessary. But the basic premise remains and the Hazop study aims mainly to identify how process deviations can be prevented or mitigated to minimize process hazards. A Hazop is capable of identifying most process deviations although all the causes and consequences of these deviations may not be diagnosed. It suggests necessary changes to a system to meet company risk guidelines or suggests procedures or changes for eliminating or reducing the probability of operating deviations.

The study is not comprehensive and any individual or corporate effort yields results directly proportional to the appropriate background experience of the participants. The system accommodates the use of standards and codes of practice yet examines their relevance in specific circumstances of plant operation.

WHY DO A HAZOP STUDY?

A Hazop study generates a list of identified problems, usually with some suggestions for improvement of the system. The results can be applied in a variety of ways. This may be as the input to a probabilistic safety assessment, as the basis for design changes, as further input to the development of operating instructions and procedures for use in training and as input to quality control and management standards. The study will obviously be much influenced by whether it deals with an existing or a new plant.

One of the most important matters to get out of a Hazop study is an intimate knowledge of the plant. Usually the act of carrying out a Hazop improves safety and quality by making people more aware of potential problems. It can also help to sort out loopholes and inconsistencies in procedures and force plant personnel to get their instructions up to date — both worthwhile achievements.

A Hazop is a means of demonstrating that a genuine effort has been made to improve the safety of the plant. Changes to the system to eliminate or reduce deviations in process operation will be recommended as a result. But this does not mean that all incidents and their causes will have been foreseen. It is necessary at some time in the future, say after three years, to go through the procedure again in order to update the study in the light of operating experience. Not only does this ensure that a thorough investigation of plant operability has been completed but it serves as a check that actual operating practice has not departed drastically from the design intent.

THE HAZOP PROCEDURE

A Hazop is carried out on a P&ID at the detailed design stage or as a check on the operability of an existing plant. The line selected usually runs from one main plant item to another, but frequently may include any heat exchanger in between. Main plant items are also considered. Other lines are noted as they join the main line. The method has also been applied to procedures.

Figure 5.1 (page 92) illustrates the Hazop process, from start to finish. This is a basic diagram which is refined later to take account of other factors. Going through each cycle the sequence is to identify the causes and consequences of each deviation. On completion of a cycle the procedure is to consider if another guide word is applicable and repeat the study. If not another parameter is considered and guide words applied again. The study continues until each node associated with the vessel currently being studied is completed. Then other sections are considered until the drawing has been totally examined.

A deviation matters if it has realistic cause and a related consequence which could create a hazard or operability problem. It is then often described as a *meaningful deviation*. In some cases it is necessary to obtain more information in order to determine whether there is a problem.

HAZOP GUIDE WORDS AND PARAMETERS

A Hazop uses parameters and guide words to suggest deviations of process variables and their causes. The deviation NO FLOW contains the guide word NO and the parameter FLOW.

The parameter selected is one relevant to the system under review. Typical parameters are flow rate, pressure, temperature, level, composition, flow quantity and a physical property such as viscosity. Flow, pressure and temperature are nearly always chosen first, often in that order.

The list of guide words and their meanings (Table 5.1, page 93) is basic and describes the intent of the vast majority of guide words. The literature contains many more.

A HAZOP EXAMPLE

In Figure 5.2 (page 93) a pressure controller is used to reduce the pressure of a mildly toxic gas from 20 bar to 5 bar over a small line running between two plant items. The relief system for the downstream section of plant is located immediately after the controller. It consists of a bursting disc which discharges gas directly to atmosphere off a short line at a height some 5 m from the ground. In view of the lack of information about this system, this example study by Hazop considers only one parameter — PRESSURE.

HAZARD IDENTIFICATION AND RISK ASSESSMENT

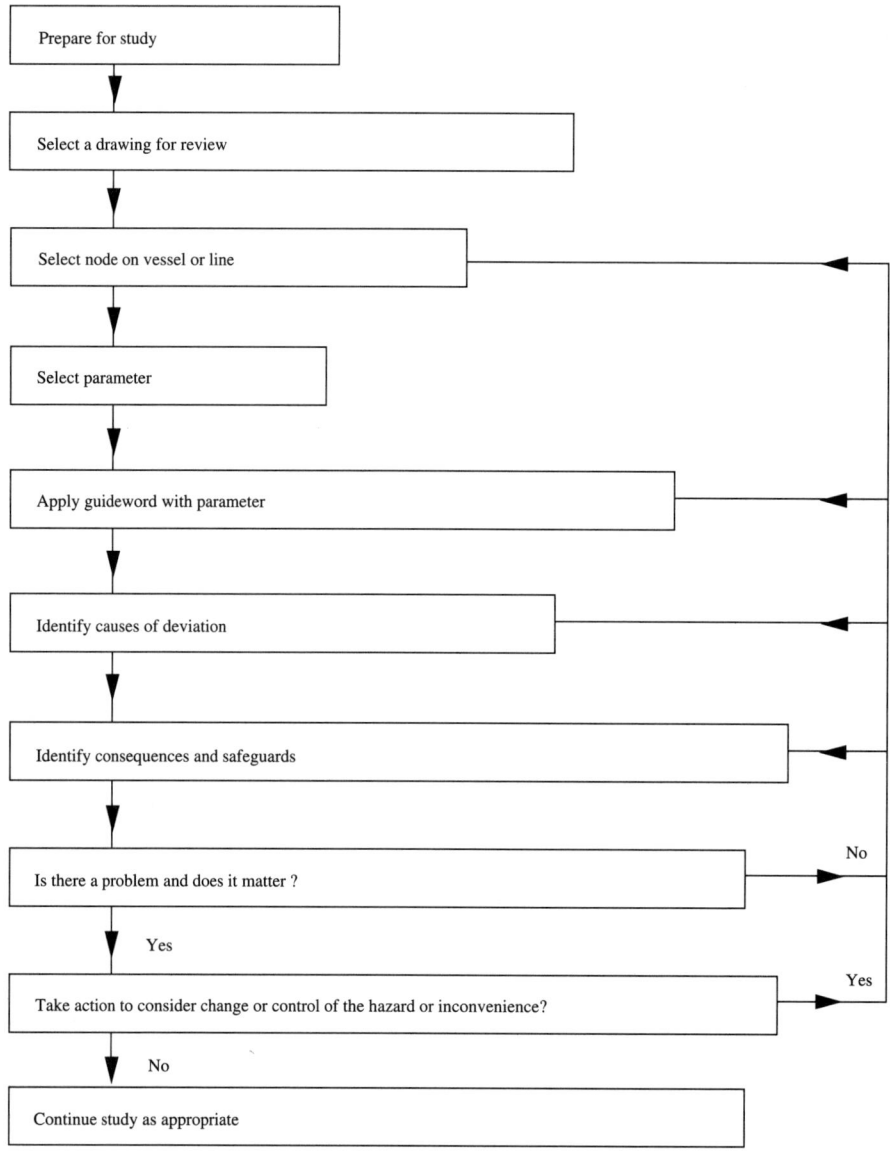

Figure 5.1   The Hazop study procedure.

IS THERE A PROBLEM?

The following account does not give a proper indication of the time and effort which will have been spent in going through significant deviations and identifying problems. When these are considered during a Hazop it is necessary to decide whether they are acceptable or whether change is required. In the large majority of cases this need can be evaluated using experience and judgement. Indeed this is one reason why a meeting format is essential. On some occasions, however, it is necessary to evaluate the risk. Further information about the technique to use is given later,

HAZARD AND OPERABILITY STUDIES (HAZOP)

TABLE 5.1
Basic guide words in Hazop

NO, NOT, NONE The activity is not carried out or ceases	No activity or operation takes place. There is no forward flow when there should be. A task may not be done, something may not be delivered or be there. There may be no action in response to an activating signal. A check is omitted. No catalyst present.
MORE OF A quantitative increase in an activity	There is more of something. More of any physical quantity than there should be — for example, of temperature, pressure, quantity or flow. More of a task can be carried out. An activity is done for a longer time.
LESS OF A quantitative decrease in an activity	There is less of something present. Less of an activity is carried out. Less time is taken.
PART OF Incomplete performance of an activity	Only part of an action is carried out. There might be a transfer of part of a load or batch. More components or an extra phase or impurities might be present.
REVERSE Inversion of an activity	Something happens backwards. A back siphon occurs. Heating rather than cooling occurs. This keyword can also be used to generate ideas as to how to recover from a situation.
OTHER (THAN)	A gas X can be sent down the line instead of gas Y. An operator might press the wrong button or open the wrong valve. This keyword is also used to identify what needs to happen other than normal operation — for example, start-up, shutdown, regeneration, maintenance.
AS WELL AS Another activity occurs as well as the original activity	Can buttons A and B be pressed when only A was meant to be pressed? Can both gas X and gas Y be sent down the line? What happens if valve D is adjusted while the system is running? What happens if the operator eats his lunch at the same time as packing cyanide ... ?
SOONER/LATER THAN An activity occurring at the wrong time relative to others	Every system has its running clock. What happens if task G is done before task K? What if the batch reaction is not completed in the normal time?

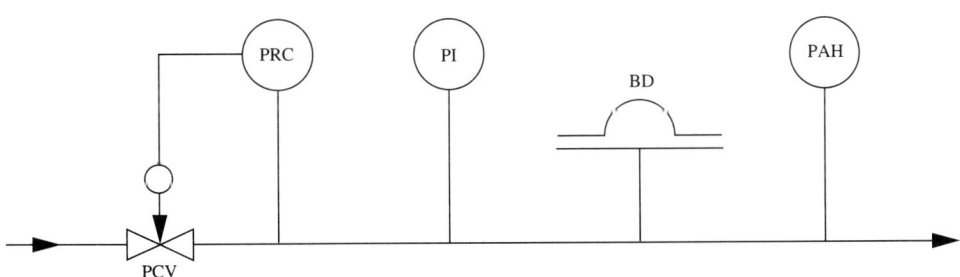

Figure 5.2 Hazop study of a pipeline. (See page 243 for instrumentation definitions.)

HAZARD IDENTIFICATION AND RISK ASSESSMENT

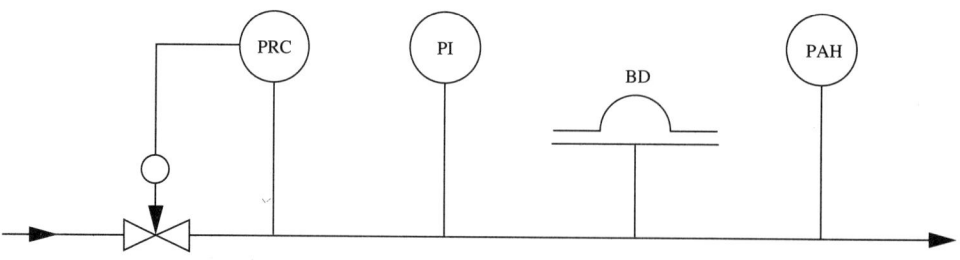

PARAMETER PRESSURE	MORE OF: PRESSURE TOO HIGH
CAUSE 1	PC system fails open
CAUSE 2	PC set incorrectly
CONSEQUENCES	Rupture of bursting disc
WARNING/PROTECTIONS	PAH, PI
Hazard/damage	Injuries to personnel with some hospitalization; classification Major (3)
Frequency	Fairly probable (10^{-2}/year)
RISK	**Unacceptable (1)**
MODIFICATION	Recommend vent disc to scrubber
CAUSE 3	Disc fails to discharge
CONSEQUENCES	Pipe bursts
PROTECTION	PAH, BD
Hazard/damage	Injuries to personnel, death of worker unlikely; classification Major (3)
Frequency	Very remote (10^{-4})
RISK	**Acceptable (−1)**
RECOMMENDATION	Suggestion of second disc and appropriate testing considered but rejected by team

Figure 5.3 Evaluation of risk in the pipeline study.

but for completeness a summary of the relevant study is given in Figure 5.3. It confirms the normal good practice that the bursting disc should be vented to a scrubber when material of this nature is discharged. Note that in the Hazop report the proposed vent scrubber is to be kept operational at all times. Spurious failure of the disc under normal operating conditions should be considered but is not noted here.

ACTION TO PREVENT THE DEVIATIONS
In this example the emphasis has been on the hazard, its mitigation and added safeguards. It may be considered important to examine the immediate causes of the deviation in more detail, considering the underlying causes such as incorrect fail-safe action, incorrect set point, instrument power failure, instrument air failure, other failures in the control loop and valve sticking. It may be necessary to examine the effectiveness of process relief through the disc and the reliability of the vent scrubber. The maintenance and testing of the different parts of the system have not been considered. Also, is further instrumentation necessary to identify when the bursting disc has

discharged, either on demand or on spurious failure? What procedures should be written to advise the operator on the action to take should the pressure be discovered to be high? Is further training required on this or on the setting of the PRC system? The Hazop team should make appropriate recommendations on these and other matters as it feels appropriate. It is not normal to identify underlying causes in the 'causes' column of the Hazop worksheet, as they may be addressed under 'comments' or 'safeguards'.

FURTHER GUIDANCE ON GUIDE WORDS AND PARAMETERS
No list of guide words and parameters can be exhaustive. Most experienced people have their own guide words which are chosen specifically for their own industry, type of process and the specific purposes of the Hazop.

The main parameters with their commonly used guide words are as follows:
- flow (no, more, less, reverse, sneak);
- temperature, pressure, level (higher, lower);
- protection, mitigation (none, too much, sooner, later, insufficient);
- composition, reaction, phase (no, more, less, as well as, part of, other than);
- reaction, phase (no, more, less, as well as, part of, other than);
- separation, addition (no, more, less, as well as, part of, other than);

TABLE 5.2
Expanded list of parameters

- corrosion, erosion (pitting, cracking, brittle fracture, thermal cracking, stress, fatigue)
- radiation (noise, thermal, ionizing)
- impact, stressing (internal impact, external impact, drop, overload, vibration, hammer)
- elevation (higher, lower)
- drawing (omission, sneak, inadequate)
- reliability, availability, redundancy, common mode failures
- integrity (mechanical, structural, civil, control systems)
- spare equipment (installed, non-installed, spares, test running)
- maintenance (inspection, monitoring, isolation, downtime, frequency, work permit system)
- services (utilities, telecommunications, heating and ventilation, computing services)
- emergency services (fire, police, ambulance)
- emergency action (protection, relief, discharge, isolation, mitigation, evacuation)
- computer system, instrumentation (control philosophy, location, response time, alarm and proof testing, auto/manual facility, fail-safe philosophy, common mode failures)
- electricity (current, voltage, abnormal operation, temperature, contamination)
- static, ignition (earthing, splashing, dust, flames, sparks, engines, stray currents)
- start-up, shutdown (initial, normal, partial, hot standby, emergency)
- checking, sampling, measurement
- labelling (secureness, lighting, climate)
- manning (manning levels, supervision)
- understanding (lack of understanding, technology transfer, training)

- viscosity, physical properties (higher, lower, emulsion);
- time, sequence (sooner, later, insufficient, longer, shorter);
- maintenance, testing (no, more, less, as well as, part of, other than);
- sampling, checking, measurement (no, more, less, as well as, part of, other than).

Further parameters which have been put forward are listed in Table 5.2, page 95. Some of these are used in every project. But if all were used on every drawing in a project, it is possible that the Hazop study would last longer than the life of the plant. Also mental fatigue and boredom are counterproductive. The Hazop team leader and the team members should decide which guide words to use in the project. It may be that team members will suggest a particular guide word to use during the discussion.

It is useful to have some information about what causes process deviations. Table 5.3 (pages 97–98) gives some typical causes of problems. Further information is given in structured form in the Hazcheck listing (Appendix 2).

A HAZOP CASE STUDY — HIGH TEMPERATURE ON A BENZENE PLANT

An exothermic reaction takes place in a fixed-bed reactor, R101 (Figure 5.4, page 99), which is used to convert toluene to benzene by hydrocracking in a stream of hydrogen. The feed to the reactor is preheated in a fired vaporizer.

The temperature controller, TRC, is set at 600°C. A high temperature alarm, TAH, is set at 650°C, and the emergency shutdown system, TZH, is activated at 700°C. Information on letter codes for instruments is given in Appendix 1, page 243.

On alarm the operator should check for a coil leak; such a leak requires the immediate manual shutdown of the plant. The alarm may also be caused by the failure of the TRC loop containing the fuel gas. If the temperature continues to rise, the flow of toluene to the plant should be stopped but not the flow of hydrogen which is mixed with the process feed just before the vaporizer. Hydrogen is provided in excess to avoid carbon formation.

If the temperature continues to rise, the shutdown system will be activated. This stops the toluene feed pump and closes the temperature control valve, TCV, on the fuel gas line. In the event of these actions not being effected, the hand control valve, HCV, on the fuel gas line should be closed and the toluene feed pump stopped manually.

The reaction inlet line has been examined for the deviation REVERSE FLOW; LESS CONCENTRATION. Table 5.4 on page 100 shows details of the study in abbreviated form.

The fuel gas line has been examined for the deviations MORE FLOW, LESS PRESSURE and MORE PRESSURE (Table 5.5, page 100). A further study of this system appears in Chapter 6 (see page 131). There is one common mode error which any competent engineer would eliminate. Can you identify it? What would you do to rectify the problem?

THE FORMAL CONDUCT OF A HAZOP MEETING

PLANNING THE STUDY

It is necessary to determine the duration of the study and the required finishing date. An average group of vessels such as those found in the examples in this chapter should take about four hours

TABLE 5.3
Some process deviations and their causes

Deviation	Typical problems
NO FLOW	Isolation in error — wrong routing — blockage — incorrectly fitted non return valve (NRV) — large leak — equipment failure (control valve, isolation valve, pump, vessel, etc) — incorrect pressure differential — delivery side overpressure — vapour lock — service failure.
REVERSE FLOW	Defective NRV — siphon effect — incorrect differential pressure — two-way flow — emergency venting — incorrect operation — pump reversed — service failure.
MORE FLOW	Increased pumping capacity — increased suction pressure — reduced delivery head — greater fluid density — exchanger tube leaks — restriction orifice plates deleted — cross-connection of systems — control faults — control set wrong — open bypass — more quantity — service failure — abnormal opening.
LESS FLOW	Line restriction — partial blockage — defective pumps — cavitation — fouling of vessels, valves, restrictor or orifice plates — density or viscosity problems — incorrect specification of process fluid — process turndown — less quantity — small leak — service failure, abnormal opening.
MORE PRESSURE	Surge problems — leakage from interconnected high pressure system — gas breakthrough — inadequate venting — thermal overpressure — failed open control valves — heating of blocked in system — explosion — fire — imbalance of input and output — external pressure — water hammer — positive displacement pumps.
LESS PRESSURE	Vacuum condition — condensation — gas dissolving in liquid — restricted pump/compressor suction line — undetected — leakage — vessel drainage — imbalance of input and output.
MORE TEMPERATURE	Ambient conditions — fouled or failed exchanger tubes — less cooling — cooling water failure — defective control — fire situation — reaction control failure — connected high temperature source — energy from machines.
LESS TEMPERATURE	Ambient conditions — reducing pressure — fouled or failed exchanger tubes — loss of heating — rain — connected source — Joule Thompson effect — auto-refrigeration.
DENSITY VISCOSITY	Incorrect material — incorrect temperature — extra phase — solids concentration
PART COMPOSITION	Phase change — incorrect feed — incorrect or reversed ratio — incorrect separation failures — change in reaction — emergency discharge — internal leaks — inadequate control — settling — lack of mixing — missing component.

TABLE 5.3 (continued)
Some process deviations and their causes

Deviation	Typical problems
CONTAMINATION	Incorrect routing — interconnected systems — effect of corrosion — wrong additives — ingress of air, water, lube oils — shutdown and start-up conditions — carry-over of solid or liquid — accumulations — inert gas failure — internal leaks.
CHANGE IN TIME OR OPERATION	Valves open too much, too long, too short, wrong duration, wrong sequence
OTHER ACTIVITIES	Start-up — shutdown — testing and inspection — relief sampling — service failure — planned abnormal operations such as purging, blowdown, catalyst activation, etc — maintenance — unusual emissions and effluents — static generation — external events.
EMERGENCY DISCHARGE	Feed change — contamination — unexpected reaction — continued reaction — incorrect ratio — duration.
LEVEL/LOAD	Overfilling — underfilling — swell — control failure — variation in loading — empty.
MEASUREMENT	Measurement error in temperature, pressure, level, flow, property.
PROTECTION	Spurious failure — protection missing or reduced — sabotage or hooliganism.
STRESSING	Loading in engines, motors, drives — excessive revolutions — bearing wear — extra phase — impact blows.
ABNORMAL OPENING	Incorrect status, failure of isolating device — opening for entry or discharge.
EXTERNAL HAZARDS	Impact — act of God — extreme weather — external interference — external event.

to Hazop; more if the process is completely new. The number of meetings should be restricted to about three a week and be no longer than five hours in duration excluding breaks. The location must have appropriate facilities for privacy, display and refreshment. Team members are selected and their availability is determined. They should include:
- an experienced Hazop team leader;
- people with knowledge and experience of Hazop;
- some people who are independent of the design;
- people with knowledge and experience of the process, plant operation and maintenance;
- specialists for specific features.

Team members are selected for their technical expertise, relevant know-how, interpersonal and communication skills, commitment and availability. Special team members may be called in from time to time according to the needs of the study.

HAZARD AND OPERABILITY STUDIES (HAZOP)

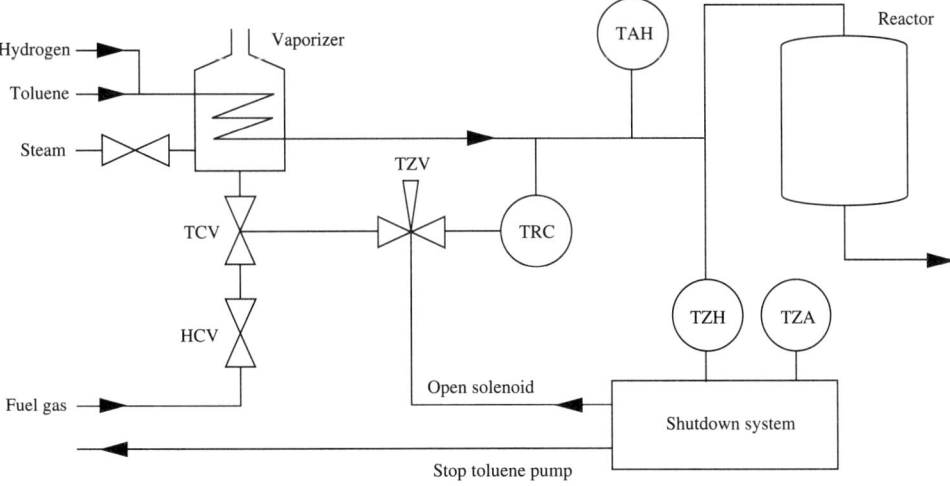

Figure 5.4   A benzene plant vaporizer and reactor. (See page 243 for instrumentation definitions.)

In any Hazop the technical expertise of the team members influences the outcome of the Hazop. It is also an opportunity to allow people inexperienced in Hazop to sit in on meetings, as this is the only valid way to learn the technique.

DEFINING THE OBJECTIVES AND SCOPE
In order to carry out a Hazop effectively, the objectives and scope must be clear. The purpose of the study and the fundamental assumptions must be clearly defined. It is essential to have a complete and accurate set of P&IDs and PFDs; for new designs they should be available at an early stage of detailed design. Other information of value is information on the process design and its intent, the process chemistry, process materials safety data sheets, plot plans and instrument control and shutdown logic diagrams. For batch plant a breakdown of the operations is needed. Data must be complete and accurate. Drawings should be the latest issue and consistent with the final design and existing site conditions.

The main objectives of a Hazop are usually to identify all deviations from the way the design is intended to function. The causes of such deviations are determined together with the potential hazards or operability problems. It must then be decided whether action is required to control the hazard or the operability problem and hence the way in which the problem might be resolved. Possible changes can then be recommended or additional information sought. Finally the actions are determined and followed up by the Hazop team.

The team leader must define precisely the scope of the study. The study may be concerned with a complex plant, a single vessel or an operating sequence. It is often important to clarify what is *not* in the study, such as streams outside battery limits or sewers. Particular care is needed to identify the mode of operation and maintenance to be considered, and when activities other than normal operation should be considered.

The Hazop leader describes the Hazop procedure briefly. Prior to the meeting it is helpful to issue guidelines, in a form similar to those described here. It must again be emphasized that

HAZARD IDENTIFICATION AND RISK ASSESSMENT

TABLE 5.4
Worksheet for Hazop on the benzene plant

Parameters:	1. Flow	2. Concentration	Node 3: Reactor inlet line	
Guide word	Deviation	Possible causes	Consequences	Recommendations
REVERSE	Reverse flow	Coil failure on vaporizer	Reactor contents feed fire	Re-locate block valves on smothering steam. Consider NRV on heater outlet. Obtain advice on best metallurgy for the coil.
LESS	Less of concentration	Reduced hydrogen flow	Carbon produced in reactor	Consider LFA on hydrogen flow and trip system turning off flow of toluene.

TABLE 5.5
Worksheet for Hazop on the benzene plant

Parameters:	1. Flow	2. Pressure	Node 2: Fuel gas line	
Guide word	Deviation	Possible causes	Consequences	Recommendations
AS WELL AS	As well as flow	Extra phase due to carry-over of aromatics and/or condensation in line	Spitting in heater and probable problems on start-up	Check layout to see if KO pot needed and trace line
LESS	Less pressure	Loss of fuel gas supply	Low temperature inlet to reactor with operator possibly unaware	Consider LPA on fuel gas. Consider diversity of fuel supply.
MORE	More pressure	Failure of PCV	High pressure gas enters heater	Recommend consulting manufacturers to establish effects

it is usually necessary to assume initially that the plant is safe as operated or designed — that is, that normal operation presents no significant hazard — and that a search is made for deviations from normal operation in order to identify causes and consequences. This assumption about the plant's initial safety is much easier to justify for an existing plant because the operating experience is available. The boundaries of the study must be defined taking into account any external threats to the plant. The validity of the results depends on having accurate information available and the P&IDs, the equipment specifications and the process, site and plant layout must be correct and up to date. Information should be provided on operating procedures when available and on company standards.

A team study is essential. Any experienced plant engineer can identify the major hazards associated with a plant. What such engineers might find difficult to do is to convince themselves, or their manager or even a judge that they had covered as much as was possible and reasonable, and that they had identified all the small credible events which might contribute to a larger (potentially disastrous) event. There are many operational inconveniences which may eventually contribute to the occurrence of an accident.

The Hazop leader must explain the overall process by appropriate means, often using block and process flow diagrams.

PROGRESS IN THE MEETING

It is essential to have an appropriate format for applying the guide words. The flow chart in Figure 5.5 (page 102) has the advantage of emphasizing the role of the operator.

The study involves selecting each drawing in turn and reviewing it in its entirety before proceeding to the next drawing. It is usual to start at the beginning of the process and follow the main material flows through the system, with utilities and minor process feeds being assimilated later.

Before the meeting the team leader should break down the process into nodes. This is normally done by identifying a main plant item, usually a vessel, and identifying first its input streams and then its output streams. A node is a point where the process parameters have an identified design intent. So a typical node section would be an input stream running from the battery limits to the first main plant item. Normally process items such as valves are located between nodes. The node assignments must be marked on the diagram and numbered sequentially.

Initially it is common to group related equipment, particularly heat exchangers and pumps. Occasionally a node might be introduced to denote a change in consequences or the impact of a deviation on other main plant items. It is not uncommon for the list of nodes to change as the study evolves in order to suit the analysis better or to correct omissions. More detail may be required in a specific area or the steps of a procedure for start-up, for example, may need improved definition.

The system under review runs between nodes, and it is usual to describe this section in terms of one node as a point of reference. This assignment must be applied consistently at the same end of the pipe throughout the process. Thus for a given line, according to the assignment, the same rupture of pipe might be described as a 'reverse flow' at the downstream node or as a 'high flow' at the upstream node.

During the study it is essential that the appropriate part of the earlier briefing on operations is repeated as each node is considered. The design intent of the section should be identified

HAZARD IDENTIFICATION AND RISK ASSESSMENT

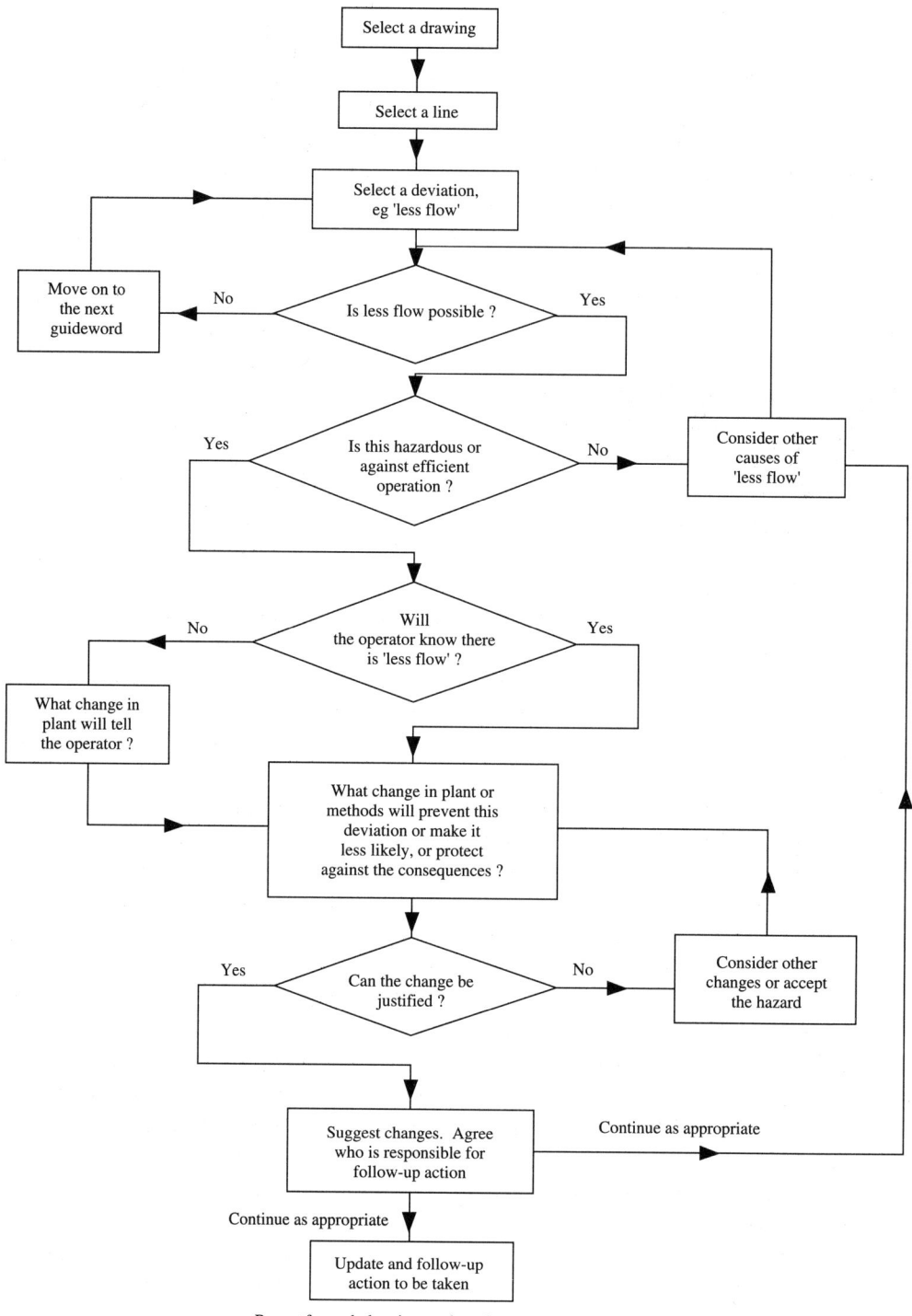

Figure 5.5   A flowchart for application of Hazop.

and noted on the Hazop worksheet. This should refer to the expected or desired operating conditions, possibly the allowable range, and occasionally reference might be made to special criteria such as particular materials of construction or an exothermic reaction.

It is important that the team leader controls the discussion. Here are some points to note:
- avoid exceeding the scope of the study;
- avoid domination of the meeting by any person, particularly the team leader;
- ensure quieter members have their say;
- control any personality clashes;
- stimulate using 'open' or 'closed' questions;
- limit the discussion and keep to the point;
- encourage the team to draw conclusions but record areas of concern or disagreement for discussion outside the meeting or later in the study;
- do not make assumptions, particularly that systems are safe;
- ensure that the notes being taken are adequate.

Avoid the perils of effecting a redesign. If a problem is noted then the solution may be readily evident and can be documented on the worksheet. Often, however, the solution is not readily conceived. In such a case avoid delaying the study by setting the problem aside for action outside the meeting.

DOCUMENTATION AND THE FINAL REPORT

The philosophy to be adopted in making records of the Hazop meeting must be determined. The most common method is to record only meaningful deviations and ensure the record is complete with respect to actions. There is some benefit for subsequent quality control in making an entry for every guide word considered. Further recording on tape gives a complete record of the meeting but a tape is very difficult because it is hard to tell who is speaking.

There are a number of computer packages available which can be used to carry out Hazop. Form-filling packages work well, and help structure a session and provide semi-automatic documentation. Table 5.6 (pages 104–105) gives an indication of the documentation of a Hazop report. The columns selected can be varied. It may be valuable to incorporate one column highlighting how the operator identifies that there is a deviation. Sometimes an estimate of risk is included using the columns Severity, S, Likelihood, L, and Risk, R. In the UK such estimates would normally be added after the meeting.

The main pros and cons of such form-filling packages are as follows:
- the packages do not make good use of past knowledge; check-lists are usually supplied but in limited, off-line format. It is unlikely that this facility would be used in a meeting;
- some consider that the form-filling is unimaginative, unattractive and difficult to follow as nodes are changed and windows filled. Others consider it to be very effective compared with older, manual methods;
- little time is saved during the group session; hence the analysis is still time-consuming, particularly in generating excessive causes;
- a better record of the meeting is usually obtained if it is documented using a computer program which provides assistance in typing in the responses on a *pro forma*. A record of the meeting can thereby be made available shortly after the meeting ends.

TABLE 5.6
Worksheet showing computer documentation of a Hazop

Company: Geoffs Industries, Inc		Facility: XYZ Plant	
Parameter: Flow	Design intent: Transfer a flow of 60 kg/h of xylene at 10 bar and 500°C		
Deviation	Causes	Consequences	Safeguards/checks
NO FLOW	(1) Loss of power to pump 103.	(1) Product flow stops and reaction ceases.	(1) Power supply is highly reliable.
		(2) Possible overheating of the material in the reactor.	
	(2) Line rupture.	(3) Potential fire and explosion due to release of flammables to the atmosphere.	(2) Line is inspected regularly by non-destructive testing (NDT).
	(3) Blockage in the inlet line.	(4) Similar to (1) above.	(3) Line too large for blockage due to debris or polymer.
	(4) Operator closes the manual block valve on the inlet line.	(5) Similar to (1) above.	(4) Valve is tagged to remain open under normal operating.
	(5) A PLC failure prematurely closes the control valve.	(6) Similar to (1) above.	(5) PLC has operator hold on critical valve functions.
MORE FLOW	(6) Pump overspeed.	(7) Excessive flow rate to the reactor and subsequent higher outlet temperature	(6) Flow rate is monitored by F12.
S = severity; L = likelihood; R = risk; P = priority (A = immediate action, B = report within two weeks); ENG = engineering; OPS = operations; LAB = works laboratory.			

A survey shows that most users prefer to have a package available, but not necessarily in meetings — where people either love them or hate them. Their documentation capabilities are really attractive and gaps in the record can always be filled in after the meeting. An alternative is to use a word processor software package such as Word for Windows.

Table 5.7 shows what might be contained in the final Hazop study report. Clearly this will vary with the nature of the plant and depend on whether full recording or recording by exception is used.

After each meeting it is necessary to develop minutes which record results and recommendations from the day's activities. Follow-up items should be ascribed to team members. The

Session: 1 21.06.93				Node: Line from storage tank to reactor			
Revision: 6				Drawing No: A 123.789			
S	L	R	Recommendations	By	P	Comments	
1	0	1	(1) Consider a spare pump with automatic switchover on loss of power to pump 103.	ENG	B	(1) This problem has occurred on the existing unit once in the past year.	
3	−2	1	(2) TIC 21 should have high temperature alarm in the control room.	OPS	A	(2) Overheating would cause a possible runaway reaction.	
3	−4	−1	(3) Check pressure rating requirements of the inlet piping.	ENG	B	(3) The piping is adequately designed for all conceivable overpressures.	
1	−5	−4	(4) No recommendations.			(4) No comments.	
1	−2	−1	(5) Eliminate valve 102 since it is unnecessary.	ENG	A	(5) Valve 103 will be adequate for reactor isolation.	
1	−2	−1	(6) Review PLC logic.	ENG	A	(6) Check for the possibility of valve closure during normal operation.	
3	−2	1	(7) Review process chemistry to determine if runaway is avoidable.	LAB	B	(7) It may be possible to size the pump to eliminate this hazard.	

TABLE 5.7
Final Hazop report

Executive summary:
- study objectives and achievements
- key recommendations

Introduction and organization of report
Scope and objectives
Study approach
Process description
Hazop results

Conclusions and action items
Appendices:
- description of technique
- study nodes and drawing list
- Hazop worksheets
- project action items
- operations action items

resolution of such items should be carried out quickly in order to answer questions before completing the study. The minutes can then be reviewed at the start of the next meeting to correct errors or misconceptions and upgrade on follow-up items. The final version of the meeting should alert management to the recommendations of the study. These are often grouped into recommendations requiring capital expenditure, those involving the change of procedures or similar, and those requiring further information.

A Hazop review action and follow-up form is generally issued which indicates the problem and the nature of the request. This can be completed with the operating or project response.

THE COMPLETENESS OF A HAZARD ANALYSIS

Some analysts place much emphasis on the measure of completeness of a hazard study. One definition of completeness is that the measure of completeness is equal to (the sum of hazards found) divided by (the sum of hazards found in the historical record).

Certainly a Task Analysis following a Hazop will find many further errors; so will appropriate further review and inspection. But this can be misleading. It allows for a lot of errors in construction and does not indicate whether the problems found gave rise to an undetected and hazardous consequence against which no safeguards had been considered. The emphasis in this calculation of completeness is on detecting initiating events. It might be more appropriate to define completeness in terms of the number of dangerous conditions against which safety barriers must be introduced. Of course there will always be ignorance about events, with some hazards remaining unknown or unanalysed due to lack of knowledge or resources. Residual accepted risks always remain and there is new accident potential which arises from things like poor management, change from design intent and new external threats.

Table 5.8 suggests some of the main causes of incompleteness.

It is absolutely certain that the plant documentation will at some time be incomplete and in error, not everyone will be trained to an appropriate level and that errors will be made by everyone working on the plant.

EFFECTIVENESS OF SAFETY SYSTEMS

There are many other parts of a system which should be subjected to a Hazop in order to improve completeness. The combination of Hazop and Task Analysis will be discussed in Chapter 8. It is preferable to apply a Hazop once the design has been frozen prior to construction. But in practice,

TABLE 5.8
The main causes of incompleteness

Documents do not completely describe the plant	Errors of judgement
Lack of knowledge of individuals	Too narrow a design framework
Errors in plant description documents	Simple oversights
Absolute lack of knowledge	Problems introduced since the analysis
Lack of resources	Masking

especially with large projects, design changes often appear after the final design. It is therefore appropriate that the revised P&ID is again subjected to Hazop. To cover the fact that this will rarely be done, a form of Sneak Analysis can be carried out to unearth construction deficiencies which have resulted in significant design or operability problems which were not part of the original design. Sneak Analysis is a technique for identifying paths down which flow of material might sneak or additional energy might enter the system. It can be accomplished by marking such paths up on the diagram.

The procedures may be subjected to a Hazop. This applies particularly to batch processes which are critically dependent on procedural activities. Further advice is given in Chapter 7.

A Hazop of the protective and mitigative systems is vital. The simplest study can merely postulate failure, spurious failure or partial failure and discuss the outcome generating a procedure to be followed. A more detailed study might examine the key factors which influence the overall probability that such a system will work and that a successful mitigated outcome will result. This study would consider the following factors:

- the detection system can measure the fault condition in sufficient time;
- the hazard size or the process deviation is large enough to be detected;
- the instrument can register the hazard as an input;
- an output is transmitted;
- the action from the output successfully operates appropriate hardware or alarms, trips, etc and achieves its desired outcome;
- the action has the desired effect.

An example of a problem needing identification is when a detection system operates too slowly and allows an ignited leak of fluid to travel from a vessel offshore into another module. If the emergency valving had closed sooner then the fire would not have escalated. So many incidents have resulted from the failure of a safety device to respond correctly that this must always be considered when carrying out a Hazop.

It is also essential to incorporate in a Hazop some study of the emergency response to a release. It should not be acceptable to complete a Hazop without such information being available.

HOW TO TARGET HAZOP EFFORT

It takes very little experience of Hazop to realize that it is a very tedious and time-consuming exercise. To Hazop every single drawing for a multi-million pound plant takes a year of full-time Hazop study. So how can effort be sensibly directed to where it is needed, without the risk of missing an important hazard/operability problem?

One way to do this comes from pre-Hazop studies — for example, Concept Hazard Analysis or Preliminary Process Hazard Analysis. If the PFD or similar for one area of the plant has been reviewed and found to contain no hazards, or to be deterministically safe, then any P&IDs arising from that plant area can be assumed not to contain any hazards. It may still be necessary to Hazop the detailed design in order to identify any operability problems but if the interest is in hazards, these P&IDs can be safely omitted. This approach shows another advantage of pre-Hazop techniques because alternative methods of screening a lot of drawings are more complicated and time-consuming.

One other screening technique which can be complementary to screening via a pre-Hazop technique can be outlined as follows:
(a) Take the number of P&IDs for the plant (say there are about 100).
(b) Divide the allowable plant risk equally between them (that is, if allowable risk is 10^{-6} deaths per annum, then for 100 drawings the allowable risk per drawing is 10^{-8}).
(c) Look at each drawing and, assuming a frequency of hazard of 1 per annum, see if the unmitigated consequence would give a risk greater than the tolerable guideline per drawing. If it does, it should be subjected to a Hazop. If it does not, it is reasonable to assume that there are no hazards of any significance to find by Hazop study of that drawing.

This may seem a simplistic approach, and it does miss operability issues so it is not always appropriate. But it has been shown to work in industry when applied with care. The difficult part is in applying the technique correctly, especially where the site/plant has other secondary criteria which are important in gaining plant acceptance.

USE, STRENGTHS AND LIMITATIONS OF HAZOP

Hazop can be used for hazard identification on a wide range of processes and plant. It is used at many stages of plant life from design through to operation. Studies at an initial design stage have not, however, always proved satisfactory as the method inhibits creative thought. Hazop is designed for use after the design is complete. At this stage for very detailed studies it may be augmented by Task Analysis and possibly by Failure Mode and Effect Analysis.

Hazop is well proven, systematic, reasonably comprehensive and flexible. It is suitable mainly for team use because it is possible to incorporate the general experience available. The use of keywords is effective. It gives good identification of cause and excellent identification of critical deviations. Hazop identifies virtually all significant deviations on the plant. All major accidents should be identified, but not necessarily their cause.

Hazop is very time-consuming and laborious, and there is a tendency for boredom to set in. It tends to be hardware-oriented and process-oriented, although the technique can be applied to human error, as will be shown in Chapter 6. It tends to generate many failure events with insignificant consequences and generate many failure events which have the same consequences.

Hazop is often used in a manner which takes little account of the probabilities of events or consequences, although approximate qualitative assessment may be made. It is poor where multiple-combination events can have severe effects.

The Hazop team leader may not make the best use of the power of the technique. The leader may neglect defects or deterioration of materials of construction, or give inappropriate weight to maintainability. When identifying consequences the leader should not encourage listing action by emergency control measures without considering that such action might fail. Hazop should not neglect the contribution which can be made by operator interventions. These mistakes can be overcome given suitable awareness by the study group and introducing appropriate comments on response in the consequence columns.

INDIVIDUAL LEARNING

For those with limited experience of the technique the following comments, based on a talk given by Steve Whitty of British Nuclear Fuels, are relevant. A Hazop exercise illustrates several

important points to the individual:
- get acquainted with the plant before the exercise (this is easier with existing plants); there must be someone in your team who knows the plant in detail;
- a team study is essential. Without a team, there can be no confidence that all the important events have been identified. Even then the study is rarely complete;
- Hazop takes time. In the time allotted it is impracticable to apply all the guide words to all possible states of the system;
- Hazop is not exciting cut and thrust. A modern Monty Python film might identify a Hazop team leader who does nothing else with the poor, boring accountant. However the leader must generate the best from the group without letting discussion drag on or be monopolized by one or two individuals;
- beware of the temptation to say 'nothing else can go wrong with this system' after one or two guide words have been applied. People with a lot of experience can pick out the guide words that are likely to reveal the most meaningful deviations. The best way to gain that experience is by applying the words and seeing what comes up. The team leader's experience is crucial here.

AN OPERABILITY PROBLEM CAN BE A HAZARD IN DISGUISE

It is often believed that if an item creates an operability problem then the plant operators will get round it somehow. Doing so may introduce a safety problem. Examples include interlocks which prevent later stages in a sequence of operations if some earlier condition is not satisfied. If the interlock is known to be faulty then the operator might try to prevent it from functioning. Similarly alarms and warnings might automatically be ignored if they give unnecessary warnings. An alarm may be always going off for high level in a stirred vessel because of splashing, so the operator automatically turns off the alarm until eventually a major overflow creates chaos. Such events should be nailed down instead of being left as accepted practice. If a plant has been running for 20 years or more, however, it should have no real operability problems. So basically be careful of any which are uncovered. Also low frequency problems can have severe effects and will not have arisen during the plant's lifetime. In such cases only make recommendations for further consideration, except when the errors are obvious omissions.

Operability problems may lead to plant shutdown. This introduces change, which is always a concern. It may introduce unusual operating and maintenance problems. Loss of quality in finished products can lead to reworking of process material. Any shutdown puts pressure on staff to get back in production.

BEWARE OF THE INCREDIBLE

It is not certain that events are incredible. Many photographs show releases of process material that were deemed impossible in any circumstances. Common cause failures may be a problem. A simple and obvious common cause failure of two pumps is power supply failure. This can be easily identified by Hazop where appropriate. If such failure modes cannot be identified this does not mean that there are none. For instance, the two pumps may have independent power supplies, but how is the power supply routed to the pumps? Do they both pass along the same trunking, and so are both vulnerable to the same event? For example, will a fire in one cable knock the other one out? A nuclear power station was once stripped of its safety systems because someone was checking the ducting which carried all the cabling using light from a candle and ... yes, it did catch fire.

HAZARD IDENTIFICATION AND RISK ASSESSMENT

Further problems may arise from factors such as common manufacture. Pumps built by the same manufacturers and to the same design probably have the same design faults. If maintained by the same team they suffer from the same maintenance practices. The use of a label such as 'incredible' is wrong. It is better to say 'very unlikely'.

There is a need to look for diversity as well as redundancy. A good example of the difference between diversity and redundancy is the requirements to be sure of lighting a camp-fire.

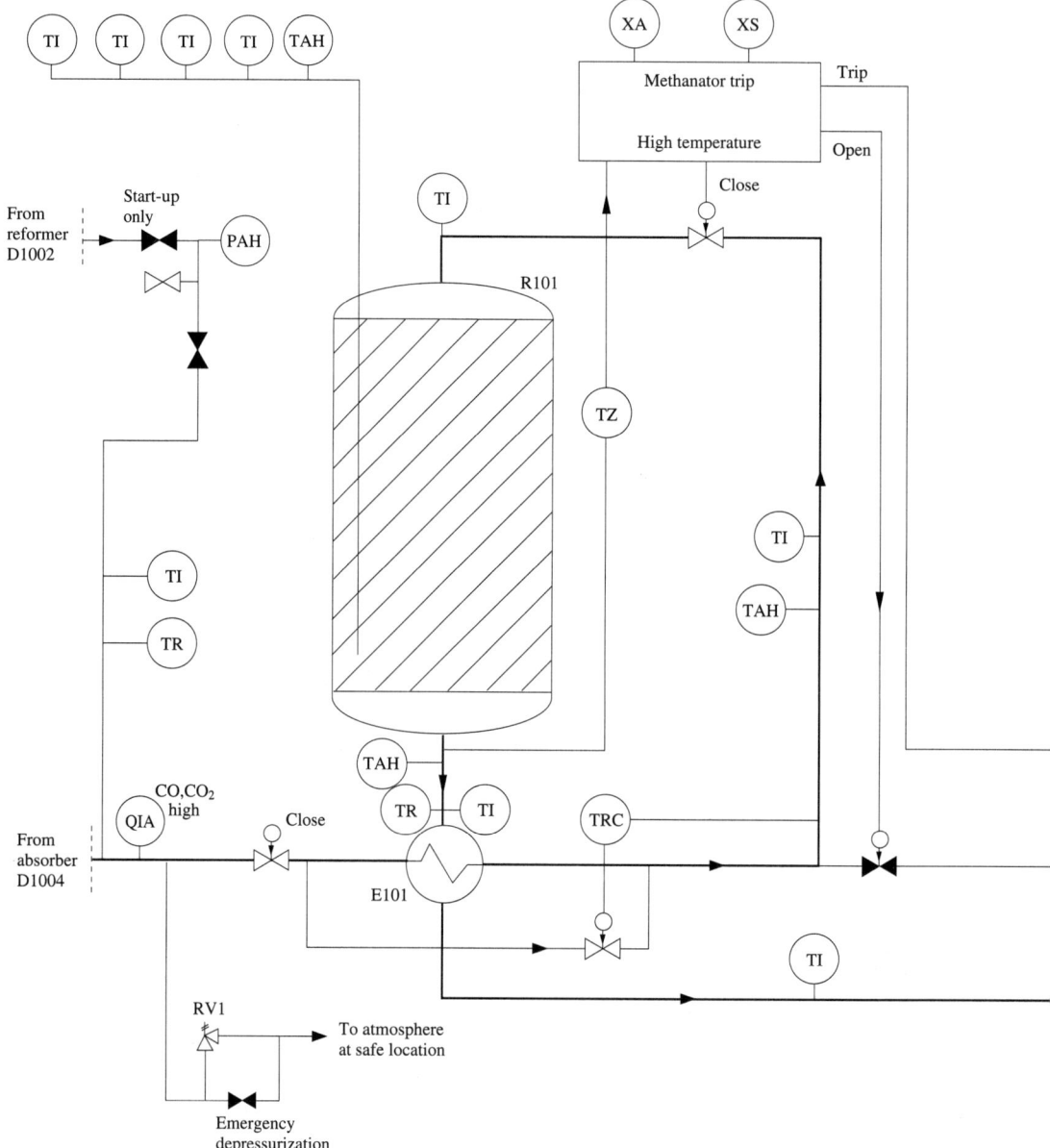

Figure 3.4   P&I diagram for methanator, Revision B. (See page 243 for instrumentation definitions.) 10 = Nitrogen, 10 bar; 11 = Cooling water supply; 12 = Cooling water return.

HAZARD AND OPERABILITY STUDIES (HAZOP)

One box of matches can get wet, or it can be very windy so they won't catch. Some redundancy can be introduced into the system by keeping two boxes of matches, but would that really help? Ideally, some other means of lighting the fire is required which relies on a different principle. A gas-fuelled cigarette lighter, for instance. This introduces some diversity, and so increases the chances of always being able to light the fire. It is still not incredible, however, that it will ever be impossible to light the fire.

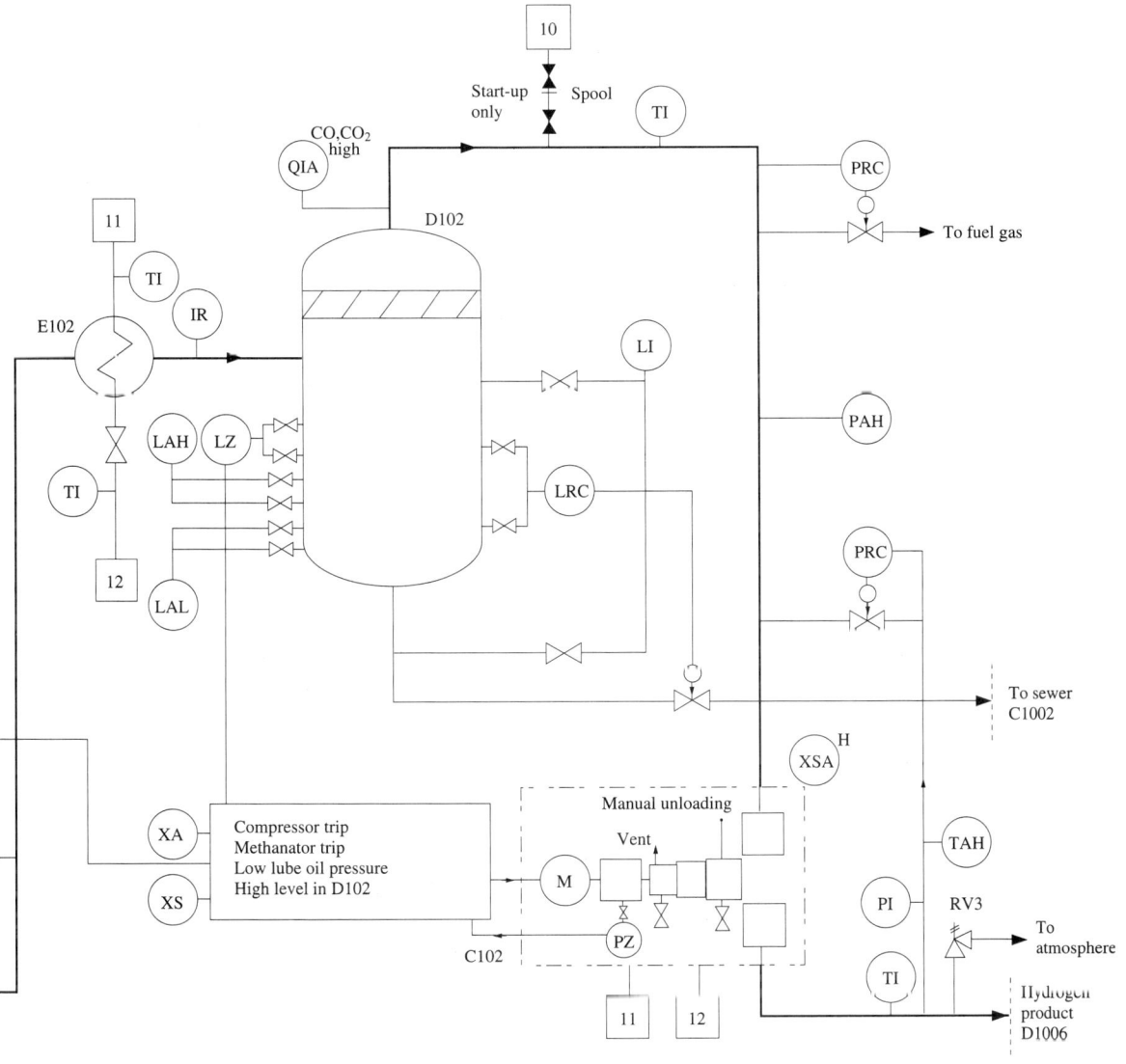

CASE STUDY — THE METHANATOR

This case study has been examined earlier and a CHA (page 30) and a PPHA (page 51) have been carried out. The purpose of the methanator is to remove oxides of carbon down to parts per million. This is because the product is used in downstream plants which can suffer catalyst damage from such oxides. A P&ID has been developed for this plant, as described earlier. Figure 3.4 is repeated on pages 110–111. It is available for a section of a hydrogen plant in which methanation is carried out as a finishing operation in order to remove oxides of carbon from hydrogen as final product. The methanation section involves the treatment of a gaseous mixture of hydrogen 98%, methane 1%, oxides of carbon 1%, which is saturated with water and at 40°C and 21 bar. The stream is heated to 400°C before passing through a fixed-bed catalytic reactor, the methanator R101. The reactor outlet stream is then cooled in the preheater, E101, before cooling to 30°C in E102. Any condensed water is removed in a knock-out pot, C102, before compression to 40 bar. The main reaction in the methanator involves the conversion of oxides of carbon to methane.

The reaction goes virtually to completion — that is, down to 20 ppm of oxides of carbon in the final product, hydrogen.

The methanator is started up from cold by preheating it with a stream of nitrogen heated in the main reformer when that unit is started up. It is then left hot until the reactor is ready to come on stream, usually a period of about two hours. The stream of nitrogen is obtained from the main process stream just below a steam superheater. Note that this stream is only available as nitrogen during the start-up of the reformer. At other times it is the untreated product direct from the reformer; it is very hot and has a very high level of oxides of nitrogen.

The reaction in the methanator is exothermic. This can cause a major problem if oxides of carbon are not removed in the absorber from a level of 20% down to at least 10%. The reaction goes to completion in any case, and consequently very high temperatures can be generated. Failure due to overtemperature occurs either in the reactor, in the outlet pipeline or in the preheater. An emergency shutdown system has been designed which, on high temperature, shuts down the methanator and the compressor. The attention of the analysts is drawn to the incident scenario envisaged for this eventuality, along with other disturbances such as gas blowby from the knock-out pot D102 which sends a considerable amount of gas to sewer, and the overtemperature at the compressor. For further information see Table 3.10 (pages 63–64). Full details of the subsequent Hazop are not given; merely a selection of notes which are of general interest.

SELECTED NOTES FROM SUBSEQUENT HAZOP STUDY

Table 5.9 on pages 113–117 shows a selection of results from the Hazop study. The following points are selected from further suggestions made by a number of experienced analysts who have studied the methanator as individuals. Some would be adopted and some are considered to be unnecessary additions. The hydrogen plant is a constantly running generator of process gas, which is not easy to shut down quickly and is slow to restart and build up production. Hydrogen reformer tubes have been known to fill with carbon if operated with no steam present. Other than that, blockages are presumably extremely rare because of the low molecular weight of all the gases involved. Interestingly, the drawings made prior to Hazop, with slight modifications (mainly extra instruments), represented an existing plant as it was in the 1960s. Incidents which have occurred since that time have suggested that the reactor should be shut down by a two-out-of-three voting system activated by probes from within the reactor bed. This is because of the speed with which the reaction takes place.

TABLE 5.9
Summary of results of Hazop

Deviation	Causes	Consequences	Actions
Feed to R101			
No flow to methanator	Catastrophic failure of hydrogen line	Fire and explosion hazard. Compressor runs dry.	Gas detection and emergency trip on push button by operator or automatic operation at 20% low flow. Vent to flare or atmosphere. Apply passive intumescent to pipework. Consider water deluge on reactors.
More flow from absorber	Excess oxides of carbon from absorber	Off-specification hydrogen due to insufficient reaction	FRAH at E101. The setting of this flow might prove difficult.
More flow to C101	Methanator bypass valve fully open	Off-specification material to downstream plant	Consider removal of bypass facility. Consider alternative trip system of main isolation valve at E101 inlet with vent to appropriate sink immediately prior. Suitable sinks are atmosphere, fuel gas or flare.
Low flow to methanator	Upstream trip of reformer.	No load at compressor.	Link methanator and compressor trip systems to total plant trip. Warn by FRA low. Check exchanger ratings.
	Blockage of methanator bed.	No load or reduced load at compressor.	Flow meter plus flow staging controls at C102. Air integrating flow recorder on hydrogen and fuel gas assists plant audit.
	Restriction in E101 due to carry-over from absorber.	Poor preheat affects oxides of carbon contamination.	Install TAL. Cleaning schedule for E101. Differential pressure measurement around E101.
	Trip activated but passing.	Runaway continues with operator unaware of cause.	Regardless of the trip system used, install valve positioners with status indicated in control room. Install TAHH on inlet.
More composition in R101	High oxides of carbon from absorber (<5%)	Catalyst affected by prolonged oxides of carbon contact	Project to advise. TAH in catalyst bed.
More temperature in R101	High oxides of carbon from start-up line after methane introduced to reformer	Runaway reaction and possible reactor failure	Isolate using double block and bleed, and PAH (also consider spectacle and removable spool). Valves locked closed and only removed under special permit. Review location of start-up line. Relocate inlet line so it goes to E101 and process analyser. Redesign E101 for increase in temperature. TA on methanator inlet.

TABLE 5.9 (continued)
Summary of results of Hazop

Deviation	Causes	Consequences	Actions
Feed to R101			
More temperature from E101 to R101	Failure of TRC or TCV open	High temperature to reactor but not runaway reaction. Maximum temperature limited by approach temperature E101.	TAH on inlet is sufficient warning
Less temperature from R101/E101	Failure of TRC	Reaction of oxides of carbon could stop	Oxides of carbon analyser and alarm downstream plus TAL on E101 outlet
	Less temperature from methanator due to less oxides of carbon in feed	No exchange of heat in E101	Problem only likely at start-up. Ensure procedure carefully lays down when methanator preheat finished and when feed can enter methanator.
R101/D102			
More temperature from E101 to E102	Methanator bypassed and initial surge of hot material	High temperature to E102 for short period	Project to check duty of E102
More temperature to E101 and E102	Runaway reaction in methanator	High temperature to E102 particularly if trip system fails	Project to check duty of E101 and E102
More temperature from E102	Poor heat transfer, increased duty, or cooling water failure	Higher moisture content to C102 and downstream plants	Project to advise if this is a problem. Consider a moisture analyser on outlet of E102.
		Higher temperature to sewer	Increased vaporization of any other oils in the sewer but less dissolved gas. Project to advise on other materials present. An atmospheric degassing vessel rejected.
		Higher duty on LCV, possibly some cavitation	Project to advise regarding specification. TAH on outlet from E102.
		Higher duty on D102. Higher duty on C102.	

TABLE 5.9 (continued)
Summary of results of Hazop

Deviation	Causes	Consequences	Actions
R101/D102			
Less temperature in water lines	Low ambient temperature	Freezing causes blockage and can burst lines	Insulate and trace cooling water and deadlegs in D102
	Poor catalyst activity	No reaction	Check catalyst initiation and monitor activity
Less temperature from E102	High cooling water throughput	More dissolved gas to sewer. Higher cooling water costs.	TRC on hydrogen outlet from E102 to regulate cooling water. Check LCV can handle extra flow.
Less temperature in water lines	Low ambient temperature.	Freezing causes blockage and can burst lines.	Insulate and trace cooling water and deadlegs in D102.
	Poor catalyst activity.	No reaction.	Check catalyst initiation and monitor activity.
Less temperature from E102	High cooling water throughput	More dissolved gas to sewer. Higher cooling water costs.	TRC on hydrogen outlet from E102 to regulate cooling water. Check LCV can handle extra flow.
Other than in E101	Tube leak	Hydrogen containing oxides of carbon leaks into hydrogen to product causing off-specification material	Oxides of carbon analyser and alarm. Appropriate checking procedure for operator prior to compressor trip.
Other than in E102	Tube leak	Hydrogen leaks into cooling water and possible fire at cooling tower	Leak sampler at high point on E102. Emergency shutdown valves may be needed to protect against excess pressure given a severe leak.
D102/Sewer			
More flow to sewer	Maloperation of LCV or controller.	Large flow of gas to sewer.	LAL and trip system or ball in pipe system. Note LAL not introduced as specified on PPHA.
	Ambient temperature.	Freezing of low bore static lines containing water causes failures.	Insulate and steam trace.

TABLE 5.9 (continued)
Summary of results of Hazop

Deviation	Causes	Consequences	Actions
D102/Sewer			
More flow down spillback line round C102	Failure open of PCV	Reduced flow downstream or increased pressure upstream	Install PAL on discharge. Check design of upstream plant and second relief by PSV. Two non-return valves to be located on hydrogen product line.
Composition phase	Flashing of dissolved gases across LCV2	Reduction of capacity of LCV102 possibly rendering it unable to control the level. Continued running without load could cause damage.	Project to check. Fit motor with low electric load alarm. Check with project if trip required.
Line from D102 to product			
More temperature at compressor	Excess compression ratio	Possible mechanical damage	TAH on overpressure
More vibration at C102	Mechanical problems at C102	Compressor explosion	High vibration trip on C102. On-line condition monitoring of C102.
Less lube oil pressure	Blocked lube oil filters	Inadequate lubrication	Lube oil filter switchover system and lube oil differential pressure alarm
Less lube oil pressure	Fouled lube oil coolers	High lube oil temperature	Lube oil high temperature alarm
High temperature in C102 recycle line	Lack of demand	Excessive circulation of gas around compressor	Relocate to put the recycled gas through E102 where it will be cooled. Project should confirm that lines are adequately sized to receive the maximum recycle flow.
High pressure at C102 outlet	Lack of demand or loss of control with PRC failed closed	High pressure followed by rupture of outlet pipe	Check location of relief valve. Additional relief valve would reduce risk and protect during proof test of other valve. Selected close to compressor rather than from knockout pot under demister pad.
D101/C102			
Less pressure from compressor	Loss of containment. Low output from compressor. Failure of PRC.	Loss of product forward rate	PAL on discharge and PAL on inlet

TABLE 5.9 (continued)
Summary of results of Hazop

Deviation	Causes	Consequences	Actions
D101/C102			
Composition phase	Cooling by ambient air will cause condensation on wall of the pipe from D102 to C102.	Liquid carry-over into compressor.	Line from D102 insulated and possibly heat traced.
	Failure/channelling of flow through demister on D102.	Liquid carry-over by entrainment from D102.	Install a nucleonic gas density scanner to detect average density changes caused by liquid entrainment.
Less flow in compressor C102	Worn/damaged discharge valves which allow reverse flow into cylinders. Catastrophic seal failure of compressor C101.	Potential to overheat compressor leading to mechanical failure. Excessive pressure/hydrogen in dead spaces, etc and possible fire/explosion.	HTA on compressor outlet line instead of recycle line (deadleg). Project should consider fitting a TAH to compressor cylinder head. Dead spaces and crankcase purged continuously with nitrogen and vented. Fit pressure alarms on high switches to activate emergency shutdown and compressor shutdown system.
Reverse flow through plant	Upstream discharge, most probably on failure of absorber LCV	Reverse flow to stripping column or atmosphere	Design trip to close on reverse flow. Consider non-return valve on inlet to E101.
More pressure to fuel gas	Failure of PRC open	Overpressure of fuel gas by high flow rate.	Ensure burn-off to flare via appropriate control and pressure relief.
		Nitrogen to fuel gas at start-up.	Line should be isolated at start-up with appropriate procedures involving lock-open/lock-close for valves.

For convenience a number of subsystems associated with the activation of the catalyst and inerting have been left out. Also no attempt has been made to allow for maintenance or examine procedures in detail.

The Hazop documentation shown in Table 5.9 is less than adequate because of space limitations. The terminology used on actions is rather terse and does not indicate that the changes are for consideration by the design team and other groups. The example in Appendix 3 gives a better idea of what an action column should look like.

HAZARD IDENTIFICATION AND RISK ASSESSMENT

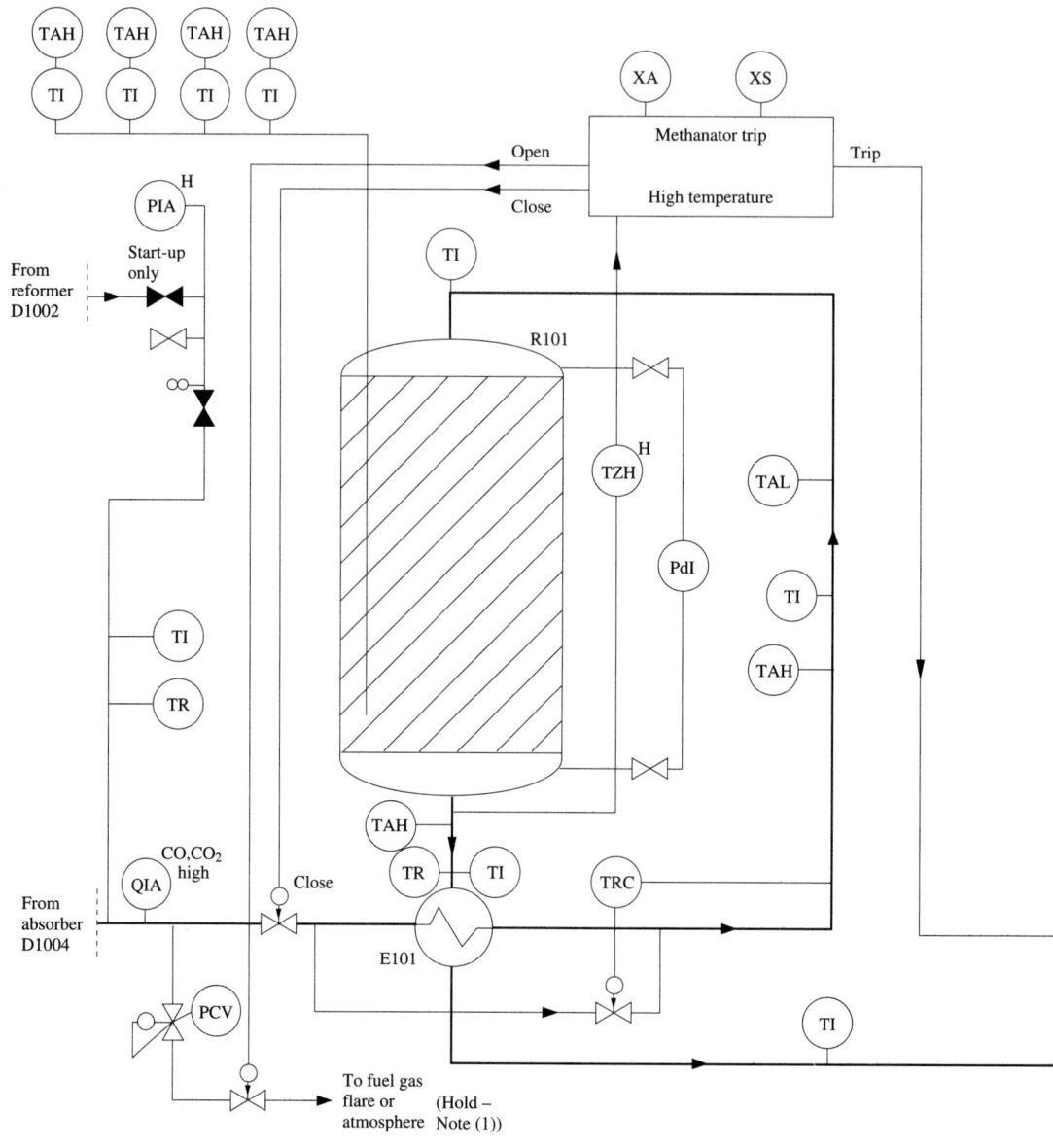

Figure 5.6   P&ID of hydrogen plant, Revision C. This drawing has been Hazoped. (See page 243 for instrumentation definitions.)

Item number	R101	D102	E101	E102	C102
Title	Methanator	KO pot	Methanator preheat	Methanator cooler	H_2 compressor
Op. temp, °C	450	30	400	100	50
Op. pres., bar	20	18.7	21	19	40

HAZARD AND OPERABILITY STUDIES (HAZOP)

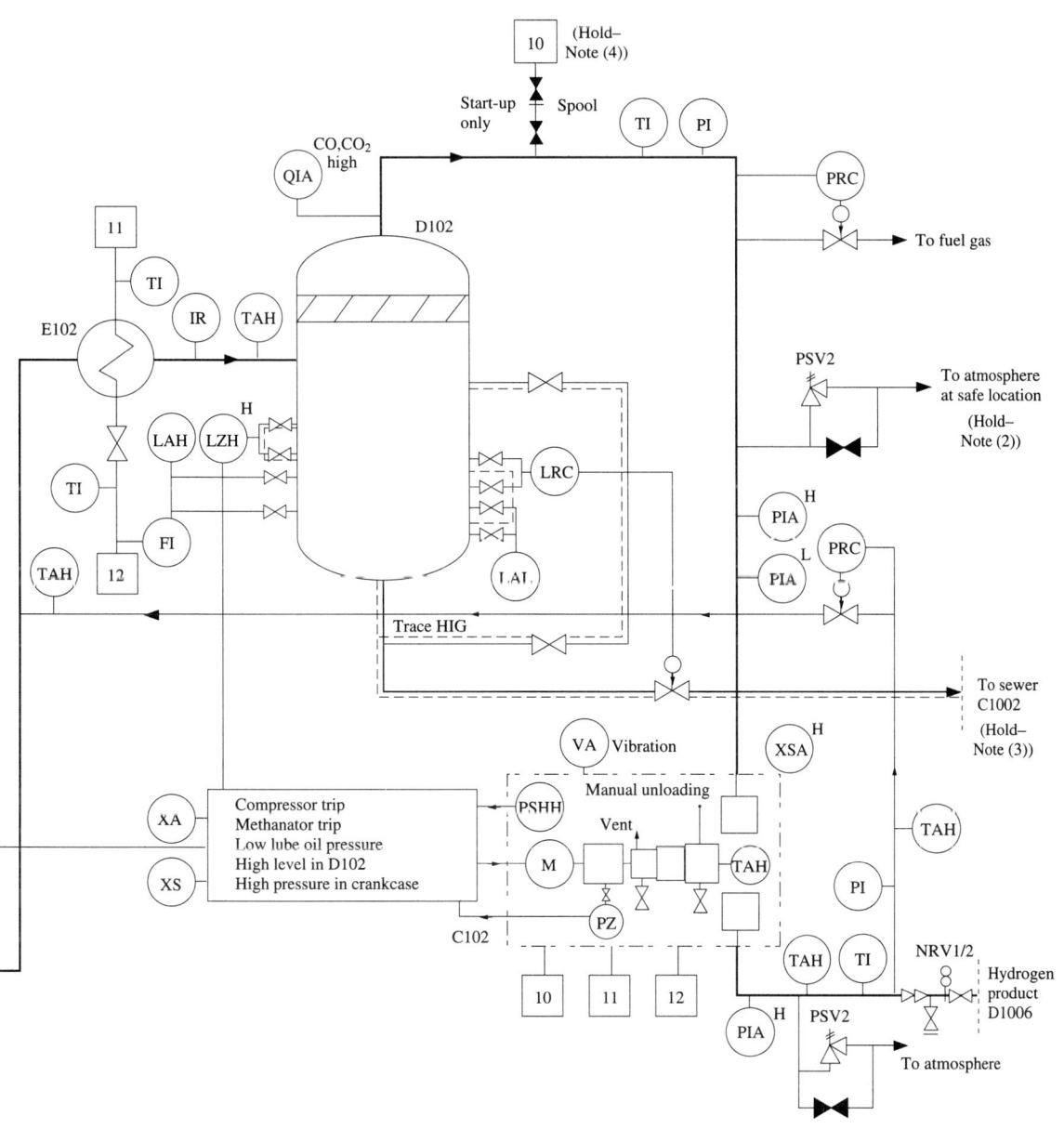

NOTES ON THE UPDATED P&ID

The drawing of the plant is now updated (Figure 5.6, pages 118–119). Four specific notes given on this drawing are worth developing:

(1) A decision has yet to be made on whether to send hydrogen to atmosphere (where it may ignite), to a main flare system or to fuel gas. Operating experience recommends passing the material to atmosphere except when sending surplus normal plant production to fuel gas. A further factor is that fuel gas needs are greatly reduced when the hydrogen plant is down. The most likely eventual decision is to vent hydrogen into fuel gas and burn off the surplus. No information has yet been obtained about whether the discharge of hydrogen to atmosphere is environmentally unfriendly. It should at least give any passing insect life a nasty surprise.

(2) Some workers have reported impressive reductions in risk by having two relief valves in parallel where only one is shown. Plant experience has never shown overpressure as a problem at this point. The sizing of the relief system is particularly dependent on correctly assessing the recycle load via the spillback line and the load for all failure modes.

(3) The sewer discharge has caused problems on the existing plant; sewage discharge caused the disaster in Guatemala City in 1992 which probably changed some attitudes. Options for improvement include a low level trip system, an atmospheric degassing chamber and further block valves. The balance of opinion suggests a trip system, although this does not solve the problem that the sewer will at all times contain flammables. High standpipes as vents have been shown to reduce the risk of serious problems on other plants.

(4) The nitrogen purge system needs resolving. Further lines are not shown in Figure 5.6 as all the plant diagrams and integration of the purge system with the reformer section is not shown. In reality nitrogen is circulated around the whole system from C102 and appropriate purge and isolation is provided. The present system still allows for the possibility of nitrogen passing to fuel gas, especially if the PRC on this line fails open. The maintainability of the system was not considered as the P&ID is still heavily simplified.

It is again emphasized that for ease of presentation the diagram in Figure 5.6 has not been upgraded to the normal standard of a P&ID. It has also not been submitted to extensive checking and the compressor section is incomplete. Full details of the Hazop study are not given here.

THE HAZOP EXAMPLE IN APPENDIX 3

The example provided in Appendix 3 contains improved documentation and is recommended for study because it is more comprehensive than the simplified examples in this chapter.

THE SAFETY SCHEDULE

The results of all the hazard identification studies should build up into a safety schedule for the plant. These assessments are integrated with other reviews of the plant used to analyse the sociotechnical system and identify consequences. Such a schedule contains specific information on the process, the control philosophy, material hazards and inventory, incident scenarios, the basic control system, the immediate causes of incidents, the engineered protection and mitigation systems, the plant emergency response and the off-site or community response. The schedule should include a note of root causes. This should refer to specific factors, and indicate how deficiencies can be measured using performance indicators and where specific engineered defences can be instigated.

The safety schedule can be built up over the life of a plant and can be used to demonstrate — in an auditable form — the quality assurance measures carried out on the project. All specific reports such as those prepared to meet the requirements of the Control of Industrial Major Accident Hazards (CIMAH) Regulations in the UK, the Occupational Safety and Health Administration (OSHA) rule on Process Safety Management of highly hazardous chemicals or the Offshore Safety Case are included in the schedule.

The following features of the plant should be identified:
- incident scenarios — the sequence of events leading to a specific undesired impact event;
- incident initiators — selected events which are basically immediate causes with possible amplification by their root causes;
- root cause performance indicators — it is not practical to identify all root causes in detail. For the key incident scenarios, however, it is important to note where their frequency can be reduced by attention to selected root causes and to the performance indicators measuring such failures;
- incident consequences — the worst expected financial, environmental and safety consequences of the unmitigated event;
- basic control system — this includes operator and maintenance monitoring/supervision, plus the alarm system and operator-initiated corrective action;
- protection systems — the engineered safety systems which are designed to detect and correct (automatically);
- mitigation systems — the (engineered) safety systems which reduce the undesired consequences of the event;
- emergency response — the emergency response both on site and off site, including post-incident response.

Such considerations fit equally well into any project stage. For example, if Preliminary Process Hazard Analysis is used to identify the scenario — initiator — consequence sequence, then the information generated can be extended to define the required functionality and integrity of the engineered protection and mitigation systems, and the events against which they are designed.

It is important to prepare reasonable descriptions of each major scenario developed. Such scenarios for the methanator are shown in abbreviated form in Table 5.10, pages 122–123.

This information can be used as a specification for the design engineers as the project continues. The information is also useful during precommissioning and production. The system is able to evolve as the project develops and different hazard identification methods are used. The safety schedule demonstrates in summary form that the plant/process is viable at the early stages of the project. It forms the specification for the detailed design and as refined later gives a summary demonstration of plant safety, environmental acceptability, etc. It can be used as a safety audit tool in which the integrity and functionality of the safety system has been defined; the audit can confirm that the actual design meets that specification.

The frequency of the worst scenario on the reactor circuit could be reduced by installing a better control system. A large base plant such as this should have programmable electronic systems both as part of the basic process control system and for the safety interlock system. The level of oxides of carbon should be monitored and the trip system should probably have some partial

TABLE 5.10
Preliminary safety schedule

INCIDENT SCENARIO AND IMPACT EVENT	IMMEDIATE CAUSES	INADEQUATE CONTROL OR ACTION	INADEQUATE EMERGENCY CONTROL/ACTION
Scenario 3 Rupture on the methanator circuit due to overtemperature leading to torch fire with possible escalation to rest of complex	High inlet temperature to reactor. $F = 0.1$ High carbon dioxide in stream from absorber or impurities in feed down start-up line or high inlet temperature to reactor. $F = 0.1$ Impurities in feed down start-up line (sneak path). $F = 0.01$	Operator fails to stop temperature trend on TAH by adjusting control (slow-acting). $P = 0.01$ Operator fails to stop trend on QHA(COx) or TAH by correcting absorber. $P = 0.1$ Operator fails to stop trend on QHA(COx) or TAH by isolation. $P = 0.05$	Hazardous temperature in reactor and high temperature shutdown system fails. Demand on system: $F = 0.01$ Probability of failure of trip system: $P = 0.05$
Notes S = severity of event; F = frequency of event; P = probability of event. Procedures on all alarms and on activation of emergency control system to be documented and training programme established. Personnel to stay clear of methanator area on trip being activated. Fire-fighting procedures to be evaluated further.			

redundancy built in to reduce the number of spurious trips, as well as improve the availability of the system to 0.995/0.999. Note that partial redundancy implies either redundancy in the sensor, the logic solver or the actuator, but not all three. A two-out-of-three voting system employing extra sensors reduces spurious trips and an additional valve system reduces the likely failure of the reactor.

The safety schedule eventually developed can be used to identify all the control and protection measures associated with each fault in a given incident scenario. In the schedule the emphasis should be on specific measures.

Attention might therefore be given to the following themes:
- failures — initial engineering and quality control; maintenance and maintenance schedule, proof testing;
- detection — normal operational measures and operator response; automatic response;
- control — operating limits; normal operational measures and operator response; passive protection systems; active protection systems;
- mitigation — constraints by management; emergency response;
- feedback — performance indicators; incident reports.

In addition general information is produced which feeds into the whole range of administrative and engineering controls designed to keep plant operating safely. The studies assist in the

FAILURE TO RECOVER SITUATION	SIGNIFICANT RELEASE	MITIGATION OF RELEASE	SEVERITY OF CONSEQUENCE
Operator fails to stop all flows into methanator by isolation of flows in and out of system. System not to be depressurized until cooled. P = 0.5	Release on overtemperature of reactor and adjacent pipework on outlet. Failure frequency of system: F = 0.0003	Release self-ignites. Reactor strengthened so that pipework most probable failure. Impingement on vessels leads to further torch fire. P = 0.1 Impingement on adjacent pipeline with pool fire and possible missiles to plant complex. P = 0.01	Major damage to plant, possible injuries whilst fighting fire. Likelihood = 3 or 4 Severity = 3 Severe damage to plant with possible injuries to plant personnel or fatality. No danger to public due to location of plant. Likelihood = 5 or 6 Severity = 4
Procedures for inspection and maintenance of emergency trip system to be documented. Note how risk is badly affected should this item fail. Action on downstream plants on failure of methanator to be determined.			

generation or review of operating and other safe work practices to provide for the control of hazards during the various operations which take place on process plants, including opening process equipment or piping and control of entrance into a facility by maintenance, contractor, technician or other support personnel. The training of employees must include emphasis on the specific health and safety hazards, emergency operations and safe work practices applicable to their tasks. There should also be due emphasis on all matters associated with maintaining the ongoing integrity of the process plant. This applies to both initial and refresher training so that such matters are understood by all concerned.

The management must ensure that the findings and recommendations from each hazard analysis are considered and resolved, and actions are documented, taken and completed. Such actions must be communicated to project, operating, maintenance and others whose work may be affected by these recommendations and actions. To this end the safety schedule serves as a source of reference and builds up the safety culture within the company.

6. FAULT TREE ANALYSIS

INTRODUCTION

Fault trees are widely used as communication aids to demonstrate system failures and their development to managers, designers and operators. The use of fault trees in qualitative analysis demonstrates the effect of system failure modes and design changes. Their study is undertaken in two parts:
- fault tree construction;
- fault tree evaluation.

TABLE 6.1
Gate symbols

Gate symbol	Gate name	Causal relationship
	AND gate	Output event occurs if all input events occur simultaneously
	OR gate	Output event occurs if any one of the input events occurs
	Inhibit gate	Input produces output when conditional event occurs
	Priority AND gate	Output event occurs if all input events occur in the order from left to right
	Exclusive OR gate	Output event occurs if one, but not both, of the input events occur
	m out of n gate	Output event occurs if m out of n input events occur

TABLE 6.2
Event symbols

Event symbol	Meaning of symbol
Rectangle	Any event: signify the top event by a double box
Diamond	Undeveloped event: event not significant or no information
Circle	Basic event that requires no further development
Oval	Conditional event used with inhibit gate
House	A switch, either occurring or not occurring
OUT / IN	Transfer symbol: 'out' signified by line from side, 'in' by line from apex

FAULT TREE SYMBOLS

A fault tree is a method by which a particular undesired system failure mode can be expressed in terms of component failure modes and operator actions. The system failure mode to be considered is termed the 'top event' and the fault tree is developed in branches below this event showing its causes.

A fault tree diagram contains two basic elements, 'gates' and 'events'. 'Gates' allow or inhibit the passage of fault logic up the tree and show the relationships between 'events' needed for the occurrence of a higher event.

A list of gate and event symbols is given in Tables 6.1 and 6.2. These are not comprehensive, but prove adequate for most purposes. In order to aid in evaluation, a combination event, drawn as a rectangle, must be used at every logical gate. Gates must not be directly connected to other gates. The fault tree is divided into levels and identified accordingly (Figure 6.1, page 126).

The following case studies are intended to demonstrate the construction of fault trees.

HAZARD IDENTIFICATION AND RISK ASSESSMENT

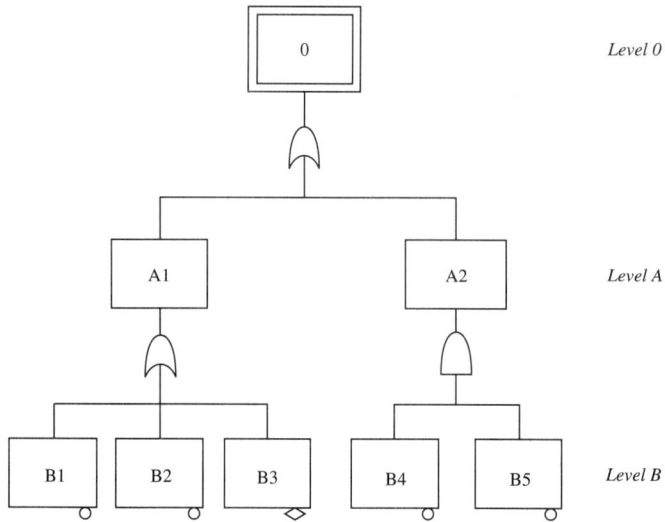

Figure 6.1   Levels in fault tree development.

A FAULT TREE CASE STUDY — A LEAK OF GAS

In Figure 6.2 a fire occurs when a gas leaks providing a flammable mixture at an ignition source. Here the use of an inhibit gate illustrates that the mix is flammable.

Obviously many other factors could have been put in, such as the effect of the wind or the distance to the different sources of ignition. This particular tree creates problems, however, not due to incompleteness but because one of the basic events — the process flame — is, in fact, not a fault. How can this 'fault' tree be drawn better?

One way is to put in an inadequate action by the operator which has a very high probability of failure. For example, it might be possible, in theory at least, for the operator to extinguish the flame. This enables the diagram to contain only faults and may also suggest an action which would mitigate against the incident.

A further alternative is to ignore all normal functions and leave them off the tree. This does not seem very appropriate in this case. If the system is quantified then the presence of the process flame will dominate matters, so maybe there is a need to allow for the wind direction and the dispersion of the gas. What seems to be a simple problem is beginning to appear extremely complex. The problem is reconsidered at the end of this chapter when event trees are described.

ANOTHER CASE STUDY — HIGH PRESSURE IN A TANK

A fault tree is to be constructed for a top event — the release of propane from a relief valve located on the top of a storage tank. One procedure used during construction is to work from the top event to the next level of events and so on, until the basic events are defined. This has been done in Figure 6.3, page 128.

The most common mistake of the inexperienced is to write the basic events in at the first level. The completed fault tree, as developed from Figure 6.3, is shown in Figure 6.4 on page 129.

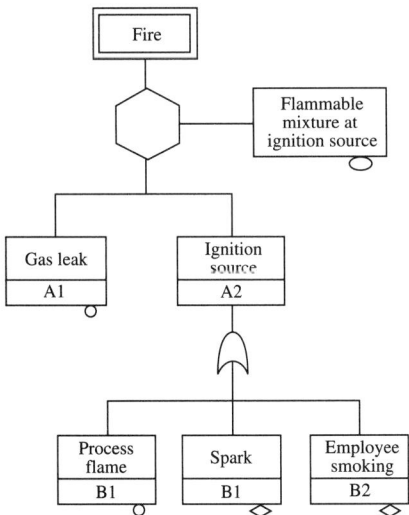

Figure 6.2 A fault tree for a gas fire or explosion.

The presence of propane in the tank could be left off if the system were defined as being in normal operation. But Figure 6.3 contains errors. Have a look in particular at the gates between the second and third levels. One of the most common errors on a fault tree is the selection of the wrong gate. Fortunately, evaluation of the frequency of the top event can generate indications that errors have been made when the results of combining events give frequencies or probabilities which do not agree with common sense estimates.

In this case study a traditional top-down approach has generated the tree. This approach often proves difficult, however, for those less experienced in using this approach. Consequently an alternative construction method based on the generic approach developed for Preliminary Process Hazard Analysis is described next before embarking on an overall methodology for Fault Tree Analysis (FTA).

GENERIC FAULT TREE FOR A RELEASE

The generic fault tree for a release of process material described in Chapter 3 is reproduced again here as Figure 6.5 on page 130.

An operating process goes through various states before unintentional release of material. These are states recognized as representing a snapshot in time, because the process is continuously changing. A fault tree is not a state diagram. In some cases equipment moves from one state to another — for example, from full operation to failed or from valve open to shut. Often, though, there is a gradual change in condition punctuated by events such as alarm or valve closure.

Basically the sequence of key events is:

IMMEDIATE CAUSE
A basic fault occurs, such as the failure of an item of plant or part of the control system.

HAZARD IDENTIFICATION AND RISK ASSESSMENT

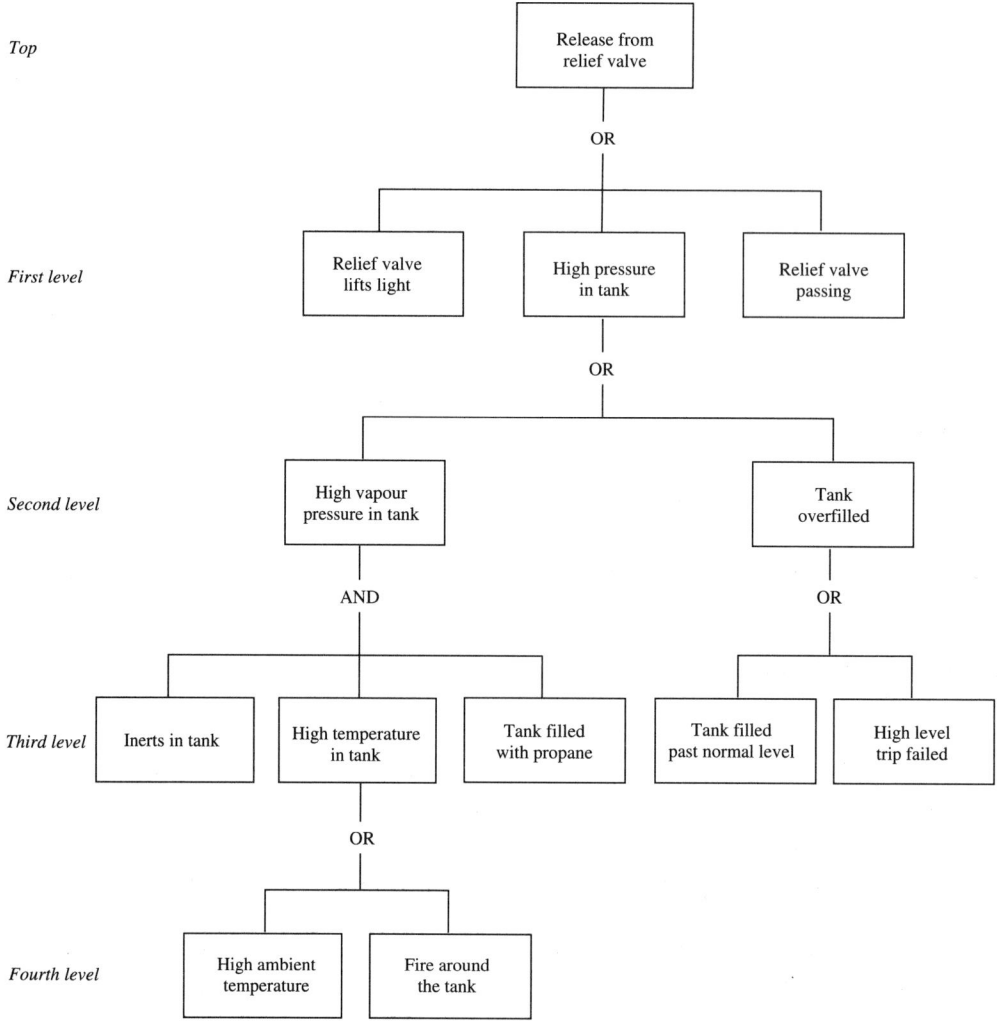

Figure 6.3 Development of a fault tree. (Note that this tree contains errors.)

INADEQUATE NORMAL CONTROL
The basic fault causes a variation in process conditions which inadequate normal control methods fail, or are unable, to correct.

PROCESS DEVIATION
The two events combine to generate a process deviation. Such a deviation corresponds to a Hazop deviation, such as more pressure or low flow.

FAILURE TO CONTROL (ON ALARM)
If the process deviation has been noted as leading to a critical deviation, it is probable that an alarm or other attention-gaining device will be installed in order to attract the operator's attention

FAULT TREE ANALYSIS

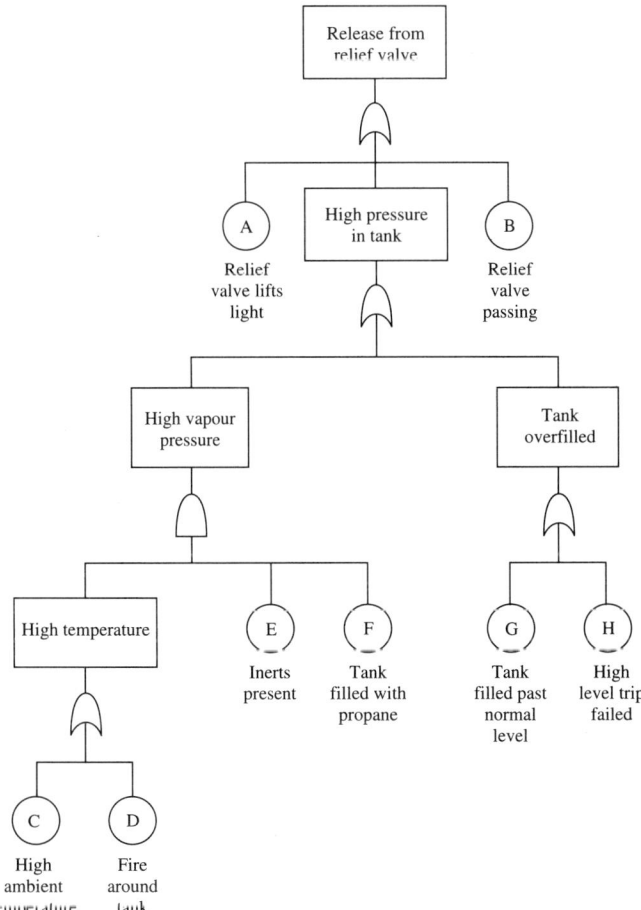

Figure 6.4   Completed fault tree. (Note that there are deliberate errors on this tree.)

to this problem. The operator may then try to correct the deviation, not necessarily in a correct manner. Ideally the procedure used should be specified earlier but even then may not succeed. Alternatively the alarm may fail or the operator may not respond either because of lack of time, knowledge, indisposition, injury or absence.

HAZARDOUS DISTURBANCE
The deviation may have other knock-on effects giving more serious effects or may itself represent a hazardous state requiring emergency action.

INADEQUATE EMERGENCY CONTROL
For many hazardous deviations action will then be taken by automatic emergency control systems or by the operator. This will prevent the deviation continuing to escalate or reduce any consequences.

129

DANGEROUS DISTURBANCE

Action on emergency control may fail, in which case it can be perceived as giving rise to a dangerous state with the deviation continuing to escalate.

FAILURE TO RECOVER THE SITUATION

There will probably be some time in which there is a last chance for the operator to recover the situation — for example, by shutting everything down.

SIGNIFICANT EVENT

Finally the top event occurs, such as the loss of integrity by the system with release of material.

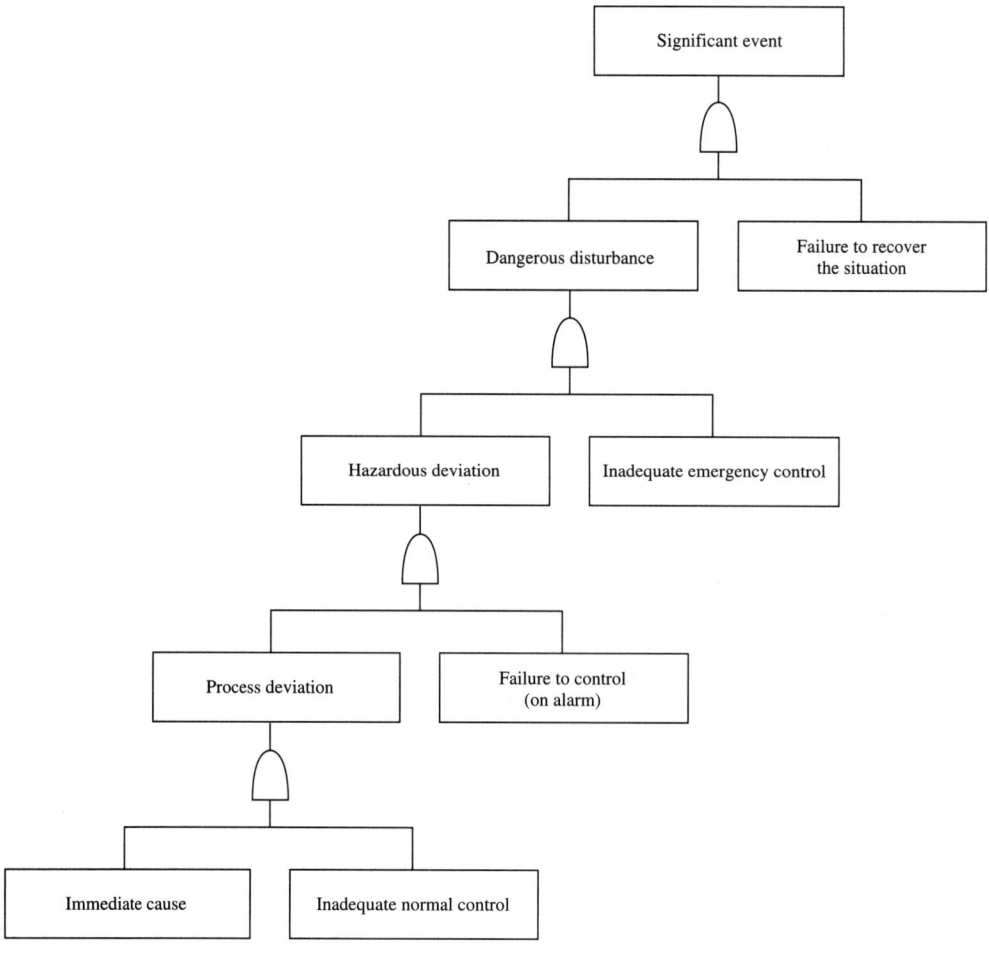

Figure 6.5   Generic fault tree representation of the general incident scenario.

FAULT TREE ANALYSIS

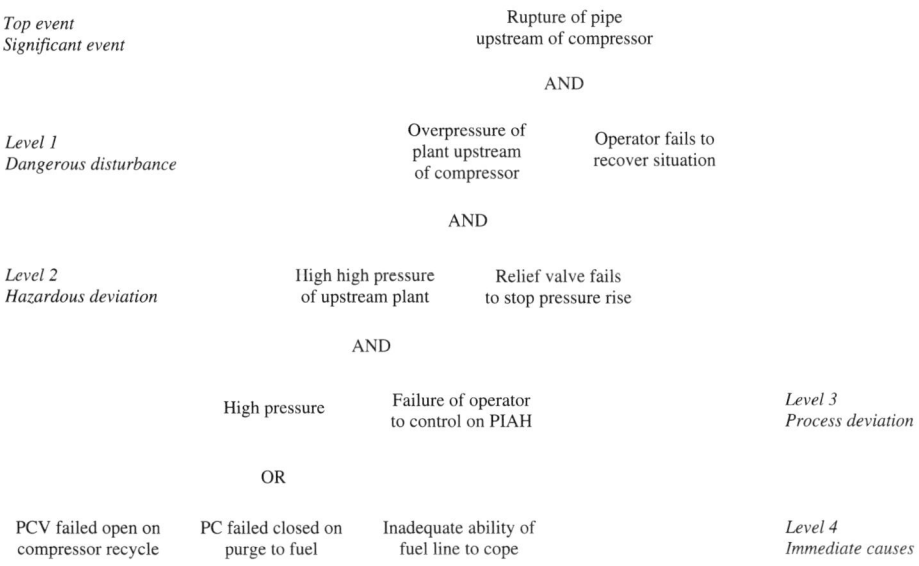

Figure 6.6   Development from generic fault tree.

CASE STUDY — HIGH PRESSURE ON THE METHANATOR

The generic fault tree can be used to develop a fault tree for the compressor system developed in Chapter 5 as Figure 5.6 (pages 118–119). To ease this task the tree presented in Figure 6.6 ignores the activation of the methanator trip system, which in any case is best studied using a separate fault tree. Figure 6.6 is not drawn in conventional form, in order to emphasize working down the levels.

This tree is simplified for ease of presentation with a number of complications associated with the immediate causes ignored. Only three such causes are included. A lack of demand for product is resolved by process relief downstream which could also be used to make unnecessary the need for recycle around the compressor. At this stage in the development of the diagram only the rise in pressure has been expanded. Each of the failures by the operator to control the situation should now be expanded. Note that the success or failure of such action will depend heavily on the experience of the operator and the availability of set and rehearsed procedures to be taken. A total plant trip might be a last resort action to avoid rupture.

The tree can now be drawn up in conventional form (not illustrated here).

CASE STUDY — HIGH TEMPERATURE ON A BENZENE PLANT

An exothermic reaction takes place in a fixed-bed reactor, R101, which is used to convert toluene to benzene by hydrocracking in a stream of hydrogen (Figure 5.4, page 99). The feed to the reactor is preheated in a fired vaporizer as shown earlier. The temperature controller, TRC, is set at 600°C, a high temperature alarm, TAH, is set at 650°C, and the emergency shutdown system, TZH, is activated at 700°C.

131

TABLE 6.3
Preliminary Process Hazard Analysis sheet for fired heater and toluene reactor

Plant: Toluene hydrocracker		**Date:** 1.4.94	
MPI: Fired heater leading to reactor		**Deviation:** Dangerous temperature	
IMMEDIATE CAUSES	**INADEQUATE NORMAL CONTROL**	**PROCESS DEVIATION**	**FAILURE TO CONTROL (ON ALARM)**
TCV1 failed open or stuck. TC loop failure or mis-set.	Control system unable to correct fault	Temperature exceeds 620°C	Failure of TAH or no response by operator. Procedure on alarm is ineffective. Incorrect intervention.
Leak on coil	Control system unable to correct fault	Temperature exceeds 620°C	Procedure on alarm is ineffective

Recommendations/comments/actions:
Scenario (1): A significant increase in the temperature of the feed to the reactor has a knock-on effect as conversion in the reactor increases and the reaction is exothermic. The events would be moderately slow to take effect in scenario (1). The failure of TCV1 acts both as immediate cause and failure of the emergency control system. The two systems should be separated.

The Hazop of this plant has already been presented. A Preliminary Process Hazard Analysis sheet has been completed (Table 6.3) and this has been interpreted as Figure 6.7 on page 134. The actual procedures used to effect recovery are not described here.

From this diagram it is possible to generate the fault tree in Figure 6.8, pages 135 and 136. This system can be modified to eliminate the common mode failure caused by the failure of the temperature control valve, TCV, on the fuel line. The shutdown system should always isolate a system using a valve separate from that of the controller which has failed to correct the situation.

On alarm the operator should check for a coil leak. Such an occurrence requires the immediate manual shutdown of the plant. The alarm may also be caused by the failure of the TRC loop containing the fuel gas. If the temperature continues to rise, the flow of toluene to the plant should be stopped but not the flow of hydrogen which is mixed with the process feed just before the vaporizer (and not shown in the diagram). If the temperature continues to rise the shutdown system will be activated. This stops the toluene feed pump and closes the temperature control valve, TCV, on the fuel gas line. In the event of these actions not being effected, the hand control valve, HCV, on the fuel gas line should be closed.

HAZARDOUS DISTURBANCE	INADEQUATE EMERGENCY CONTROL	DANGEROUS DISTURBANCE	FAILURE TO RECOVER SITUATION	SIGNIFICANT EVENT
Temperature exceeds 650°C	Emergency control fails as TCV open or stuck. TZV fails closed. TZH not activated.	Temperature exceeds 700°C	Failure of TZH or no response by operator. Procedure on alarm is ineffective. Incorrect intervention.	Temperature exceeds 700°C
Temperature exceeds 650°C	Emergency control system fails to correct the problem	Temperature exceeds 700°C	Incorrect intervention	Temperature exceeds 700°C

Recommendations/comments/actions (cont):
Scenario (2): It is possible that the flow would be backwards from the reactor. The resulting fire would require complete closure of the plant.

THE EVALUATION OF GATES

The evaluation of a fault tree is intended to evaluate the failure rate of the top event as a frequency or probability from basic data on failure rates. The basic combination of gates can be demonstrated for both AND and OR gates using Venn diagrams. The diagrams represent the probability of occurrence of each set in a system. This is fault tree parlance for the occurrence of a fault. The area of each set is made equal to the probability of the occurrence of this fault, and the total area of the universal set is equal to 1. The probability of compound events is represented by the area of the combined set.

The union of mutually exclusive or disjoint pairs is shown in Figure 6.9 on page 137. The probability of occurrence of each event A and B corresponds to the area within the circle.

The intersection of two events A and B occurs when the probability of A occurring and B occurring overlaps, as shown in Figure 6.10 on page 137. If these events are failure events, the result corresponds to the AND gate on a fault tree.

The union of independent events A and B occurs when the probability of A occurring and B occurring is represented as shown in Figure 6.11 on page 137. If these events are failure

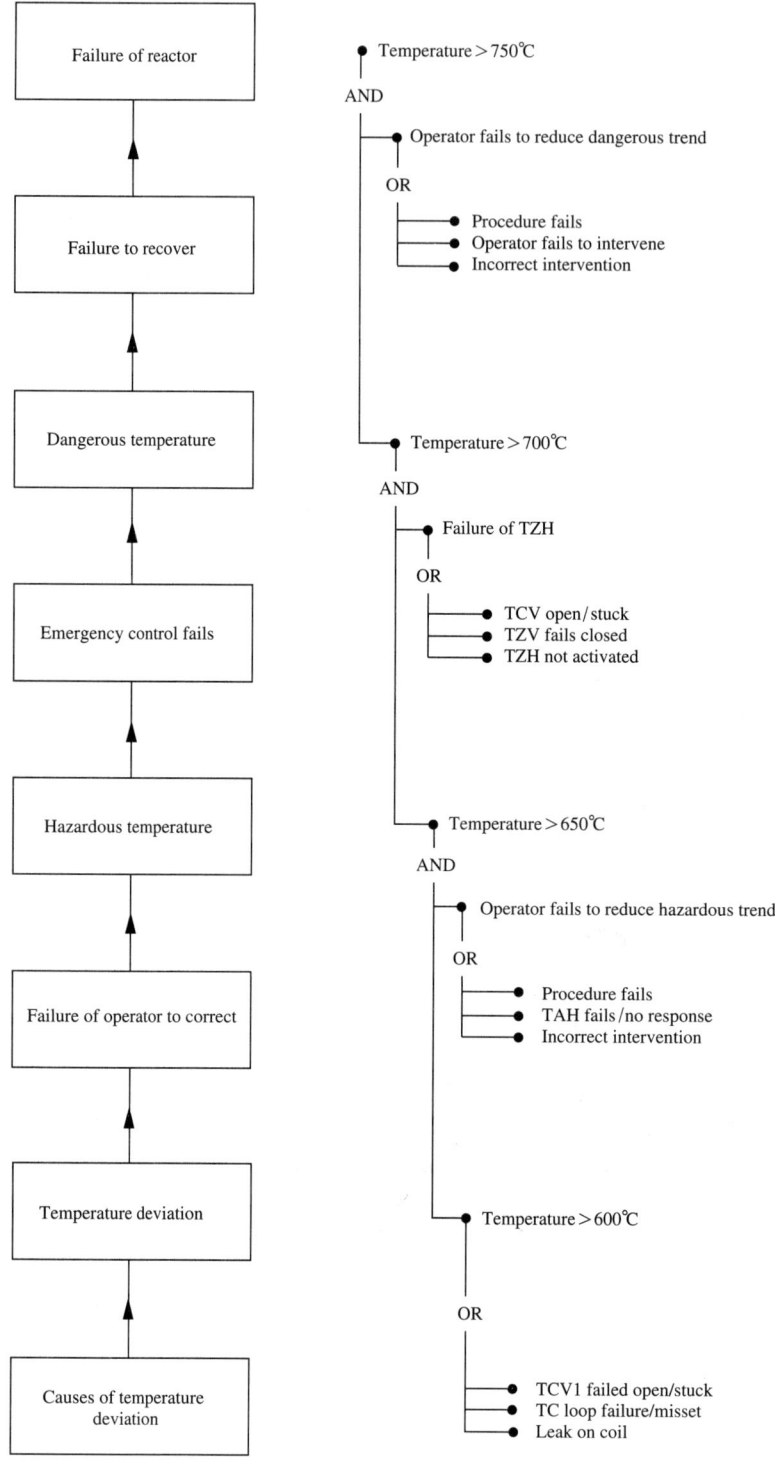

Figure 6.7 Incident scenario — toluene reactor.

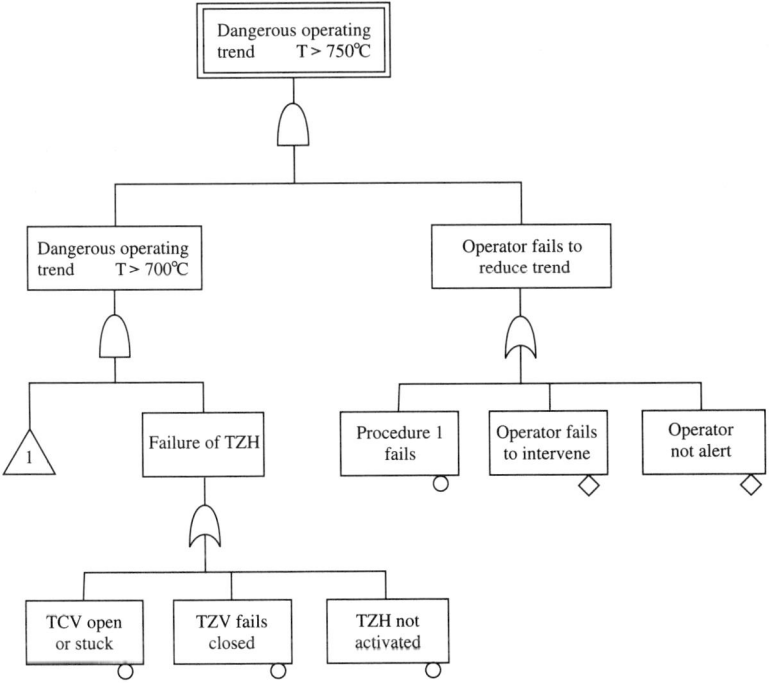

Figure 6.8   (Part 1) Fault tree for the toluene reactor.

events, the result corresponds to the OR gate on a fault tree. Note that the product of A and B must be subtracted from the sum of the probabilities. Otherwise the total space occupied within the shaded area is evaluated incorrectly.

Similar results occur when more independent events overlap. Figure 6.12 on page 138 shows the Venn diagram corresponding to the combination using an OR gate of the three failure events A, B and C.

The notation used for the union and intersection of sets is cumbersome and is frequently modified to **+** and **.** respectively. These signs are identified by bold type and this is appropriate because these operations have many but not all of the properties of addition and multiplication. Hence it is possible to write the sets as follows:

$$P(A\mathbf{+}B\mathbf{+}C) = P(A) + P(B) + P(C) - P(A).P(B) - P(A).P(C) - P(B).P(C) + P(A).P(B).P(C)$$

$$P(A\mathbf{.}B\mathbf{.}C) = P(A).P(B).P(C)$$

THE USE OF FREQUENCIES AND PROBABILITIES

A common problem arises from confusion in the use of frequency and probability units. Information about events may be measured in units of frequency or probability. Some programs work out everything by converting all frequencies to probabilities, but this makes major assumptions about the distribution on which the information is based.

HAZARD IDENTIFICATION AND RISK ASSESSMENT

There is no difficulty in using both sets of units as long as care is taken that the units are always correct after their combination. Probabilities are dimensionless; frequency is measured as the product of the probability per operation and the number of operations per year, or other time interval. This means that certain rules apply (Table 6.4, page 138).

When working with mixed units few problems arise if deviations of parameters such as immediate cause, hazardous trends and dangerous trends are measured in frequencies while failures to control the situation are measured in probabilities.

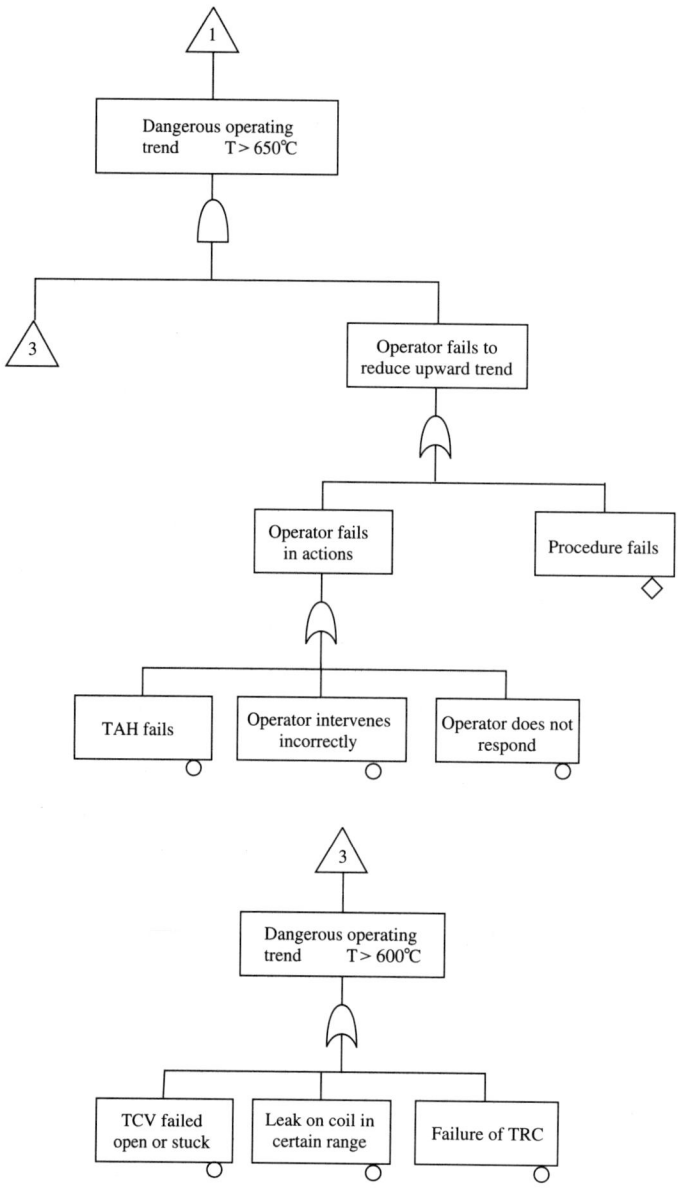

Figure 6.8   (Part 2) Fault tree for the toluene reactor.

FAULT TREE ANALYSIS

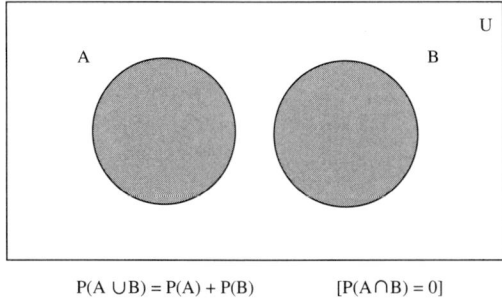

$P(A \cup B) = P(A) + P(B)$ $[P(A \cap B) = 0]$

Figure 6.9   The union of mutually exclusive or disjoint pairs.

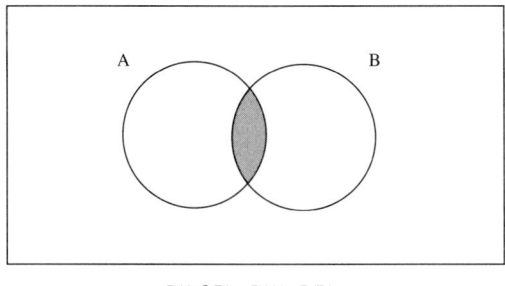

$P(A \cap B) = P(A) \cdot P(B)$

Figure 6.10   The intersection of independent events.

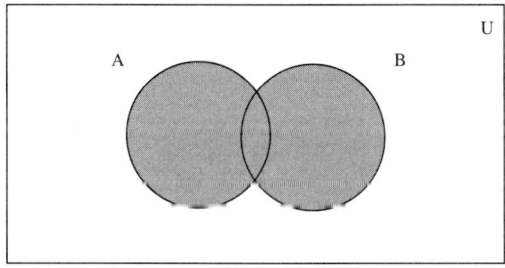

$P(A \cup B) = P(A) + P(B) - P(A \cap B)$

Figure 6.11   The union of independent events.

137

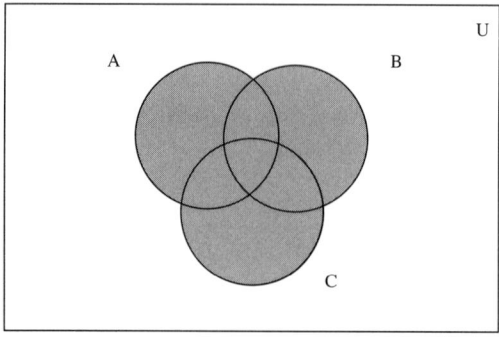

$P(A \cup B) = P(A) + P(B) + P(C) - P(A \cap B) - P(A \cap C) - P(B \cap C) + P(A \cap B \cap C)$

Figure 6.12 The union of three independent events.

TABLE 6.4
Rules for the combination of gates

AND gate

$P(A.B) = P(A) \times P(B)$	result in units of probability
$F(A.B) = F(A) \times P(B)$	result in units of frequency

Not allowed: $F(A).F(B)$ as the result is in units of (frequency)^2

OR gate

$P(A + B) = P(A) + P(B) - P(A).P(B)$	result in units of probability
$F(A + B) = F(A) + F(B)$	result in units of frequency

Not allowed: $F(A) + P(B)$ as the result is meaningless

Note that when the values for the probability of the occurrence of faults are small, the value of the coproducts is small and can be ignored as a correction factor in OR gates.

APPLICATION TO A SIMPLE PUMP SYSTEM

In a pump system (Figure 6.13) it is assumed that the supply tank is always full and the pipe system and the valves never fail. The only failures that are perceived as being significant have the probabilities of failure as follows:

Probability of mechanical failure of pump X $P(X) = 0.25$

Probability of mechanical failure of pump Y $P(Y) = 0.25$

Probability of electrical supply failure $P(G) = 0.1$

FAULT TREE ANALYSIS

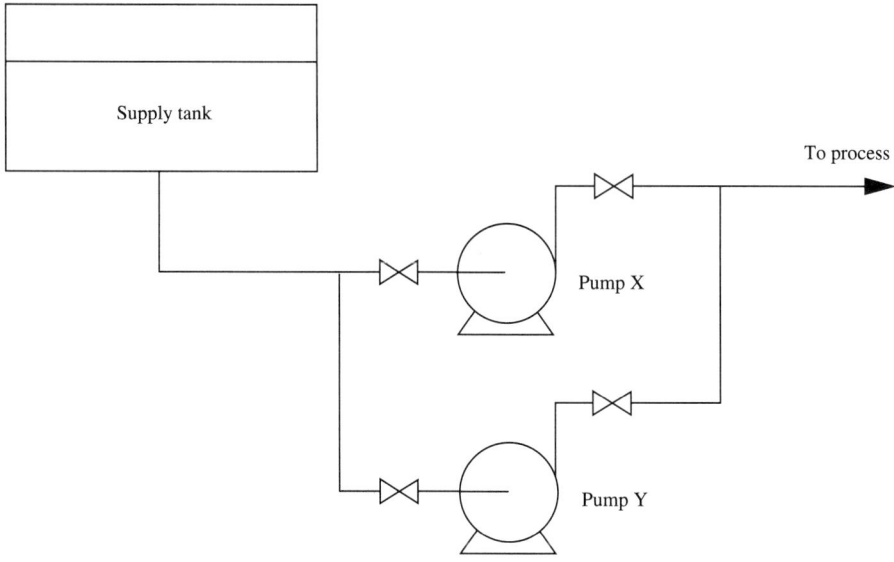

Figure 6.13 A simple pump system.

The fault tree for this problem is first written as Figure 6.14 — see page 140. It is evaluated as follows:

P(PUMP X FAILS) = P(G) + P(X) − P(G).P(X) = 0.325

P(PUMP Y FAILS) = P(G) + P(Y) − P(G).P(Y) = 0.325

Hence the probability that both pumps fail is given by:

P(TOP) = P(PUMP X FAILS).(PUMP Y FAILS) = 0.106

The fault tree has been evaluated in a manner that obeys the rules already described, but is this answer correct? Before responding, first consider the rules of Boolean algebra.

BOOLEAN MANIPULATION

Boolean algebra is widely used to evaluate fault trees. However, its use will not necessarily be appreciated when using programs to evaluate fault trees. Its concepts may at first seem odd but are easy to master. It then enables the analyst to identify inconsistencies in the program logic, particularly where a basic event appears more than once on the diagram. For convenience the probability sign P will not be used here when evaluating Boolean algebra, but the following notations are used instead for the union and intersection of the sets.

Union: A + B   Intersection: A.B

The rules of Boolean algebra are conveniently divided into those which follow normal numerical manipulation and those which follow the specific rules of Boolean algebra.

139

HAZARD IDENTIFICATION AND RISK ASSESSMENT

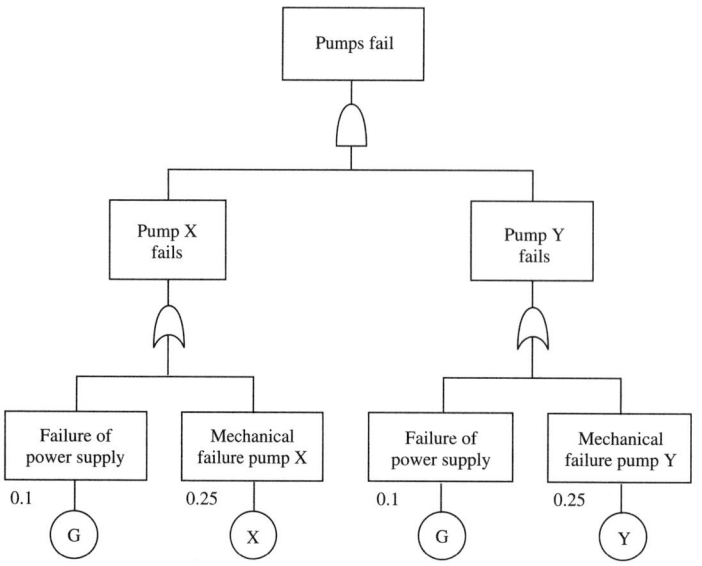

Figure 6.14 The fault tree for a simple pump system.

RULES SIMILAR TO THOSE USED IN NUMERICAL MANIPULATION
- The order in which operations are performed is • before + unless overwritten by brackets. In (A+B•C) first evaluate B•C; call the result X; then evaluate (A+X).
- Operations are commutative:

    A + B = B + A and A•A = B•A

- Operations are associative:

    (A + B) + C = A + (B + C) and (A•B)•C = A•(B•C)

- Operations have identities:

    A + 0 = A and A•1 = A

RULES DIFFERENT FROM THOSE USED IN NUMERICAL MANIPULATION
- The union (•) and the intersection (+) are idempotent: their values to any power are the same.

    A + A = A and A•A = A

    This means that duplicates can be eliminated. If the Venn diagram is drawn for this system, all that appears is a single circle representing A.

- Events can be combined using the absorption law.
    For example, suppose a fault tree was evaluated as:

    TOP = A + A•B = A

FAULT TREE ANALYSIS

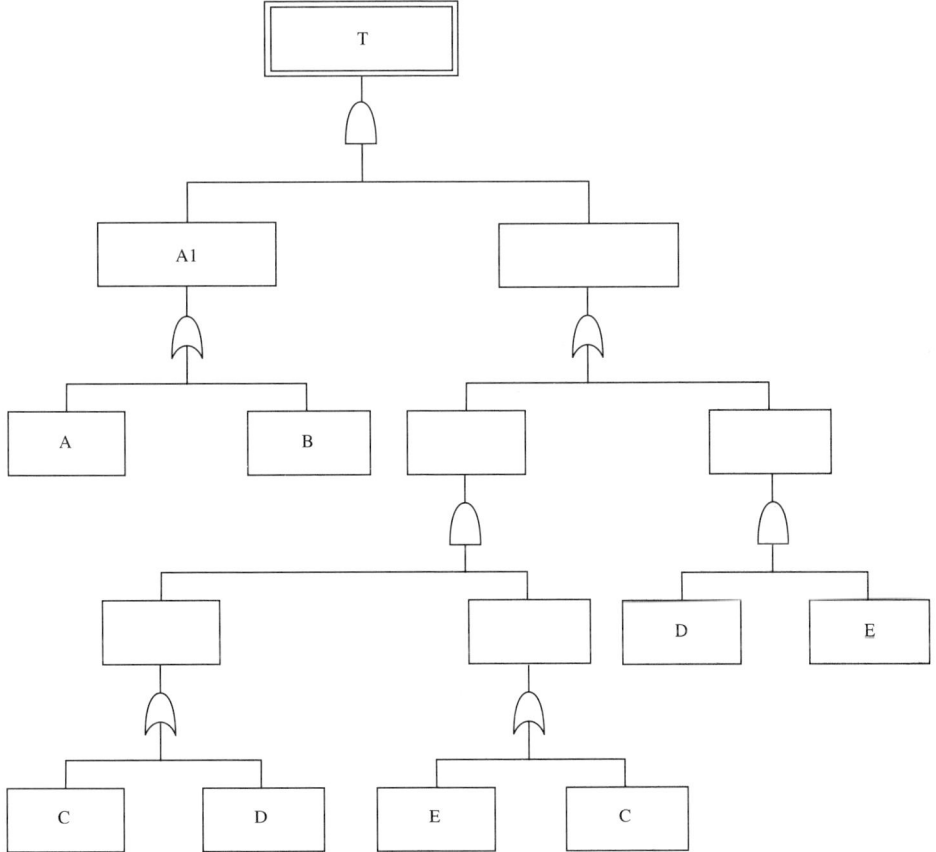

Figure 6.15 Fault tree for a processing system.

The Venn diagram containing an overlap of A.B is completely masked by the circle representing A alone.

Similarly the expression below can be reduced as indicated.

$$X = A + A \cdot B + B \cdot C + B \cdot C \cdot D + A \cdot D = A + B \cdot C$$

It makes no difference in this expression if event D occurs, as whenever B and C both occur this has the same effect on the top event X as if B.C.D had happened.

BOOLEAN ELIMINATION OF REDUNDANT TERMS

Redundant terms in fault trees may be eliminated using the following procedure:

• The initial fault tree for a processing system shown in Figure 6.15 can be written down in Boolean algebra as follows:

$$(A + B) \cdot [(C + D) \cdot (E + C) + D.E]$$

141

- This expression can be expanded using the multiplicative distribution law to eliminate the bracket.

$$(A + B) \bullet (C.E + D.E + C.C + D.C + D.E)$$

- All duplicates can be eliminated in this set:

$$(A + B) \bullet (C.E + D \bullet E + C + D \bullet C + D \bullet E)$$

All sets are placed in length order:

$$(A + B) \bullet (C + C.E + D \bullet E + D \bullet C + D.E)$$

These are rearranged in alphabetical order to give:

$$(A + B) \bullet (C + C \bullet D + C \bullet E + D \bullet E + D.E)$$

Subsets are checked and eliminated.

(a) First use the idempotent law: $(A + B) \bullet (C + C.D + C \bullet E + D \bullet E)$.

(b) Then use the absorption law: $(A + B) \bullet (C + D \bullet E)$.

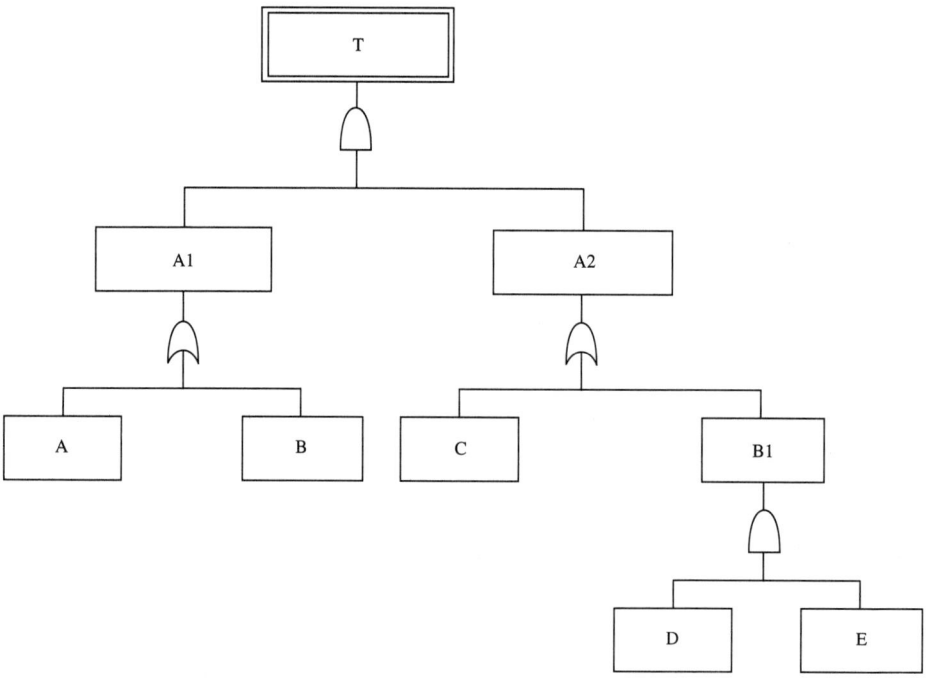

Figure 6.16 Revised fault tree for a processing reactor.

The revised fault tree can then be redrawn as Figure 6.16. Comparing Figures 6.15 and 6.16 it is clear that the second tree shows considerable improvements over the first. The redundant terms on the left-hand side of Gate A2 have been eliminated and reduced simply to event C. In retrospect it would have been easy to do this by inspection. But in a more complex tree this would have been less simple to accomplish.

Different answers for the frequency or probability of the top event arise when the same values for basic events are substituted in the two cases. Clearly the second diagram is correct and the first diagram contains a number of redundant terms which introduce errors.

In order to study the problem further we will revise the answers from the pump problem using the above technique.

THE PUMP EXAMPLE REVISITED

The fault tree for the pump example given in Figure 6.14 (page 140) can be reduced in order to eliminate duplicate events. The top event is written as follows:

$$\text{TOP} = (G + X) \bullet (G + Y) = (G \bullet G + G \bullet X + G \bullet Y + X \bullet Y)$$

After removing duplicates and absorption this becomes:

$$\text{TOP} = G + X \bullet Y$$

The new fault tree is given in Figure 6.17. Probabilities must still be converted into conventional mathematics. The equation:

$$\text{TOP} = G + X \bullet Y$$

can be evaluated as:

$$P(\text{TOP}) = P(G) + P(X).P(Y) - P(G).P(X).P(Y)$$

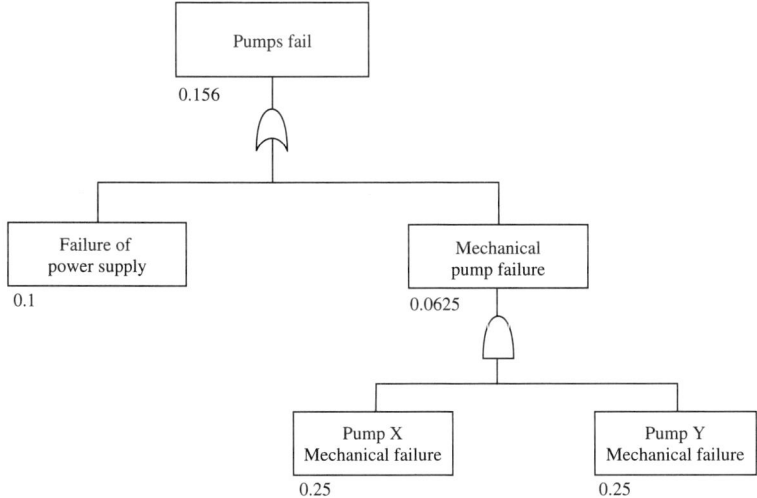

Figure 6.17 Revised fault tree for the pumping system.

COMMON MODE FAILURES

Systems employ redundancy and voting techniques to achieve high reliability. These techniques are often applied at a subsystem level or a major component level rather than at a fundamental component level. Redundancy can either be standby or active, uniform or diverse. Systems using redundancy techniques can tolerate a certain number of types of independent failures while continuing to maintain the required relationship between input and output conditions. These systems are prone to common mode failures which are caused by influences on the system from some external source that is common to the redundant components.

A *common mode failure* is the result of an event which, because of dependencies, causes a coincidence of failure states of components in two or more separate channels of a redundancy system, leading to the defined system failing to perform its intended function.

The general policy in the design and operation of redundancy systems must be the prevention of common mode failures, or at least the minimization of their frequency of occurrence and their effects upon the system. The extent to which this policy is applied depends on the reliability requirements of the system and the financial, operational and safety needs of a particular project.

The various causes of common mode failures can be classified in the same way as root causes of incidents. They include engineering design failures, manufacturing installation and commissioning faults, inadequate maintenance and operating procedures and practices and external causes including normal extremes, energetic and toxic events. Their prevention is directed through appropriate administrative control of the engineering and operational processes. Feedback can be arranged through design and safety reviews, independent reliability assessment and performance monitoring of reliability. Appropriate actions must be taken during the operational life of the plant to ensure that the system reliability does not deteriorate.

BOOLEAN REDUCTION OF NON-INDEPENDENT EVENTS

Boolean reduction is required when redundant terms appear in the logic equations describing a fault tree. Such redundancy can occur from several sources. One of the more frequent is the need to account for practical non-independence when systems comprising theoretically independent components are assessed.

COMMON EXTERNAL CAUSE

A typical case arises where the signal processing elements of several independent protection devices are located adjacent to each other in the same cabinet or rack. A temperature rise in the cabinet due to a power unit fault might cause the simultaneous failure of several of the processing elements. Where such a common cause is explicit, it can be included on the fault tree as a separate event. But because many common causes can be postulated, albeit at low hazard rates, their inclusion in the analysis can lead to diminishing practical returns.

A system consists of N nominally identical subsystems, all of which must fail to give system failure. The subsystem failure probabilities are $P1$, $P2$, etc and the system failure probability, $P(S)$ is given by:

$$P(S) = P1 \times P2 \ldots \times PN$$

An event of probability PQ can occur which simultaneously results in the failure of all the subsystems. The system failure probability becomes:

$$P(S) = (P1 + PQ)(P2 + PQ)(P3 + PQ) \ldots (PN + PQ)$$

and

$$P(S) = (P1 \times P2 \times P3 \ldots \times PN) + Q$$

COMMON INTERNAL CAUSE

The presence of a faulty batch of subcomponents could merely lead to an increased random failure rate for the production run so affected. But if the subcomponents are of such poor quality that they are suffering from early wear-out failure, the associated failures may become non-independent. A more likely case of non-independence is where a faulty batch of subcomponents make the whole batch unsuitable for the design duty. This whole batch may then be susceptible to an otherwise external event which would normally cause no damage. Faulty maintenance can cause such a situation, as when a fitter consistently reassembles items in an incorrect manner or uses non-standard replacements or lubricants. In such circumstances the event cannot be meaningfully specified or included in the fault tree; indeed, even in retrospect it may not be known.

Interaction

The failure of one system can either directly cause a system to fail or cause degradation so that its failure rate increases. Another form of interaction is when a failure causes an explosion and the subsequent missiles strike another plant item. The effects of interaction are difficult to examine analytically unless gross simplifying assumptions are made.

The beta factor model

A common term has been postulated for the three cases of non-independence considered. It is believed that basic generic failure data already include a degree of non-independence. This is because the compilers of such data have, in many cases, an incomplete knowledge of the conditions under which the failures occurred.

The generic failure rate quoted by a source is $F(S)$. This is assumed to contain a random element, $F(R)$, and a common term, $F(C)$.

$$F(S) = F(R) + F(C)$$

This common term is assumed to be proportional to the random term, thus numerically:

$$F(S) = F(R) + \beta.F(R)$$

Unfortunately the value of $F(R)$ is not known, but it is known to be a little less than $F(S)$. Hence its replacement by $F(S)$ can be justified as giving a slightly pessimistic approximation. Hence the value to use for the frequency, $F(A)$, becomes:

$$F(A) = F(S) + \beta.F(S)$$

This is known as the beta factor model for common mode failure. Typical values of beta are between 0.2 and 0.02, usually towards the lower end of the scale. In practice a fault tree is constructed under the assumption of independence. It is then examined to determine where dependencies are likely to exist in practice. Then an appropriate common term calculated in this way is added to the tree.

OVERALL METHODOLOGY FOR FAULT TREE ANALYSIS

Here is the overall methodology for Fault Tree Analysis (FTA) when it is used to estimate the frequency of unwanted events in multicomponent systems:

(i) Define the top event of a fault tree.

(ii) Choose the events identified by the hazard identification exercise which can lead to this top event. A selective approach is often desirable, including only those events which are likely to be significant. Do not assume remote coincidences. If normal functioning propagates a fault sequence, assume the component functions normally.

(iii) Decide on the hierarchical construction of the fault tree.

(iv) Construct the fault tree; it is preferable that all inputs to a particular gate should be completely defined before further analysis of one of them is undertaken. Gaps can be filled in later; the use of the generic tree makes this task easier.

(v) Quantify the basic events. It may be possible to eliminate repeated basic events by analysis of the fault tree.

(vi) Quantify the top event.

(vii) Analyse the results to determine the significance of particular basic events or combination events. Also check if the results seem sensible.

(viii) Carry out a sensitivity analysis on the fault tree to test the following factors:

- the uncertainty of basic data;
- the effect of improving the reliability of plant and control systems;
- the effect of varying the method of operation on the plant. Has the role of the operators in recovering the situation been taken into account?
- the effect of plant modernization;
- the effect of improved training of operators.

(ix) Report the results. Try to avoid reporting to an absurd number of decimal places or very low powers of ten.

Clearly the evaluation of the results can give useful information to the analyst. If the results seem inaccurate, it is usually because one or more wrong gates have been selected. As well as providing information on the frequency of significant events, this data serves as an input to risk assessment. The note on the accuracy of the results is important. If the frequency of a major event is reported as 10^{-30} times per year, one begins to lose faith in the analysis. This frequency might be correct for death by tidal wave when standing in the middle of the Sahara desert but not for many seriously credible risks.

Further information will be given later about the development and construction of event trees. Fault trees are often useful for generating information on significant releases of material, but event trees are better able to express the different outcomes of incidents stemming from this event. This is because of the wide diversion of scenarios from this event onwards.

LIMITS OF RESOLUTION AND ACCURACY

The top event and the system boundary must be selected with care. The objective at the end of the study is not to have identified every failure scenario, but to have included the vast majority of those which are significant (Figure 6.18). The choice of system boundary determines how comprehensive a study is made. The external boundary is defined by the feature of system performance which is of interest. The level of resolution to which the system is developed must be established. It may not be necessary to extend the analysis to the level of certain subsystems or component levels. The choice of a limit of resolution therefore sets the detail of the analysis, and

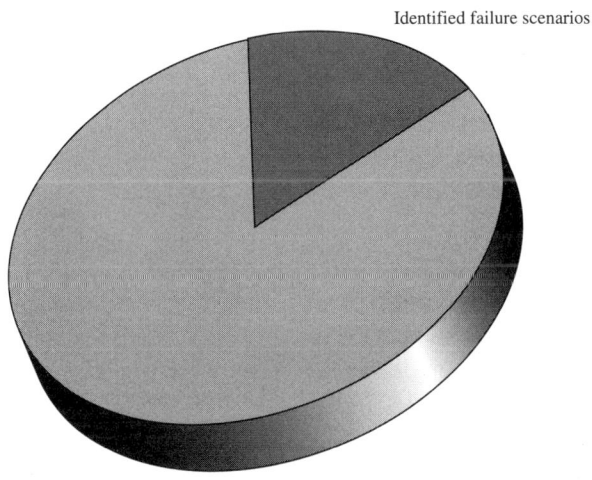

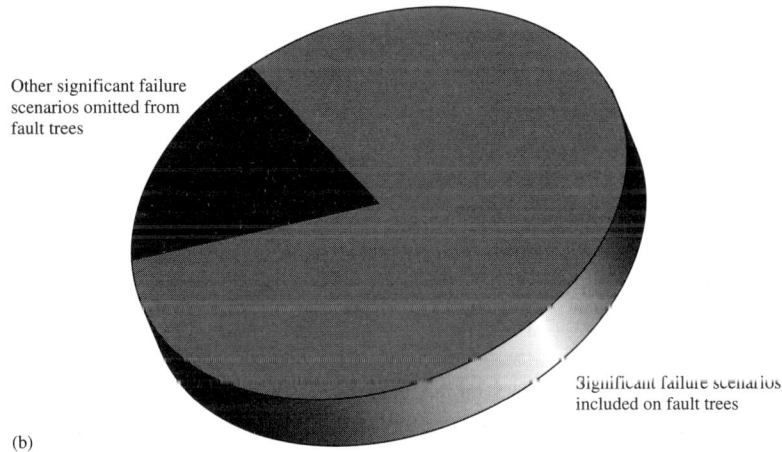

Figure 6.18 The need for completeness of the fault tree.
(a) Possible failure scenarios. (b) Significant failure scenarios.

base events indicate the limit of resolution of the fault tree. It is determined, amongst other factors, by considerations of the objective and feasibility of the analysis. For example, if the failure probability of a system is to be calculated, the limit of resolution is affected by the availability of data on component failures.

The most common problems encountered in Fault Tree Analysis are:
- the inadequate specification of the system boundary;
- failure to incorporate all significant failure modes;
- poor construction of the fault tree and selecting the wrong gate;
- inconsistent units on fault trees;
- poor quality of failure data and weak treatment of uncertainty in the analysis;
- common mode effects.

It should be noted again that a fault tree is merely a snapshot in time. It does not reflect system dynamics. Components can exhibit degraded behaviour as well as an on-off working state. Also the reliability for repairable components is not amenable to exact solution.

EVENT TREES

Event trees can be used to analyse the probabilities of different outcomes which have been identified during a safety study. From the event which initiates any specific accident sequence, normally a significant release, the functioning and failure of safety subsystems are analysed using forward logic. This results in a tree structure with the branches developing from left to right. By assigning a probability to each branch the probabilities of every possible outcome following the initiating event can be determined.

A particular construction is adopted and it involves expanding branches. Occasionally branches are drawn in which recombination occurs; a cause-consequence diagram is one such example. Such diagrams are rarely used and are not considered here.

The steps in an event tree analysis are as follows:

STEP 1: IDENTIFY THE INITIATING EVENT
Normally this will be a failure event corresponding to the release of hazardous material.

STEP 2: IDENTIFY THE DEVELOPMENT OF THE INCIDENT
Consider further attempts to control the situation and mitigation.

STEP 3: CONSTRUCT THE EVENT TREE
At each failure mode, the alternative events that materially affect the outcome are shown. For an example of likely tree outcomes, see the various illustrations in this chapter.
- The event tree is constructed from left to right.
- At each failure mode, the alternative events that materially affect the outcome are shown.
- The event headings are indicated above the node. A 'success yes, or did occur' response on the tree is made to branch upwards; a 'failure no, or did not occur' response is made to branch downwards.

Sometimes each heading is labelled with a letter identifier so that each sequence can be identified by a unique letter combination. A bar over a letter indicates the response to the heading was 'failure or did not occur'.

FAULT TREE ANALYSIS

STEP 4: CLASSIFY THE OUTCOMES OF THE INCIDENT
The development of the event tree should only be so far as to meet the goal of the analysis. It may be reasonable to develop the tree only as far as a major incident.

STEP 5: ESTIMATE THE PROBABILITY OF EACH BRANCH IN THE TREE
Each heading in the event tree, other than the initiating event, corresponds to a conditional probability of some outcome if the preceding event has occurred. Thus the probabilities associated with each branch must sum to 1.0 for each handling.

STEP 6: QUANTIFY THE OUTCOMES AND REVIEW THEIR ACCURACY
The frequency of each outcome is determined by multiplying the initiating event frequency by the conditional probabilities along each path leading to that outcome. The calculations assume no dependency among events or partial success/failure. The results should be checked against the frequency of occurrences in the historical record.

EVENT TREE FOR A LEAK OF GAS
A leak of gas was considered in Figure 6.2 (see page 127). This is now presented as an event tree in Figure 6.19. Here it has been possible to consider the probability of ignition by the process flame or other source and, according to their location, an estimate is made of the likelihood of delayed or immediate ignition and the development of the incident into a torch fire or a vapour cloud explosion (VCE). It will be noted that the tree expands and the frequency of a torch fire must be evaluated by summing two chains of events.

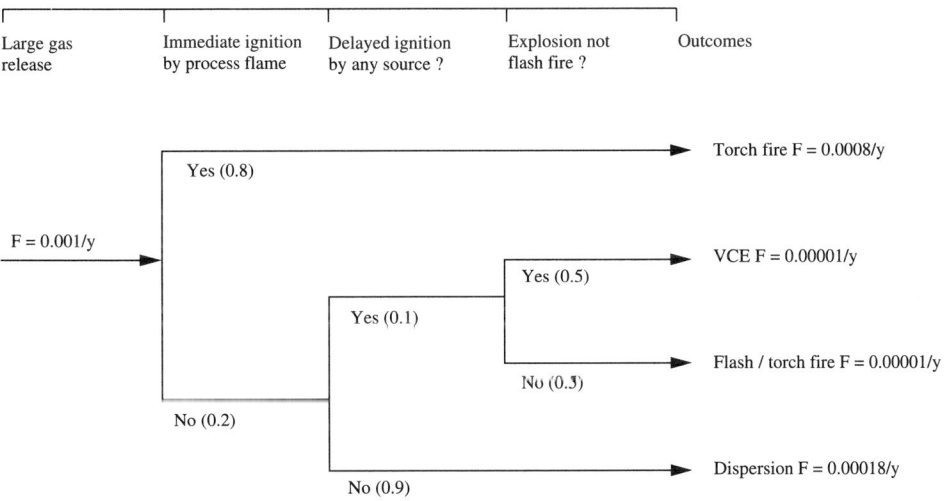

Figure 6.19 The event tree for a gas release.

149

EVENT TREES: A SIMPLE DESCRIPTION OF AN LPG RELEASE

A further event tree is given in Figure 6.20 which represents the energetic events stemming from a large release of LPG and considers immediate or delayed ignition, Pascal weather type of D5 or F2, and the likelihood of a vapour cloud explosion. The section where flame impinges on an LPG tank and possibly gives rise to a BLEVE is grossly simplified as many factors are involved and what happens at source following a flash fire needs to be described in more detail. The diagram can be expanded to allow for the frequency of an individual fatality, considering such factors as the weather type, outdoor or indoors, present or not at the location, the effect of the emergency response and the probability of survival of a given impact.

USE OF EVENT TREES TO CHECK FAULT TREES

An event tree can be drawn out to check that the fault tree has been drawn correctly. For example, the fault tree in Figure 6.1 contained errors (see page 126). In Figure 6.21 the event trees corresponding to a section of the fault tree have been sketched and the section revised. In the upper diagram there is only one undesired outcome indicating one AND gate between all the actions. Propane added was left off the fault tree as a normal event. However it would be included should its frequency be intermittent. In the lower diagram there are three outcomes indicating two (or one) OR gates with an AND gate linking in to indicate the presence of propane in the tank. The latter is a normal state but is left in for the purpose of illustration. It would have a value close to one.

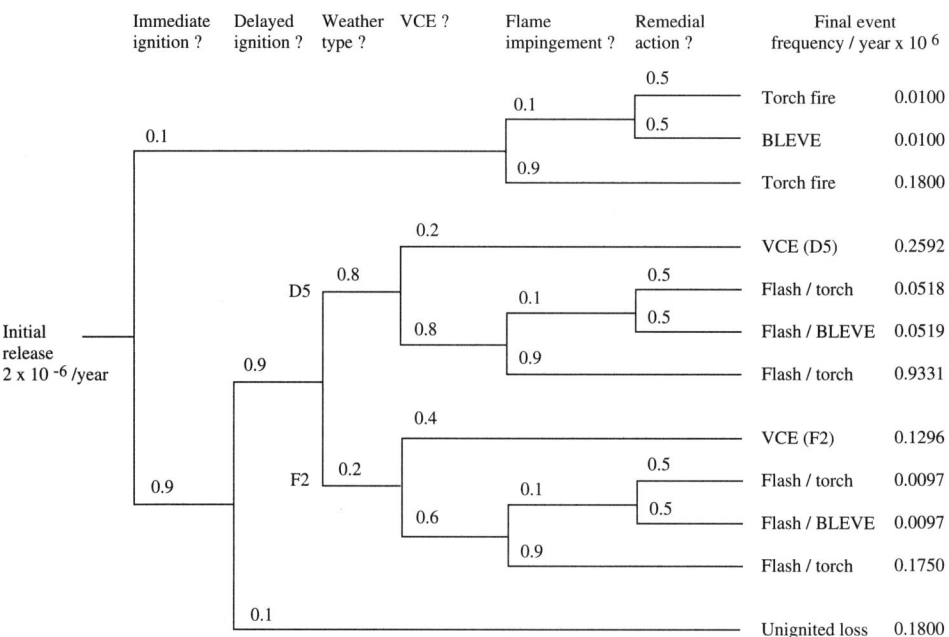

Figure 6.20 Event tree for an LPG release.

150

FAULT TREE ANALYSIS

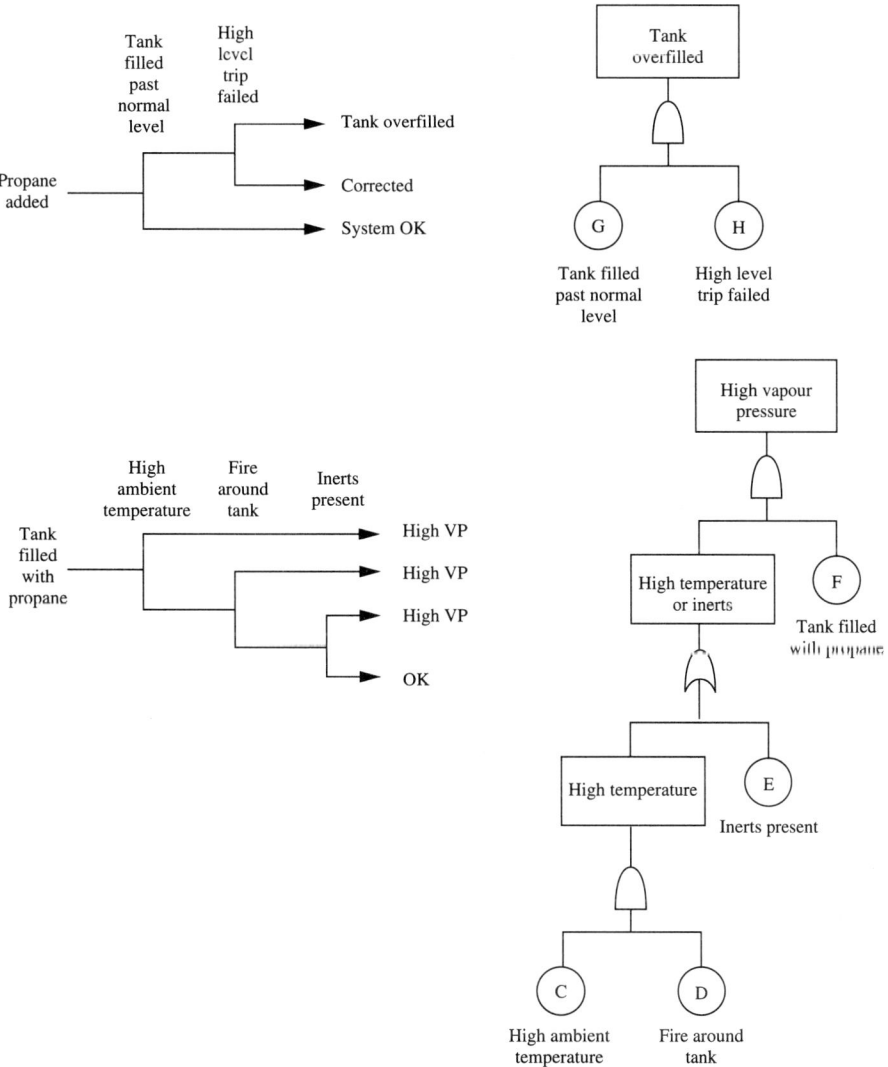

Figure 6.21   Comparison of fault and event trees.

151

7. TASK ANALYSIS

INTRODUCTION

The errors made by operators can be classified in a similar manner to hardware faults. Although human error may be perceived as the immediate cause of incidents, this does not automatically imply blame. An error is often the end-point of a complex series of events involving failures associated with the process and plant, the procedures and practices and communication systems. The fault may arise in the failure to provide adequate resources and training. The management and supervision may be inadequate. Clearly an error is more likely if the working conditions are not optimal, or if pressure on the operator to make a correct decision is high due to any cause. Unfamiliar conditions and lack of technical back-up and information can also increase the difficulty of the task.

All these factors may be described as 'performance shaping factors' which — along with a number of problems associated with the operator's cognitive processes incorporating features such as psychological mechanisms — affect the likelihood of making an error. For example, owing to a strong habit actions are diverted along a familiar but unintended pathway. Analysis of this helps in the prevention or reduction in the frequency of occurrence. At this stage in the analysis the focus is not the likelihood of error but rather the immediate cause of the accident and reasons for failing to control or recover the situation. The main emphasis is on identifying how the cause should have been recognized by analysis of the proposed or existing plant.

VIOLATIONS, SLIPS AND MISTAKES

A *violation* is an intentional departure from accepted practice. There can be many reasons which lead to a violation; an example is not wearing a hard hat because a hat is hot to wear, because it is getting in the way, because it is less comfortable than being bare-headed or because a hat may not be readily available.

Work may be done consciously or unconsciously (automatically). Conscious behaviour is being used whenever a task is being learned. There is an intermediate level of rule-based behaviour as rules are developed about how to do the task. Finally automatic behaviour, generally known as skill, takes over once complete proficiency is attained.

At work people try for as long as possible to control demands on their resources using skill-based behaviour. In this way they can work for several hours without unacceptable fatigue. When the situation requires more analytical reasoning, as when incidents occur, they switch to rule-based or conscious behaviour.

Slips are errors in automatic skill-based behaviour. *Mistakes* are errors in conscious behaviour which may be either rule-based, if the task merely involves following a set of procedures, or knowledge-based when the task involves evaluation of a new situation. An *error* may be defined as the failure of planned actions to achieve their goals. Slips occur when actions are not as planned and mistakes arise when planning is inadequate.

EXTERNAL ERROR MODES

There are basic external error modes (Table 7.1) which afflict the performance of humans. Here the performance under scrutiny is mainly that of plant operators. The descriptions of why, when or how the error occurred are merely indicative of a vast range of problems; the net effect is the same — the omission or commission of an error. It is unusual to put in the operator's condition at this stage, but an analogy can be drawn with modes of equipment failure.

An external error mode is one that can be observed, as opposed to the psychological error mode which may underlie it. Identification of it is a first step in determining appropriate ways of reducing the likelihood of errors occurring and determining their probability of occurrence.

EQUIPMENT FAILURE MODES

There are many similarities between equipment failure modes (Table 7.2, page 154) and human errors. To call one a failure and the other an error is not necessarily helpful, with the inference that 'to err is human; to forgive, divine'.

Actions by equipment are normally subdivided into passive or active actions as follows:
- active — physical position or activity in the performance of an equipment's function;
- passive — equipment not physically actuated in order to perform its function.

TABLE 7.1
Basic external modes of human errors

Omission of a task step or substep	An error involving: • lack of attention (fatigued, high workload, distracted) • lack of response (absent, incapacitated, time pressure, equipment malfunction) • failure to support or retain • unaware of need for action (no signal, wrong reading, incorrect information processing)
Commission of an extra step or action	An error involving: • a selection mistake (wrong object, wrong action) • wrong action (too much, too little, too long, too late) • wrong direction (insertion, misalignment) • wrong timing (delay or premature action) • wrong duration (mistiming, equipment problems) • performance out of sequence (bad procedure, miscommunication) • replacement of correct action (drop, lift, close, open) • use of excess force (tightening, fitting, closing)
Change in operator's physical or mental condition	An error involving: • operator working under non-optimal conditions • operator not in optimal conditions • involuntary action due to fall • operator absent or unable to act

TABLE 7.2
Modes of equipment failure

Omission	• no change on demand (failure to actuate, close, open, start, stop, insert, withdraw) • failure to respond to command • improper response (partially open, closed) • failure to assume a fixed position (oscillate) • failure to operate
Commission	• a spurious action (actuation, closing, opening, starting, stopping, insertion, withdrawal) • actuation out of sequence (premature actuation, delayed actuation) • failure to stay open or closed
Change in item or equipment condition	• failure to retain (loss of integrity, fracture, distortion, leak, breach of pressure or static fluid boundary) • failure to support (loss of structural support) • failure in energy transmission (loss of heat transfer, blocked or stopped flow, loss of transport or exchange)

Equipment which is well constructed and operating in the correct environment normally has a high reliability and gives a quality performance. So too does a well-trained and able operator.

There are differences between the lists in the two Tables but they seem to be, on the face of it, insignificant. Information on the basic modes of failure enables the consequent effects on the system to be identified. But to get more information about the mode of failure of both systems, it is necessary to probe deeper into the underlying mechanisms which affect the system and the error- or fault-inducing mechanisms stemming from its environment.

There is also one major difference between the equipment and human failure modes. Common mode failures in technical systems are the exception rather than the rule. But in human systems, common mode failures are more usual — as when action due to fatigue or stress affects general performance, or a bias generates actions affecting the whole sequence of events.

SOME EXAMPLES OF ERRORS LEADING TO INCIDENTS
Trevor Kletz[4,5] has discussed some of the external error modes in Table 7.1. His remarks are amplified here with further process examples.

MISTAKES DUE TO LACK OF KNOWLEDGE
People may not know, or may think they know, what to do. They may lack elementary knowledge of the properties of the materials, the process or the equipment. They may believe they must always follow rules, and be unable to react correctly when flexibility is needed. It is not possible to foresee every situation that might arise and people must be trained to diagnose and handle some unforeseen problems. Without an appropriate safety culture such action becomes difficult to effect in the safest manner.

People may be given contradictory or ambiguous instructions. Also instructions may have implied contradictions. For example, there is pressure on staff to achieve a certain output, or complete a repair, by a certain time. It may be difficult to do this without relaxing one of the normal safety instructions. What do they do? It is very difficult to avoid accepting the pressure, as in the Grangemouth flare incident — in which a flange was opened incorrectly releasing flammable material. The line had not been properly isolated and the task was continued despite noting the release of vapour. People get into a 'heads I win, tails you lose' situation. If there is an accident, they are in trouble for breaking the safety rules. If they stick to the rules and the output or repair is not completed in time, they are in trouble for that reason. The correct action may be difficult to take, particularly when people are inexperienced and keen.

At Chernobyl the operating staff were asked to carry out a series of experiments, and seem to have assumed that these instructions overrode the normal safety instructions, which were disregarded. Probably no-one actually told them that the normal instructions were suspended, but they may have got that impression from a lot of talk about the experiments without any mention of the need to follow the normal safety instructions.

SLIPS

The errors made in operating beverage vending machines are slips because the instructions are clear. Yet sometimes people press the wrong button despite having the necessary ability and motivation to carry out the task. The probability of this type of error has been studied and can be estimated from the number, size and distance apart of the buttons and the quality of the labelling. Some people make more errors than expected. Is this because they are more prone to make slips than the average person? It is more likely that their high error rate is due to not giving attention to everyday details.

An operator pressed the wrong button and opened an electrically operated valve leading to equipment under repair. Flammable gas was expelled and exploded. Four men were killed. The operator's error was a slip but to prevent the accident a better method of working was needed. The line should have been blinded and the valve at least defused.

A similar error occurred when a group was being shown round the emergency service facilities at ICI. The guide left the ambulance garage by pressing the button to open the garage door. No-one in the garage noticed that the door would knock down a man working overhead as it opened. Fortunately he noticed the danger and alerted the guide to the problem. In this case the error arose from not isolating either the area or the door-opening mechanism.

Many physical errors are linked to slips — that is to say, to skill-based performance errors. Unfortunately rules do not reflect very well how errors are made under pressure and in haste. People often have to make decisions on the spur of the moment, on the telephone and so on. It is easier for a study to be made after the event by others to show how the decision was wrong. Out-of-hours decisions place supervisors in difficult situations, as in the Grangemouth flare incident where the decision to continue opening the flange was made after the fitters had noted an initial leak.

In the Kegworth air disaster, when an aircraft crashed on the M1 motorway in the UK, the wrong engine was turned off when smoke entered the cockpit. The pilots were under pressure. They got it fixed in their minds early on in the incident that a specific engine was to blame and the inquiry, in retrospect, seemed equally 'mind-set' that the men were to blame. Yet clearly there was a lack of training on the specific aircraft and on the design of the engines, and the instrumentation was inadequate.

The Kegworth incident shows how people still make occasional errors even when they are basically well-trained and well-motivated, and capable physically and mentally of doing everything asked of them. Commercial airline pilots are very well trained. If the aeroplane crashes they are unable to bale out and likely to die or suffer injuries along with their passengers. Yet, like ordinary mortals, they may forget to close or open something or make a slip in calculation. Such errors are part of everyday life, although the consequences differ according to circumstances.

Slips and lapses do not necessarily occur because people are poorly trained. They can occur because they are well trained. Routine skill-based tasks are delegated to the lower levels of the brain and are not continually monitored by the conscious mind. People would never get through the day if every action required their full attention. They put themselves on auto-pilot. Errors are likely to occur whenever the smooth running of the programme is interrupted for any reason, particularly if conditions generate stress or distraction.

Since these slips cannot be prevented, it must be accepted that they will occur from time to time. Plants and methods of working should be designed to remove opportunities for error, or provide opportunities for recovery, or install devices to guard against both their occurrence and consequences. A simple error should not lead readily to dire consequences.

Training remains vital. Many incidents have escalated because the defences of the system were out of action. Both the Bhopal and Piper Alpha incidents are examples. In evaluating the reliability of the system, note that there is often the probability that operators will deliberately neglect or isolate the system because they are not convinced of the system's importance or because it is important to maintain production.

EXAMPLES — INCORRECT CONNECTIONS

In 1989, in a polyethylene plant in Texas, a leak of ethylene exploded, killing 23 people. The leak occurred because a line was opened for repair while the air-operated valve isolating it from the rest of the plant was open. It was open because identical couplings were used for the two compressed air connections and they were interchanged. As well as this slip, there was a violation. This was a failure (authorized at a senior level) to follow company rules and industry practice and fit a blind flange or double isolation valve (see Reference 6).

In 1990, a baby died when a nurse connected a nitrous oxide supply to the baby instead of an oxygen supply. The couplings were identical.

EXAMPLES — FORGETTING TO CLOSE A VALVE

In 1967 on a benzene plant the line was drained when it was not in use. On start-up the operator failed to close the line and benzene at 30 bar pressure spilled out, caught fire and caused damage valued at £20,000. The operator was untrained and no information was made available about the valve being open.

The liquid in a pair of pumps gave off gas when the pumps were standing. So the pumps were drained when not in use with the drain valves left open. One day an operator, in a hurry to start up the spare pump and prevent a plant upset, forgot to close the drain valve and corrosive liquid spilled out and caused a burn. The accident was said to be due to 'human failing' and the operator was told to take more care. However, a small change was made to the 'work situation': a reminder notice was placed near the pump. A better solution would have been to fit small relief valves to the pumps.

VIOLATIONS

A violation with intent to sabotage is rare. But other accidents have occurred because operators, maintenance workers or supervisors did not bother to carry out procedures that they considered troublesome or unnecessary. Indeed such practice may be condoned by management. For example, correct protective clothing not worn or the full permit-to-work procedure not followed. People need to be convinced that procedures are necessary. Training is required to convince people, and one way is to discuss accidents that occurred because the procedures were not followed. Supervisors should not allow the design intent to change in normal practice without revision of the procedure. Better motivation and development of a safety culture is an important response.

At Zeebrugge in 1987, it had allegedly become normal practice for cross-Channel ferry boats to depart sometimes without closing the doors, in order to maintain the schedule and clear exhaust fumes from the decks. This has to be considered as a violation because it was carried out as a deliberate practice. The subsequent failure to close the doors on increasing speed was an error in control and represented a slip rather than a violation.

EXAMPLE — PROTECTIVE CLOTHING NOT WORN

Many violations occur because protective clothing is deliberately not worn. A valve had to be changed on a line carrying corrosive chemicals. The line was emptied but a few drops of liquid remained. The permit asked for goggles and gloves to be worn. The fitter did not wear them and was splashed in the eye by a drop of the chemical. At first sight this seems like a violation, a deliberate failure to follow clear, written instructions. A look at the permit book, however, showed that every permit asked for goggles and gloves, even for jobs on low pressure water lines in clean areas. The maintenance crew therefore ignored the instruction. The instruction to wear goggles and gloves on such water lines is an example unfortunately far too common of instructions which are written to protect the writer rather than help the reader. The manager concerned should have spotted that more protective clothing than was necessary was being requested for jobs on these water lines and stopped the practice.

ERRORS WITHIN INCIDENT SCENARIOS

Errors can occur at any time during the life cycle of the plant. They are particularly dangerous when made by plant operators during maintenance when the containment of the plant is impaired in order to allow the contents to escape or a vessel to be entered. Many errors also occur when trying to control a dangerous situation or mitigate against the consequences of a release. Much is heard about analysing tasks leading to an immediate cause of an incident caused by such errors. Less frequently are studies undertaken on what to do when emergency control and other actions fail. Yet it is at times like this that guidance and rehearsal of the situation scenario is vital.

Amalberti[7] has studied the strategies of fighter pilots to ensure flight safety. For a high-speed low-altitude mission the pilots spend more time in preparation than execution. Owing to the high levels of risk and time constraint, each pilot develops various theoretical strategies to cope with a specific stage of the flight. All pilot activity on the flight is then directed towards keeping the flight in the domain — normal and abnormal — which has been envisaged before the flight. Moreover pilots organize their activity in this domain to avoid transient situations in which the prepared responses to events would not be applicable.

The initial response to a given situation requiring response does not aim at solving the problem, but just escaping from the risky situation. This first response is prepared beforehand or by deductive reasoning in flight. A later response involves the definitive treatment of the incident according to the prevailing circumstances. Should an incident arise which is totally unexpected, the initial response is produced more slowly and is devoted only to escaping danger. This response is often imperfect and leads to a series of short-term corrections which may become a series of poor decisions.

Pilots are therefore concerned with anticipation and action. Because of resource limitations the pilot's active behaviour forces the planned model to suit the anticipation mode — applying the prepared response and escaping, thus freeing very limited time, some 20 seconds, in order to optimize the response.

These pilot studies may at first glance seem inappropriate for slow-responding situations. But it is noticeable that in such situations the first response by operators is action to converge towards a stabilized or safe situation and only subsequently do they return to analysing the causes.

The immediate causes of an incident are not necessarily readily recognized. A human error may precede an equipment failure and vice versa. In the case of the Kegworth air disaster, there was an engine failure and the wrong engine was shut down in error. Many equipment and instrument design faults continued to affect the incident.

In the US Three Mile Island nuclear plant failure, a relief valve stuck in the open position. But the operators assumed it was shut because it was indicated on the panel as having been commanded to shut. The operators gave the component status the highest priority. Operators tend to postpone tasks which are judged less important, and these may be neglected as a consequence. There is also a tendency in transient situations for operators to jump to the end action without verifying conditions or finishing preparatory actions. Arguably this was a case of an immediate cause related to equipment being compounded by human error. Yet further study of the incident reveals that other human errors during and following maintenance created the original problems leading to this incident.

EXAMPLE — THE EXPLOSION AT KOKKOLA
On 21 February 1989 an explosion occurred in Kokkola, Finland, on a meta-aminophenol plant[8]. The process is illustrated in Figure 7.1; the incident occurred in the section involving sulphonation

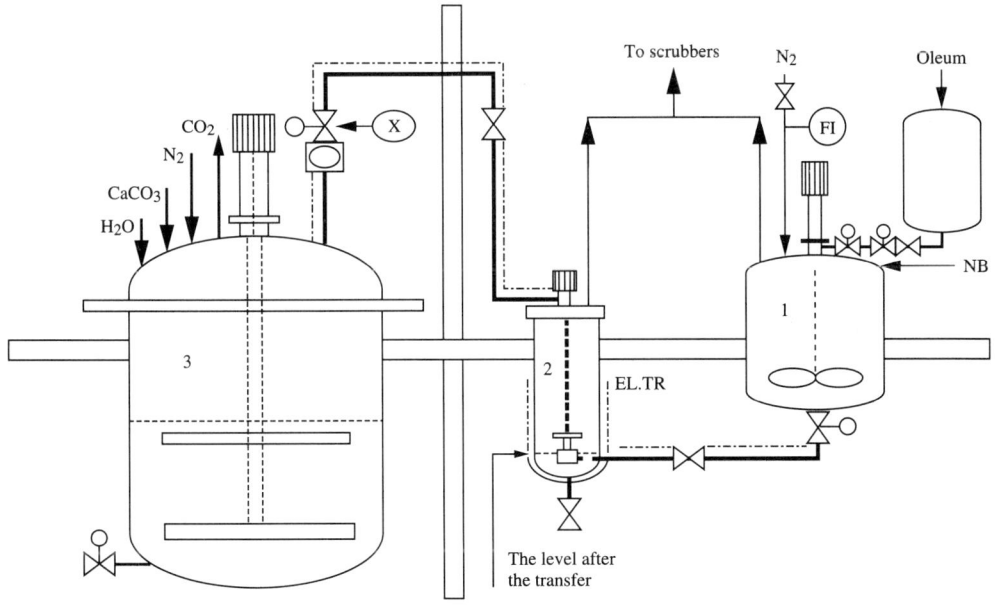

Figure 7.1 A sulphonation process. (Source: Bergroth[8].)

of nitrobenzene (NB) with oleum. The reaction mixture from sulphonation (vessel 1) was neutralized in the following step with a slurry of calcium carbonate (vessel 3).

The sequence of events was as follows.
- The automation system failed to open a valve, X, in the transfer line between the sulphonation reactor 1 and the neutralization reactor 3.
- The foreman opened valve X manually from the instrument cross-connection room.
- The transfer operation was activated and the pump (in vessel 2) was switched off by a flow switch when the sulphonation reactor was empty.
- The sulphonation reactor still contained a small amount of liquid.
- Attempts were made to empty the reactor, but the pump could not transfer a small amount of liquid.
- The manually opened valve would normally close after the pump had been shut off but had remained open for an additional 15 minutes.
- A temperature alarm from the pump tank 2 was activated soon after the terminated transfer operation. The temperature had exceeded 120°C which was unusual but had happened before. The operators did not react to the alarm.
- An exothermic reaction in the pump tank occurred. This was followed by an explosion which caused severe damage to the factory.

At first glance this explosion appears to be largely due to a lack of action by the operators in a difficult situation. They have some responsibility for making an inadequate recovery but most of the blame lies elsewhere.

The immediate cause of the incident was the exothermic reaction which took place between oleum and water. The explosion was the result of thermal decomposition of a nitro-compound which occurs at temperatures above 150°C. The trigger of the reaction was reverse flow of steam from the neutralization reactor which reacted with a small excess of oleum and sulphuric acid in the reaction mixture from the sulphonation reaction.

A Preliminary Process Hazard Analysis should have identified that such a reaction was possible. Sections of plant where this might occur should have been identified and studied. Such action is a basic design requirement.

If this had failed to identify the problem, then a Hazop study should have identified the problem. A non-return valve would have helped to reduce the likelihood of the occurrence. Better still would have been to eliminate the pumping tank and to transfer material by gravity or pressurized nitrogen.

In this case there was a major problem in trying to recover the situation once the unexpected occurred. The operators faced great difficulties in diagnosing a particular problem, given the lack of the necessary process knowledge or back-up from technical staff and a prepared contingency plan. A similar situation arose in an incident on a plant run by Dow when a drying operation went wrong some years previously. The incident at Bhopal also involved an exothermic reaction. Operators and management cannot be expected to deal with exothermic reactions when difficult situations develop unless they are particularly well trained and some analysis of the situation has been prepared beforehand.

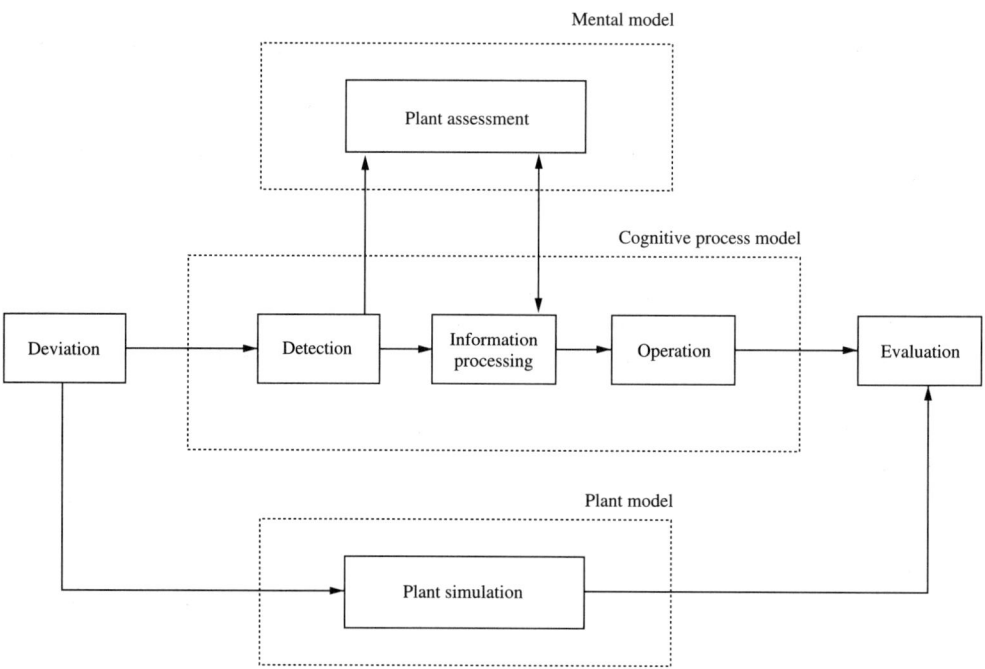

Figure 7.2   Cognitive model for an operator, based on Ujita[9].

SYSTEM-RELATED CAUSES

The list of human errors in Table 7.1 (page 153) is helpful for identifying likely problems. But to identify how likely these are to occur it is necessary to probe more deeply. Such a study must refer to a more comprehensive list of modes of human error and consider the system-related causes influencing why things go wrong.

The basic cognitive process is illustrated by the model in Figure 7.2. Each node of the network corresponds to an element of the cognitive process as follows:
- the detection phase with activation by a change from normality, maybe stimulated by alarm;
- an information processing phase involving observation, identification, interpretation, goal evaluation, task definition and procedure formulation;
- an operation phase involving execution and confirmation. This phase can include error detection and recovery.

The links and components within the network are far more complex. Operators carry out many parallel tasks and use information from a single reading of an instrument in many ways. There are many feedback routes from confirmation. The response time varies between operators according to event experience and their diagnosis differs with their level of skill.

It has also been noted that, whenever possible, the brain uses the method involving the least effort in the skilled mode. This means that the brain will try to apply learned skills to a set of conditions if they appear similar to the set of conditions for which the skills were learned. It has been observed that, after an incident occurs, the operators' cognitive process level becomes deeper as time passes, based on reconfirmation. It is skill-based for about one minute, rule-based for about five minutes and knowledge-based for about thirty minutes.

INDIVIDUAL MODES OF HUMAN ERROR

Various classifications of the modes of human errors have been made. One of the simpler classifications is based on looking at inadequacies in physical and mental ability and the circumstances in which the decisions are carried out. Some of these are indicated in Table 7.3 (page 162).

Human error modes are affected by many direct system-related causes, the most immediate of which may be those shown in Table 7.4 on page 162. Some people consider that the human operator works as a fallible machine. This is doubtless a simplification but undoubtedly a helpful concept.

Much work in process operations within a chemical plant is skilled or unconscious, and errors in this type of behaviour are slips or lapses. Slips and lapses can be readily (if time-consumingly) predicted and steps taken to reduce the predicted frequency if necessary.

Mistakes are much less predictable, less likely to be detected and potentially more dangerous. A mistake is likely to be the result of an inadequate or inappropriate theory.

PSYCHOLOGICAL MECHANISMS FOR ERRORS

It is helpful to have a list of psychological mechanisms which affect cognitive processes. The following list is based on the work of Embrey[10]:
- Failure to consider special circumstances. A task is similar to other tasks but special circumstances prevail which are ignored, and the task is carried out inappropriately.
- Short cut invoked. A wrong intention is formed based on familiar cues which activate a short cut or inappropriate rule.

TABLE 7.3
Individual modes of human error

Incorrect motor adjustment	• a simple/discrete error in holding, moving, pushing, pulling, twisting, turning • a complex/continuous error in adjusting, positioning
Incorrect searching and receiving of information	• an error in detecting, inspecting, monitoring, observing, reading, recognizing
Incorrect identification of actions, events, objects	• an error in identifying, locating
Incorrect information processing	• an error in calculating, comparison, interpolating, remembering, verifying
Incorrect problem-solving and decision-making	• an error in diagnosing, matching, selecting, use of rule, misjudgement
Incorrect communication	• an error in answering, directing, informing, receiving, recording, requesting, understanding, remembering

TABLE 7.4
Some system-related causes

Incorrect or absent information	When information is:		
	• not provided	• not accessible	
	• not existing	• not readily available	
	• ambiguous	• too detailed	
	• incomplete	• not used	
	• incorrect format	• illegible	
Incorrect or absent interface	When interface is:		
	• not provided	• out of action	
	• inappropriate	• inadequate	
	• confined	• hazardous	
	• indiscriminable	• difficult handling	
	• inaccessible	• unprotected	
Adverse environmental conditions	When conditions are affected by:		
	• chemicals	• condensation	
	• dust	• lack of visibility	
	• noise	• temperature	
	• radiation	• vibration	
	• ventilation	• weather	
	• emergency chaos	• stress	

- Stereotype take-over. Owing to a strong habit, actions are diverted along some familiar but unintended pathway.
- Need for information not prompted. Failure of external or internal cues to prompt need to search for information.
- Misinterpretation. Response is based on wrong apprehension of information such as misreading of text or instrument, misunderstanding of verbal message.
- Assumption. Response is inappropriately based on information supplied by the operator (by recall, guesses, etc) which does not correspond with information available from outside.
- Forget isolated act. Operator forgets to perform an isolated act or function — that is, an act or function which is not cued by the functional context, or which does not have an immediate effect upon the task sequence, or is an item which is not an integrated part of a memorized structure.
- Mistake among alternatives. A wrong intention causes the wrong object to be selected and acted on, or the object presents alternative modes of operation and the wrong one is chosen.
- Place losing error. The current position in the action sequence is misidentified as being later than the actual position.
- Other slip of memory.
- Motor variability. Lack of manual precision, too big/small force applied, inappropriate timing. Includes deviations from 'good craftsmanship'.
- Topographic or spatial orientation inadequate. In spite of the operator's correct intention and his correct recall of identification marks, tagging, etc, he unwittingly performs a task/act in the wrong place or on the wrong object. This occurs because he follows his immediate sense of locality where this is not applicable (not updated, surviving imprints of old habits, etc).

SYSTEM-INDUCED ERRORS

The information on error modes and tendencies gives no indication of the causes which induce error and the problems generated by operation in an environment which punishes error.

A young seaman, armed with a spanner, was sent elsewhere in the engine room to read a gauge. The location was dark and the gauge covered with condensed moisture. So he removed the gauge to read it. The resulting oil leak caught fire. Seamen could not escape in the confined space and fatalities ensued.

In addition the equipment can induce human errors, as in the case when it does not react to an action yet the operator carries on the operation as if it had responded. Furthermore an indication or signal can be misleading, as when a valve had not actually closed although it was signalled as such at Three Mile Island. Similarly hardware failures can be human-induced, particularly by overload of equipment or when incorrect maintenance leaves defects.

MODELS OF INCIDENT CAUSATION

The simplest models for human reliability are based on the classical stimulus—organism—response paradigm. The response is a function of the stimulus and the current state of the organism. This can be developed further to emphasize the input from the organism as corresponding to that from an information processing system using a fallible machine which often produces correct results but may also produce incorrect results. The stimulus-based system does not give an exact representation, because behaviour is not simply a function of the input and output mechanisms. Nevertheless the classical approach remains useful, providing that attention should focus on the overall performance rather than on the mechanisms.

The task of an operator can be identified by appropriate characteristics and broken down into task elements. Essentially a list of pre-established, unequivocal, unambiguous actions which the operator must perform is drawn up. Performance-shaping factors are identified. Finally the operator's performance is characterized by a failure probability of this action under normal, standard and usual conditions. The influence of the performance-shaping factors is analysed, the probability is corrected by expert judgement and the outcomes aggregated. There is much argument over the exact way in which performance-shaping factors should be used but, given the often linear or simple continuous functions used, the most that can be hoped for is a rough appraisal. Embrey[10] classifies errors with significant consequences as stemming from a combination of error-inducing conditions, intrinsic error tendencies and an unforgiving environment. Some of these factors are amplified in Table 7.5.

Skill-based errors are particularly influenced by the recency and frequency of previous use, environmental control signals, shared properties in a plan and similarities between concurrent plans.

TABLE 7.5
Factors contributing to incidents

Intrinsic error tendencies:
- inadequate information — for example, monitoring, memory or file lapses
- inadequate information processing/decision-making — for example, mental, computer or control lapses, deliberate violation
- inadequate physical and mental attributes — for example, slip, lack of ability, lapse of attention

Error-inducing conditions:
- resource overload/pressure/haste
- production/safety trade-off
- inadequate communication/instruction
- inadequate procedures/practices
- difficult operational interface/communication channels
- inadequate supervision/technical back-up
- inadequate control/definition of responsibilities/command
- inadequate servicing/training
- low motivation/co-operation
- personal stress/illness/tiredness/medication

Unforgiving environment:
- high risk situation
- inadequate situation for recovery
- inadequate emergency preparedness
- inadequate active protection systems
- inadequate passive protection systems

Rule-based errors stem from features caused by mind-set ('it's always worked before'), availability (first come best preferred), matched bias (like relates to like), over-confidence ('I'm sure I'm right') and over-simplification ('the halo effect').

Knowledge-based errors are affected by bounded rationality as instanced by selectivity, overload of the working memory and out-of-sight-out-of-mind. Other factors include thematic vagabonding and encysting, memory-cueing and reasoning-by-analogy, matching bias revisited and having an incomplete or incorrect mental model.

EXAMPLE — BOIL-OVER IN A REACTOR

An exothermic reaction occurred when the stirrer blade dropped off in the reactor. This was not noted by the operators and the temperature did not increase immediately. When it did, the reaction was 'unstoppable' and the vessel boiled over releasing noxious fumes. This affected members of the public who bombarded the works complaining of the smell. No-one was apparently hurt by the incident. The incident had happened before but on that occasion no runaway occurred.

The works investigation showed that an alarm indicating low current was inoperable. The agitator blade and impeller shaft had not been serviced since the vessel had been repaired. The incident cause was reported as an equipment failure and inadequate design. It was noted that human error by the maintenance staff arose, as they failed to use set procedures and critical actions were omitted. The company stopped making the product. This effectively ensured that complaints from the public were stopped. The maintenance personnel were disciplined and a directive was issued regarding the use of procedures. No-one informed either the process licensors or the equipment suppliers about the problem.

A good deal can be learned from the conditions which induced error:

Design

It is not acceptable to have an alarm as the only 'protective' device, if ignoring failure can cause an event of such magnitude. Recovery from failure must always be considered and adequate protection provided against known hazards.

Training

The operators were unfamiliar with the importance of the alarm and the effect of loss of the blade on operations. No exercises had been undertaken simulating the problems. It had not been appreciated that a runaway reaction was a problem. Operating procedures and recovery action were not provided.

Procedures

The maintenance procedures were obscure, difficult to read and had never been examined by the company. The procedures were part of a manual developed by the suppliers. The maintenance department did not participate in their development and were unsure of why actions were required.

Resources

The maintenance department generally worked on demand, except at annual shutdowns. The company had cut back on staff and generally the department was overworked and underpaid. The motivation of employees was low and the level of supervision had been greatly reduced.

Culture
The maintenance department generally considered procedures as being required only by beginners. The general atmosphere on the site was that of a low reliability culture where mistakes were tolerated.

This analysis could obviously be continued to identify the general policy failures such as inadequate work scheduling, ineffective training, inadequate process and mechanical design, ineffective design of procedures and no effective learning from a previous incident at the works or from previous, related accidents reported in the literature.

In high reliability industries there exists a culture which is strongly averse to failure. The studies described in this overview show how good design combined with Task Analysis can greatly reduce the probability of simple human error giving rise to an incident. As well as leading to a better basic plant, it results in better procedures and practices. This in turn improves job orientation, skill training, task instruction and related activities.

TASK ANALYSIS

Task Analysis has been developed as a systematic method for analysing a task into its goals, operations and plans. A task is a specific work assignment. It represents the set of operations/actions required to achieve a stated goal. A goal is what the person doing the job is aiming to achieve. It is a specific work objective. The operations represent the stages involved in the task. Plans describe the methods and conditions under which the operations are carried out. A procedure is a step-by-step description of how to proceed from beginning to end in order to perform a given task correctly. A practice is a set of guidelines helpful to the performance of a specific type of work which may not always follow a set procedure. Some tasks may not be described by a rigid procedure.

Task Analysis is a process of sorting out what people might do or actually do when carrying out operations. In general terms the analyst must respond to questions such as:
- what actions do the operators carry out?
- how do operators respond to different cues in their environment?
- what errors might be made and deviations caused in plant operations?
- how might any error be recovered from, or any deviation be controlled?
- how do operators plan their actions?

Task Analysis can provide input to such features as:
- the specification of equipment, controls, emergency controls and process interface;
- hazard identification and analysis;
- reliability evaluation and risk assessment;
- the preparation of procedures and operating instructions;
- the development of written practices;
- the specification of training requirements and development of skills.

Task Analysis is widely used for the analysis of existing tasks. But it can also be used to check the safety of features of a design and the effect of deviations. A close link with Hazop can be developed and other techniques such as Failure Mode and Effect Analysis and Sneak Analysis.

DATA FOR TASK ANALYSIS

A Task Analysis requires data from sources of information such as:
- documentation process diagrams, functional models, job descriptions, working practices and permits, instructions, operating manuals, design specifications, existing documentation of task;
- output from hazard reviews;
- study of plant records, computer output, logs, etc;
- debriefing of operators following completion of the task;
- discussion or interview with management, actual or potential users, design engineers, safety specialists and human factors experts;
- observation, recording and inspection during operation and user trails including use of video, audio, transcripts, coding schemes.

It may be necessary to derive or seek specific information on desired or set performance criteria and standards, hazards and exposure limits, previous incidents, modes of operation and control and the failure modes of activities.

The task should be placed in context and factors influencing performance considered such as task similarities, frequency and duration, level of training and experience of personnel carrying out the task, team structure, co-operation and supervision, communications and information processing, job aids and technical back-up, the working environment, pressure and time constraints. Pressure and time constraints will be considered later.

STEPS IN CARRYING OUT A TASK ANALYSIS

GOALS OF THE ANALYSIS

The overall goal of the task to be carried out should be stated. A task is sometimes called an operation. Examples of an operation include:
- the start up of a process section such as a fired heater;
- a stage activity such as the processing of a specific batch or activation of catalyst;
- the operation of an item such as an analyser or breathing apparatus.

BREAKDOWN OF THE TASK INTO STEPS OR ACTIVITIES

The task is then broken down into a set of subordinate tasks or operations which must be performed to achieve the goal at that level. For example, a process operation may be divided into appropriate stages. The diagram is expanded down to a level appropriate for the study being undertaken, often the level of the individual steps in the task. The basic guideline normally adopted is to list all the steps which are critical for performing the task correctly but to exclude details of those tasks which are unlikely to cause problems if not fully identified.

CREATING A PLAN

A statement is produced of the conditions under which each of the subordinate operations either is carried out or should be carried out. A plan should refer to each of its subordinate operations. It should not refer to operations that are not present. Plans are important in defining when to carry out various operations, their sequence, the status of valves, the duration of an activity, the process conditions and so on. The plan might take the form of a hierarchical diagram or a tabular list. Plans are discussed later in this section.

ANALYSING THE PLAN

There may be many features of the system which require change. An initial analysis should identify hazards from equipment and materials, difficulties created by the environment, the lack of appropriate controls and protections, etc. The plan should be subjected to Critical Examination.

The analysis of the plan or system should examine possible deviations from the system, determine the likelihood of those deviations and determine any deviation at the start and end of the procedure.

MODIFYING THE PLAN

Modifications should be made to improve the method of working and to reduce the effect of deviations and appropriate controls, precautions and mitigation introduced. The analysis of the

TABLE 7.6
Changing a flat tyre on a car

Fixed sequence	**0**	**Change flat tyre on a car**
		Do 1, 2, 3, 4
	1	**Jack up tyre**
Decision plan		Do 1.1, if front then 1.2, if rear then 1.3, exit
		1.1 Identify flat tyre and location at front or rear
		1.2 Jack up front until tyre just on ground
		1.3 Jack up rear until tyre just on ground
	2	**Remove tyre and fetch spare**
Time sharing		Do 2.1 and 2.2 together
		2.1 Remove tyre
Fixed sequence		Do 2.1.1, 2.1.2, 2.1.3
		2.1.1 Get wheel spanner
		2.1.2 Undo nuts and remove
Contingent fixed sequence		Do 2.1.2.1, then if nuts loosen do 2.1.2.2
		2.1.2.1 Loosen nuts
		If wheel turns lower jack and retry
		2.1.2.2 Remove nuts
		2.1.3 Remove tyre
		2.2 Get tyre from boot
Fixed sequence		Do 2.2.1, 2.2.2
		2.2.1 Open boot
		2.2.2 Take tyre to location
	3	**Fit tyre and place flat tyre in boot**
Branching plan		Do 3.1 and 3.2
		3.1 Fit tyre and replace jack
		3.2 Place flat tyre in boot
	4	**Continue journey**

plan leads to improvements in such features as working methods, procedures and practices, work environment and exposure frequency, communication and information processing, skill and capabilities of people, and management, supervision and control.

Some action will involve capital expenditure and be, or become, part of a project. Others involve a change in control procedures and practices. Appropriate control action must be developed and actions and precautions specified to prevent inefficiency and loss. All require efficient management of change.

The plans should also give advice on action to take when deviations arise and recovery as necessary including such features as:
- what to do if the valve status is different at the start of the procedure;
- how to effect immediate recovery from human error or equipment malfunction;
- what to do in the event of a release of process material;
- what to do if the transfer of material is to the wrong place or if it is off-specification;
- implementing the plan.

The plan must be properly documented with appropriate performance standards, practices and procedures clearly identified. Task statements should indicate what is done and relate this to a given performance standard. Procedures must indicate the sources of information, control measures and different modes of operation. The risk control measures must be implemented with adequate inspection, maintenance and monitoring procedures to secure continued operation. Appropriate review and appraisal is essential during operation. The emphasis is on compliance through safety culture and effective management.

A procedure should therefore start by indicating the task purpose and features of importance. The major steps should be outlined in their proper order and a diagram is helpful in showing the sequence as numbers. It is usual to indicate key features of the procedure which should be remembered.

Practices should present positive guidelines for correct performance. They should emphasize features such as hazard controls, clothing and personal procedures, emergency response and procedures. Any special problem areas should be highlighted. The most important rules and regulations should be included.

THE STRUCTURE OF PLANS

A plan might involve several different types of method:
- in a set or fixed sequence, the person follows a set of activities in a specific order;
- in a timesharing sequence the operations are carried out in co-ordination with each other or at the same time, and more than one individual may be involved;
- in a branching sequence the person will do a task, then depending on the outcome from that task, the individual will carry out a particular option;
- a plan may allow for the selection of the task which is most appropriate to the situation.

EXAMPLE — CHANGING A TYRE

Table 7.6 shows a plan for changing a flat tyre on a car. It can be assumed that the ground is level, the spare tyre is properly inflated and the car is not in a hazardous location.

TABLE 7.7
Notes on plans for diluting caustic

	Sub-task or plan	Notes
0	ENSURE CAUSTIC CONCENTRATION IS WITHIN LIMITS SPECIFIED BY MANUFACTURING INSTRUCTIONS PO: → 1 → 2 → 3 → 4 After 30 mins └── 9 ← 7 ← ── → 5 → 6 → Exit └── 9 ← 8 ← ──┘	*The 'notes' are at this stage merely suggestions that the analyst is making to improve sub-tasks. They may be taken up during instruction or they may be suggestions to a plant manager for modifying the task.*
1	1. Put on protective clothing. 2. Collect sample. 3. Test sample. 4. Compare caustic concentration with specification and determine the necessary action. 5. Empty caustic solution. 6. Remove gloves and goggles. 7. Add water to correct concentration. 8. Add caustic to correct concentration. 9. Circulate mix.	Emphasize the dangers of handling caustic without proper protection. Instruct using laboratory training materials. Emphasize the importance of selecting the correct action after comparing the concentration with the specification option.

Each task could be expanded with respect to action if problems in performance occur. For example, it might not be possible to undo the wheel nuts. What if the jack was missing? In every case there is a limited source of labour available. All such factors may require appropriate planning.

EXAMPLE — DILUTING CAUSTIC SODA

Preliminary plans given for diluting caustic soda in Table 7.7 are based on Embrey[10]. Here the plan is extended by providing notes on suggestions to improve sub-tasks. Each of the items can then be developed using hierarchical step analysis.

This development of the plan is noteworthy because the different steps have been well structured. It is possible to see at a glance the method used for the process. It can readily be developed, with the help of the operators, into written plant operating instructions.

It should also be noted just how much work has gone into structuring a very simple task. This is a problem in itself as the further analysis of the topic has yet to be explored. The plan needs considerable further development. Initially it should be developed by noting all the preconditions for carrying out the task such as getting protective clothing, seeking a permit if necessary and familiarizing with instructions. Then it is necessary to note if the plant is not in the normal state and what action then to take. Each step of the task is analysed further, taking care to note what to do if matters go wrong. Finally there must be notes on what to do when the job is completed.

ANALYSIS OF TASK DEVIATIONS
There are many ways of analysing task deviations. The traditional method study approach employs What? When? How? Where? Who? questions to probe the task. The objective is to improve safety, production and quality with possible savings in costs.

HOW CAN WE DO IT BETTER?
Carry out a Critical Examination which challenges the purpose of an individual step and seeks to alter, eliminate or avoid the step. Such radical change to the process is reasonable if redesign is possible and an appropriate structure for this is discussed in a subsequent section.

The actions can include many features of total quality control but arguably most emphasis is likely to be placed on actions to:
- improve operability, safety, quality of working life, use of resources and management;
- increase production and reduce lapses in quality;
- reduce costs;
- use fully the employee's knowledge and abilities.

HOW CAN WE DO IT WORSE?
Deviations arising from the existing system can be probed using the structure in Table 7.8 (page 172). It may still be possible to change where the task is carried out and who does it. The emphasis is on identifying deviations in the way the task is carried out and trying to prevent such occurrences or reducing the consequences. In general a task study is focused on a specific small area of the sociotechnical system. Attention is given to materials, equipment, people and the operating environment with both people and plant being affected by performance-influencing factors.

CASE STUDY — FILLING A BATCH TANK
The steps in filling a batch tank (Figure 7.3, page 172) with flammable liquid are used here to illustrate the typical sub-operations making up a task. The tank is 5 m high. Note that the further protection and operating aids provided are a LAL at 0.5 m, LAH at 4.1 m and a LHH trip system closing valve FV2 at 4.2 m.

STEPS TO LOAD BATCH TANK
Plan: Do 1, 2, 3.
1. Check the tank is empty and supply pressure is adequate.
1.1 Check level in tank below 1 m.
1.2 Check supply pressure is greater than 1.3 bar.
2. Open valve FV1 and start to fill tank.
2.1 Check FV2 is open.
2.2 Open FV1 fully.
3. Monitor tank filling and stop flow when level reaches 4 m.
3.1 Monitor tank filling OK.
3.2 Close FV1 when level at 4 m.

TABLE 7.8
Method study questions for task

Proposal	Alternatives	Conclusion
What has to be done?	What if omitted? What else might be done as well as or instead? What else can go wrong?	Does the deviation matter and what needs to be done?
Why: for what purpose?		Why?
When is it done?	What if done earlier or later, before or after, or out of sequence?	What might happen and then what needs to be done?
Why then?		Why?
How is it done? How much is done? How fast is it done? How often is it done?	What if done some other way? What if more or less done? What if quicker or slower? What if done more or less often?	What might happen and then what needs to be done?
Why that way?		
Where is it done?	What if done elsewhere?	What might happen and then what needs to be done?
Why there?		Why?
Who does it?	Who else can do it?	Who should do it?
Why them?		Why?

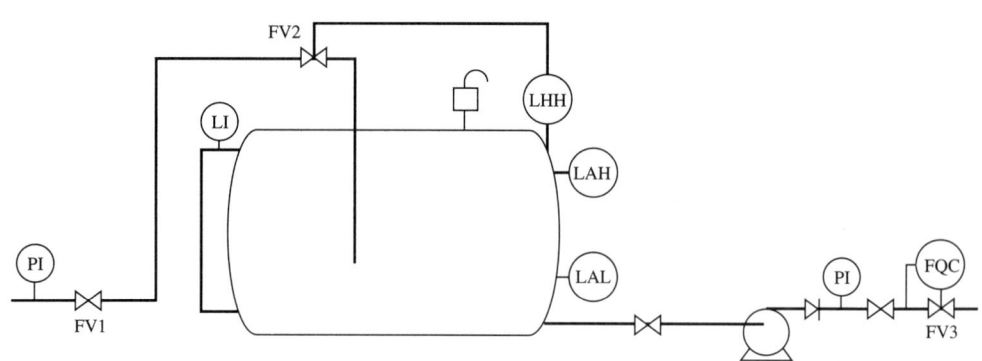

Figure 7.3   Batch tank. (See page 243 for instrumentation definitions.)

TASK ANALYSIS

ANALYSIS OF TASK DEVIATIONS

Note that in this plan there is no instruction about what to do if matters go wrong. Certainly there should be a note on what action to take if either of the level alarms goes off. If the tank is in use to put in the next batch downstream, the operator's task could take a long while. This is of some concern as it would encourage the operator to leave the task in an uncompleted state. There should be some advice about what action to take in the event of an overflow of flammable liquid. There is no indication of whether the tank is surrounded by a bund. Table 7.9 gives a partial analysis of task deviations when filling the tank and recommended actions.

The design shown here is not a recommended one and would be unlikely to be installed as such today. However such a design without the trip system would not have been uncommon on plant a decade or so ago. Note also that operational plant often cannot readily be changed, as this itself is an activity which can be fraught with danger.

TABLE 7.9
Analysis of task deviations on filling a batch tank

What has to be done?	Operator monitors tank filling and closes FV1 when level reaches 4 m.
What if omitted?	If the operator does not stop the flow the alarm should sound. The operator should then stop the flow of material. If this action is not carried out then the trip system stops the flow of material when the tank flow reaches 4.2 m.
What else can go wrong?	If the operator does not respond to the alarm and the trip system fails, material overflows through the vent onto a concrete surface and slowly passes down the surface drains and thence to the works effluent system. A loss of material would probably be detected by flammable gas alarms in the area. It is also likely to ignite causing a pool fire.
When is it done?	If done late the scenario is as noted above. If done early too little material is available for the next stage which is reaction. The operator should detect the smaller batch size. The only effect is loss of production, with no effect on safety and quality.
Who does it?	Normally it is done by the operator assigned to the task but intervention on alarm can be carried out by any operator. All operators must be trained in the activity.
Does the deviation matter and what should be done?	An overflow from the tank should be avoided. Estimates of the reliability of the operation are given in Figure 7.4 (page 174) and appear satisfactory. However, given the loss of production and the operator time involved from low level it is recommended that the installation of a level control be considered. Training must emphasize the need to complete this task avoiding any overflow. The operators carrying out this task should be warned that any occurrence of the trip system must be reported as a near-miss incident. The number of occasions the alarm is activated will be monitored as a performance indicator until further notice. The proof test interval of the trip system must be advised to maintenance. Consideration should be given to installing a bund.

HAZARD IDENTIFICATION AND RISK ASSESSMENT

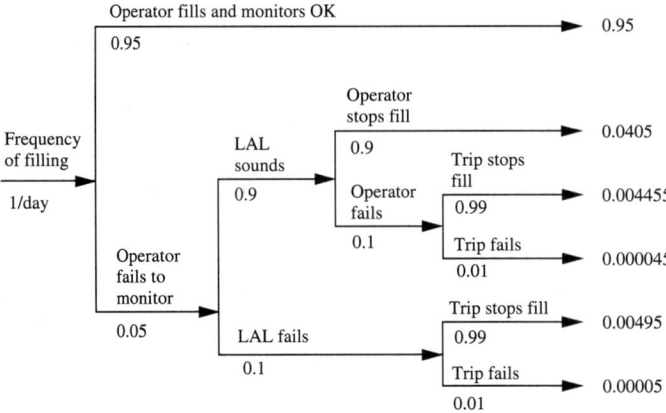

Figure 7.4   Event tree for filling a batch tank.

Plant reliability

The probability of the tank overfilling is assessed using the likelihood of each activity as shown in Figure 7.4. Assuming the plant is operating 250 days/year, then the number of failures causing overfill is estimated as:

250 (0.000045 + 0.00005) = 0.024 times per year.

In practice this system failed within two years. Why was this the case? The operator got used to controlling the level by relying on the trip system to cut-out the inflow of material. It became a routine violation until the day the trip failed.

This plant was very badly run. Few plants operate exactly as designed. Actual practice is established and this may become accepted procedure ... until the day of the incident. Such misuse could have been easily dealt with by treating every operation of the trip system as a near-miss incident.

RECOVERY

Assistance in normal operation is usually carried out by automation to various levels. This reduces the workload of repetitive actions on the operator. Earlier reference was made of the observations of Amalberti[7] on the activities of pilots and operators. He notes that the main concern for adapting to events is to anticipate and continually adapt the plan before events occur. Once the automatic control system is on, the system rigorously keeps the system within bounds — often for long periods, particularly given full automation with computer control. The operators can neither apply their own safety strategies of adapting to future events by accommodation nor can they realistically ignore the action by the control systems and conduct systems manually. Their experience and performance given certain future events is thereby weakened. The operators may try to get round the rigidity of the automatic control system by deactivating some protections in order to give more (illegal) flexibility to the system. Alternatively the operators merely monitor the actions of the controls and check that systems are on. The result is that an operator waits too long before

174

reacting. This means that action is taken without anticipation and a situation has developed in which performance is known to be poor. Fortunately there have been developments which have led to the improvement of workplace displays which can provide operators with more and better quality information, thereby stimulating rule-based behaviour. There still remains the problem of lack of manual training which may result in lack of competence and less experience.

Assistance in abnormal situations is to some extent dependent on the effect of this situation on safety or production. Often the immediate action is one of abort or escape to stabilize the situation. This can be followed by a period of reflection in which other alternative solutions can be tried. If the operators have no experience of such a style of analysis from normal operation, they are going to flounder. The consequences are the same if they are expected to follow procedures at all times in an unswerving manner, rather than building up experience of anticipation or accommodation of deviations. Problem areas should be regularly discussed with management and the experience of the operators taken into account. In the section on Task Analysis, note was made of the development of procedures. It is vital that it should be the operators who actually write the procedures with appropriate trained assistance from management and other competent people.

8. TASK ANALYSIS AND HAZARD IDENTIFICATION

INTRODUCTION

This chapter identifies the links between Hazop and Task Analysis in order to carry out a thorough study of linked process steps. This technique is particularly necessary for batch processing and the example used throughout the chapter is a batch process.

The basic steps of the process are identified. A survey is then carried out on incidents from the past. A full study of the literature is not necessary because of the availability of a general study on batch plants. This would usually be backed up by a further study of incidents involving the specific chemicals to be used.

A Critical Examination of the system can reveal which tasks to study as identified by considerations of safety, product quality and possible loss of production. A method study style analysis can then be carried out by asking What? Why? Where? When? questions about the steps. This study can be adapted to examine the hazards more thoroughly. Alternative approaches include using a Hazop alone or combining the Hazop approach with that of Task Analysis.

THE BATCH PROCESS

A batch process produces an ethylene derivative. The main problems are overpressure and overtemperature if the reaction runs away. The aim is to identify causes of incidents and determine ways of reducing the likelihood and extent of any accident.

Ethylene is contacted with a highly toxic mixture of other materials, here called M. In nature this material has properties similar to vinyl chloride. The product, P, is flammable but not particularly toxic. The reactor is first charged with ethylene and then a batch of materials M is added. Further material is added at intervals controlled by reactor temperature.

The outline P&ID of the batch process is shown in Figure 8.1. It is to be developed by introducing process relief and considering improvements to the control system, with the use of interlocks, trip systems and other emergency control systems.

Ethylene is produced continuously on-site and an intermediate storage of five tonnes is located close to the autoclaves. The reactor holds a total batch size of 1000 gallons, including 800 gallons of ethylene. M is stored in a 5000 gallon tank at atmospheric pressure and two batches of 100 gallons can be pumped into the autoclave over one hour in two distinct batches. Supplies of M are obtained locally. The temperature is allowed to recover to 25°C between batches. Each batch takes four hours to process and, after reaction and testing, the liquid product is stored in two 5000 gallon tanks.

For this reaction ethylene is provided in excess. If at any time the ratio of ethylene to M gets above a critical value, the pressure of the system begins to increase rapidly. If the agitator stops it is possible that layering will occur. If the temperature increases then the reaction rate is increased and hence the pressure.

TASK ANALYSIS AND HAZARD IDENTIFICATION

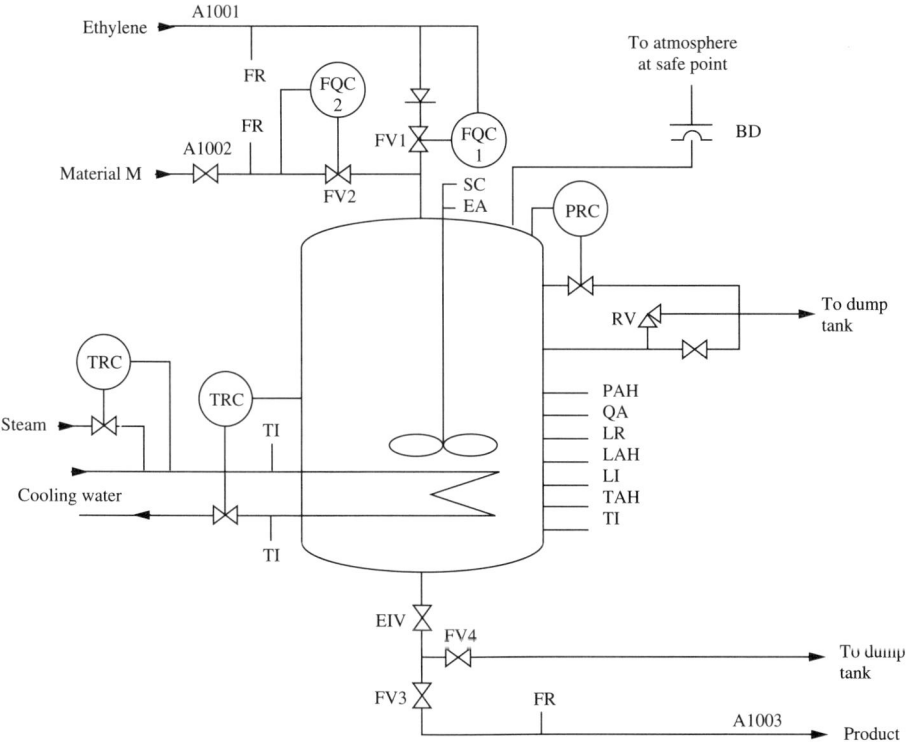

Figure 8.1 Process diagram of ethylene derivative batch plant.
EIV — emergency isolation valve. (See page 243 for other instrumentation definitions.)

LESSONS FROM PAST INCIDENTS IN BATCH REACTORS
Rasmussen[11] has made a study of overpressure incidents in batch reactors producing the data listed in Table 8.1 (page 178). Although such incidents rarely give rise to harm outside the works boundary, they nevertheless can cause major accidents which might escalate to serious or catastrophic events in certain locations. The list of initiation mechanisms in Table 8.1 suggests many top events which need to be studied further by Task Analysis or other methods.

The results suggest that a batch reactor which can give rise to an incident is unlikely to kill anyone outside the plant. But it can harm members of the workforce and result in loss of equipment and materials. Apart from the problems associated with the reaction itself, it is clear that attention needs to be given to charging, cooling, heating, agitation and control.

CRITICAL EXAMINATION OF THE PLANT
Table 8.2 (page 179) shows the results of a Critical Examination of the plant. This has not been done in great depth, as the intention of this example is to demonstrate different studies on the same plant throughout. The main feature of note is the importance of using trained and skilled operators, preferably with assistance of computer control.

TABLE 8.1
Information on batch reaction, selected from Rasmussen[11]

Total incidents studied: 66	Incident rate in 10 year period: 2.6×10^{-3}/reactor/year

Proportion of pressure excursion turning into incidents estimated as 0.5

Immediate cause of incidents:		Consequences to materials/environment:	
Incorrect charging	17.2	No danger	12%
Inadequate cooling	13.1	Danger to equipment/production	14%
Inadequate heating	9.6	Loss of production/damage to equipment	33%
Inadequate agitation	10.1	Loss of equipment and production	41%
Incorrect batch control	9.1	Damage to ecological system	1%
Recovering from fault	2.0	**Consequences to humans:**	
Regular reaction inadequate	3.5	No danger	37%
Regular reactant unknown, decomposition	8.1	Near miss	9%
Water ingress and vaporization	0.5	Injuries	41%
Exotherm of unknown type	3.5	Loss of lives inside plant	17%
Cause unknown	6.1	Loss of lives outside plant	0%
Effect on reactor:		**Physical state involved:**	
Vessel open, hazardous release	27.3	Solid	9%
Glasswork shattered and release	24.2	Solution (water slurry or solution)	29%
Vessel ruptured and release	28.8	Liquid	64%
Vessel ruptured	1.5	Gas	12%
Explosion	7.5	**Where accident arose:**	
Hazardous release	7.6	Continuous process	11%
Catchpot ruptured	1.5	Batch operation	57%
Catchpot fire	1.5	Holding/storage	24%
Inititation mechanism causing accident:		Other	8%
Wrong substance mixed	19%		
Incorrect mixing conditions	19%		
Stirrer stopped	13%		
Impure/contaminated chemicals used	6%		
Contaminated vessel used	7%		
Stray catalyst	6%		
Accumulation of reactants or intermediates	9%		

TABLE 8.2
Critical Examination of the whole batch plant

Plant: Ethylene derivative **System:** Batch plant	**Date:** 1.1.94 **Page:** 1 of 1
What is to be done and how is it to be done? Ethylene and material M are passed from storage into a batch reactor to produce a product P. The reaction is carried out at 14 bar gauge pressure and 30°C. Material P passes to storage. Any material not up to specification, or material transferred out of the system in emergency, passes into a system involving a dump tank, a direct quench and a scrubber.	
Why do it? Basic route adopted to produce product P	**Why not?** Produce by another route using ... Reduce pressure, so less dispersion on release.
Why this way? Alternative continuous route being studied, see separate report. The pressure is at 15 bar to ensure liquid phase reaction at a temperature of about 40°C.	
Who should do it? Experienced operators	**What are the main potential hazards?** Rupture with release of process material. Fire and explosion on batch plant. Runaway exothermic reaction in reactor or dump tank. External threats from existing plant and transport. **What knowledge and skills are required?** Technical knowledge and safety awareness with specific information about runaway reactions. Operating experience with hazardous materials and high pressure plant. Detailed procedures and training required.
Where should it be done? Location of plant at Sheffield site, see plot plan. Ethylene manufacturing facilities available on site. Plant location at appropriate distance from other plants.	**What can be done to improve safety?** Improve the means for stopping any runaway reaction as by material transfer or injecting a reaction suppressant. Improve the control and emergency control systems by computer control. Provide fire-fighting facilities and water curtains. Prepare emergency plans with safe havens. Monitor environment for release of material.
When should it be done? Processing seven days a week	**Why not?** Other options give better segregation and emergency cover but are not economic

HAZARD IDENTIFICATION AND RISK ASSESSMENT

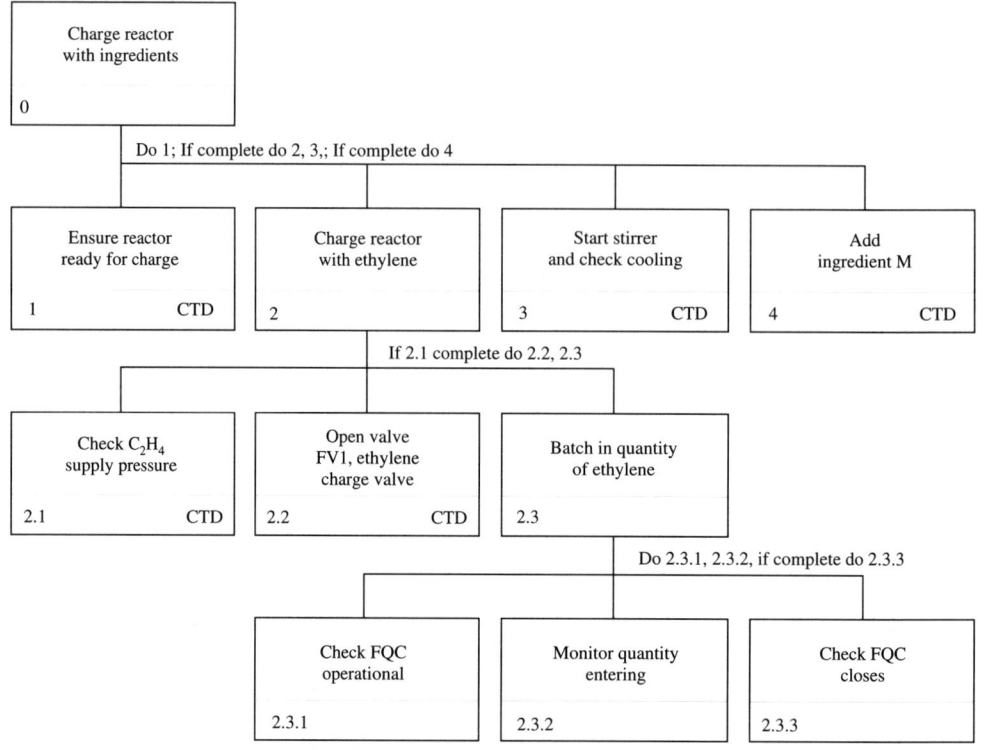

Figure 8.2 Hierarchical steps for charging the batch reactor. (Items marked CTD are to be expanded.)

HIERARCHICAL TASK ANALYSIS

The hierarchical steps in processing material are given in Figure 8.2 and further listed in Table 8.3. Each step was then analysed to identify critical steps.

IDENTIFICATION OF CRITICAL TASKS

Task Analysis takes a great deal of time and effort. Consequently it may be worthwhile to identify which tasks give rise to a safety risk, a quality lapse or a production lapse. These are then summed to indicate a critical task. For the simple scales used here, a value greater than three should be given high priority. In order to emphasize safety, the value of safety risk used here is double the normal risk calculated by the Short-Cut Risk Assessment Method (SCRAM).

The assessments were made as follows:

Safety risk, SR

Each severity, S, is rated from 1 to 5 and each likelihood by the power index, L, normally from 0 to −5. The sum of severity and likelihood (S + L) is multiplied by two in order to give the safety risk, SR. A positive value gives a ranking of the safety criticality. Any negative values are treated as zero when evaluating the critical tasks.

TABLE 8.3
Hierarchical task steps to process a batch

Charge reactor with ingredients and process batch
Plan: Complete each step in turn. Do 1, 2, 3, 4, 5, 6, 7
1   Ensure reactor ready for charge
Do 1.1, 1.2, 1.3
      1.1   Check empty using log and LR
      1.2   Check FV1, FV2, FV3, FV4 closed
      1.3   Check cooling water pressure and cooling water on
2   Charge reactor with ethylene
If 2.1 complete, do 2.2, 2.3
      2.1   Check ethylene supply pressure by PR
      2.2   Set FQC and open valve FV1
      2.3   Batch in ethylene
      Do 2.3.1, 2.3.2. If complete, do 2.3.3
            2.3.1   Check FQC operates
            2.3.2   Monitor quality
            2.3.3   Check FQC closes FV1
3   Start stirrer and check cooling
Do 3.1, 3.2, 3.3
      3.1   Start stirrer and check current amps
      3.2   Check TRC functioning
      3.3   Check PRC functioning
4   Add other ingredients
Do 4.1. If complete, do 4.2, 4.3, 4.4
      4.1   Check storage level satisfactory for batch
      4.2   Start pump
      4.3   Set FQC and open valve V2
      4.4   Batch in M
            4.4.1   Check FQC operates
            4.4.2   Monitor temperature and pressure stopping flows of M if temperature exceeds 35°C. Cool to 25°C
            4.4.3   Add second batch of M
            4.4.4   Monitor all M added
            4.4.5   Check FQC closes FV2
5   Cool the reactants
Do 5.1, 5.2, 5.3, 5.4
      5.1   Monitor temperature and pressure in reactor
      5.2   If temperature alarm sounds stop all inlet flows and prepare to dump contents if temperature goes above 40°C
      5.3   Monitor pressure and stop flows if pressure rises above 16 bar
      5.4   Cool batch to 25°C
6   Check product quality
Do 6.1, 6.2, 6.3, 6.4, 6.5, 6.6
      6.1   Check mix cooled to target temperature
      6.2   Analyse product quality. If satisfactory go to 6.6
      6.3   If ethylene level high continue agitation and test again after 15 minutes
      6.4   If ethylene level high carry out special procedure 1 after checking with supervisor
      6.5   If ethylene level low carry our special procedure 2 after checking with supervisor
      6.6   Tank ready to empty
7   Empty vessel
      To be continued

TABLE 8.4
Identification of critical tasks: batch plant

Task/activity	Lapse	Evaluation					
		S	L	SR	QL	PL	CT
(1) Ensure reactor ready for charge.	Mixing of ethylene with other chemicals in storage or destruction of batch.	1	−2	0	0	1	1
(2) Charge reactor with ethylene.	Inadequate or excess batch size leads to quality lapse. There could also be a highly incorrect mix in the reactor leading to fast runaway.	3	−2	2	1	0	3
(3) Set stirrer and check cooling.	Runaway reaction if malfunction followed by inadequate response. Cooling water failure also affects dump tank.	3	−2	2	0	0	2
(4) Add other ingredients.	Adding M too quickly could result in excess reaction with possible reactor runaway. Incorrect addition results in a quality lapse.	2	−1	2	1	0	3
(5) Monitor the reactants.	Monitoring the system allows recovery without the need for emergency action. It should enable better control of quality. See also item (3) on reactor cooling.	1	−1	0	1	0	1
(6) Check product quality.	Great care must be taken that the correct procedures are used to adjust quality.	2	−2	0	1	0	1
(7) Empty vessel.	Main problem likely to be spillage but emptying when product impure could cause quality and production loss.	1	−2	0	1	1	2

Abbreviations:
S = severity; L = likelihood/year; SR = safety risk; QL = quality lapse; PL = production lapse; CT = critical task.

Quality lapse, QL
A quality lapse is rated according to the power index of its likelihood per year. Values normally range from 2 to −2 — that is, 10^2 to 10^{-2} times per year.

Production lapse, PL
A production lapse is selected on a similar basis to a quality lapse. A significant loss of material is deemed to be a batch.

CRITICAL TASKS ON THE BATCH PLANT
The evaluation of critical tasks on the batch plant has been carried out in Table 8.4. The study suggests that the design is adequate if run by very able and skilled operators but that if such capabilities are not available then additional control, probably by computer, is necessary. As expected, the key steps involve the addition of the ingredients.

JUSTIFYING SAFETY STUDIES
The principles behind the safety studies are important. It is often seen as costing money to include add-on safety features. Of course if it is not done and matters go wrong then costs can be horrendous.
Improvements in production output and quality must be considered when carrying out a safety study. Quality control and safety are two sides of the same coin and savings in production are good news. Similarly it is hard to distinguish between safety and environmental issues for a given accidental discharge. If the discharge can be avoided or made very unlikely, a major improvement has been made.
It is necessary to ask how can things be done worse and this is now tackled using Hazard Analysis.

HAZARD ANALYSIS
As noted earlier, this study has solely considered reactor equipment problems. The specific problems considered are those giving rise to an exothermic reaction (generating a dangerous temperature or a dangerous pressure or both). They are inadequate cooling, inadequate agitation, inadequate charging of ethylene, inadequate addition of ingredients M and inadequate heating.
The study for adding ingredients M is given in Table 8.5, pages 184–185. It is not as comprehensive as a Preliminary Hazard Analysis and is only intended to highlight the sections noted. The discussion is intended to consider whether it can be modified or avoided and what must be done if it arises. The relief system has subsequently been developed as shown in Figure 8.3 on page 185.

TASK ANALYSIS OF DEVIATIONS
Each deviation of the task which might physically affect the process directly can be examined. First let us tackle it using the what-if approach. This will be followed by a Hazop.
The what-if questions selected for this type of study include the following:
- What has to be done and how is this effected?
- When is it done?

TABLE 8.5
Hazard analysis of the batch system

Plant: Ethylene derivative System: Reactor Type: Batch	Node: 4 Date: 1.1.94 Page: 1 of 2

Process step: Add ingredients, M (see step 4.4 in Table 8.3, page 181)	
What is to be done and how is it to be done? Mixture M is added to a batch of ethylene already present in the autoclave with agitator and cooling water on. The material is added in two batches. The reaction temperature is carefully monitored and between batches the process material is cooled to 25°C. Flow of M is stopped if the temperature exceeds 35°C. The contents are dumped if the reaction temperature exceeds 40°C.	
Why do it? Basic route adopted to produce product P	**Why not?** Produce by another route using ...
Why this way? Alternative continuous route being studied. Ethylene is added first to reduce maximum temperature rise. The operational temperatures have been selected following consideration of the reaction isotherm.	See report on alternative route. Why not batch both materials in continuously with excess ethylene? Why not semi-batch? Why not provide means to stop the reaction — for example, using an inhibitor or recycling product P.

What can go wrong affecting safety, production and quality?

Dangerous deviation	Discussion
Dangerous temperature and pressure due to excess reaction	Given a dangerous temperature, discharge the material into the dump tank. Temperature and pressure will continue to rise so also depressurize to the same tank. The bulk of material will flash over into the quench tank filled with product P. Pressure continues to rise and the relief system must cope with high backpressure. Consider relief to a blowdown system with back-up from a large bursting disc (problems if spurious failure) as shown in Figure 8.3. Consider pressure resistant construction of the batch reactor. Consider adding an inhibitor providing the stirrer is operational. Consider a high quality process control system which would reduce the likelihood of incorrect charging and help make decisions when to dump material on high temperature/pressure. The process control system might include a separate hard-wire system. Consider emergency cooling: unfortunately a lot of heat is to be removed and the stirrer must be operational. Install automatic back-up of the cooling water system.
Loss of cooling	This creates a major problem as it also affects the dump tank and the direct condenser. A back-up cooling water pump is essential with a drilled port in the control valve to avoid complete closure.

TABLE 8.5 (continued)
Hazard analysis of the batch system

Immediate cause of deviation	Discussion
Loss of agitator	Layering problems could arise. The stirrer must be on and intact. Interlocks and cut-out could prevent addition of M if the stirrer is not on or the blade is off. In the event of failure no more M should be added and the batch gradually discharged to the dump tank.
Inadequate charging	Batch M in two parts into the total quantity of ethylene. On no account mix ethylene into M. Any errors contrary to the procedure must be corrected by set procedures involving transfer to dump tank. Study of problems associated with the dump tank is a high priority. The system is overdependent on the action of the controller FQC1 and FQC2. Back-up emergency controls must be provided to cut out this material. The same systems are essential to isolate plant in the event of leaks on FV1 or FV2.
Inadequate heating	Admission of steam to cooling water could prevent cooling. Redesign as a preheater of limited capacity is preferred. Note that if the coil fails ethylene will end up at the cooling tower. Appropriate block valves might be considered to reduce this danger. Redesign.

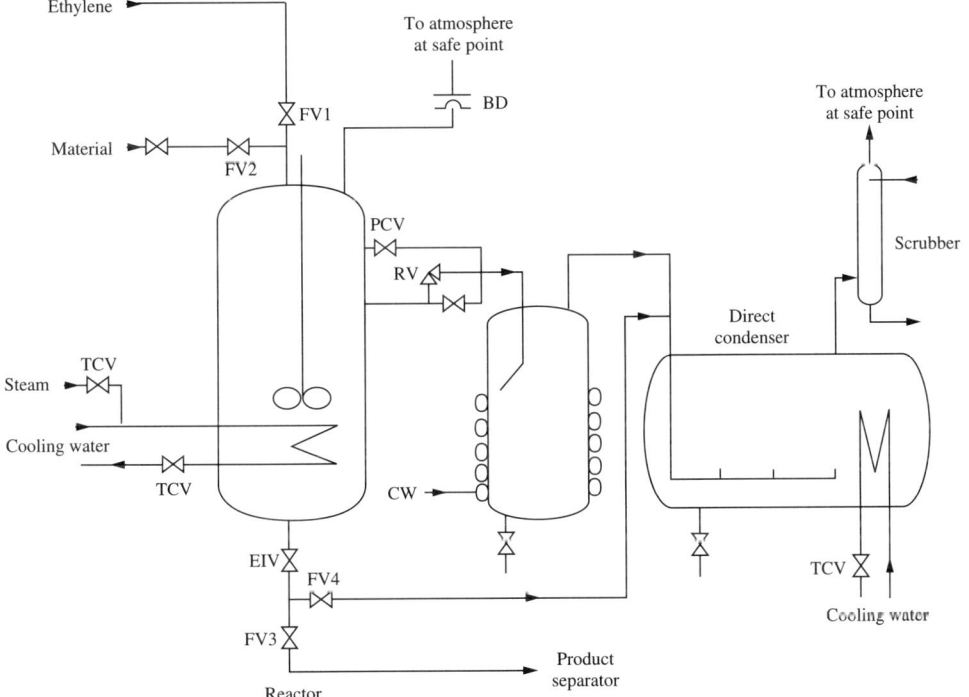

Figure 8.3   Relief system for batch reactor. BD — bursting disc; CW — cooling water; EIV — emergency isolation valve. (See page 243 for other instrumentation definitions.)

- What if an error is made? Consider the following error modes:
  What if omission of task, step or substep of the task?
  What if error of commission involving wrong selection of valve, indicator, control?
  What if wrong sequence, duration or timing?
  What if a qualitative mistake by excess, insufficient or default?
  What if task not completed or done incorrectly due to mix of errors and equipment difficulties causing actions too much, too little, not closed, not opened, not aligned, too long, too short, excess force, too little force?
- What else can go wrong?

   This approach is very general and does not aim to answer other important questions such as Does the operator know an error has occurred? An omission can occur because the equipment does not respond or because the operator knows it has occurred and does not correct it — say, because of insufficient time.

IDENTIFICATION OF CRITICAL TASKS IN A SPECIFIC TASK

This method will now be used to examine a specific task, step 2.2. The objective is to charge the autoclave with ethylene via valve FV1 (see Figure 8.1, page 177) using the flow controller FQC1 activated from the control room. The task activities involved in Step 2.2 are described in Table 8.6.

   It will be noted that many of the actions recommended will be occasional actions carried out by the operator with a high conditional probability of failure in some cases. Consequently appropriate add-on safety measures are called for, particularly interlocks and alarms. Appropriate procedures must be drawn up, both identifying the correct way to operate this valve and emphasizing appropriate recovery action if things go wrong. Clearly an on-line computer control system is beneficial in such circumstances. Some of the required interlock systems are suggested for valve FV2 in Figure 8.4.

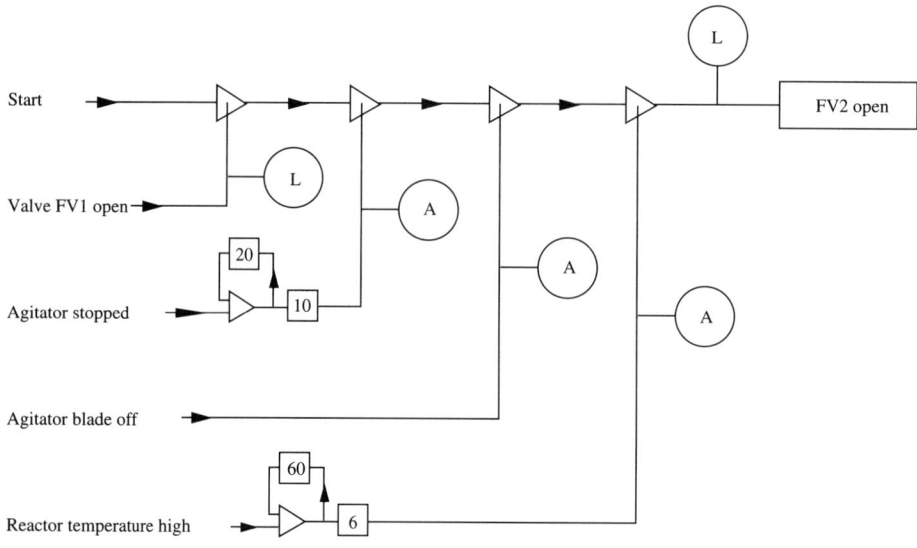

Figure 8.4   Interlocks before FV2 in Figure 8.1 can be opened.

TASK ANALYSIS AND HAZARD IDENTIFICATION

The study could be continued to consider what happens in the event of such failures. But for a plant of this size and hazard, it is unlikely that a full quantified risk assessment involving detailed estimation of reliability would be required.

HAZOP OF THE BATCH PLANT

Has the previous exercise been carried out in a satisfactory manner? Could other methods serve the same purpose? A conventional Hazop study is carried out on line A1001, the charge line for

TABLE 8.6
Task analysis of step 2.2 — open valve FV1 (see Figure 8.1, page 177)

Question and answer	Deviation	Action required
What has to be done and how is it effected? Charge autoclave with ethylene via FV1 using FQC activated from control room		
What if omission of task? Autoclave will contain no ethylene.	If M added can create difficult situation and incident unless discharged safely	Operator to discharge M to dump tank provided this is empty
What if too little ethylene added? Valve open but insufficient or no ethylene available.	If M added then can get runaway or reverse flow to ethylene if NRV fails and valve not closed	Interlocks are necessary on FV1/2. PAL on ethylene supply. TAH on ethylene supply tanks.
Valve jammed open or FQC failure	Autoclave overfilled	On alarm by LAH close all valves
When is it done? Valve opened after ensuring autoclave empty according to LR and FV2, FV3 and FV4 closed		
What if opened when reactor full?	Reverse flow remotely possible. Warn by TAH in storage. See also next item.	On alarm by TAH close all valves
What if more done?	If autoclave still full adding further material creates problems but not danger	LAH required to warn operator. Discharge to dump vessel. The design of this LAH will be difficult.
What if done out of sequence?	If FV3 is open, ethylene passes into product. If FV4 is open it passes into dump. If FV2 is open and NRV fails, C_2H_4 flows into M storage with runaway reaction.	Interlocks on FV1/2/3/4 (no valve can open if others are open). Status of valves (positioners) to be shown on control panel
What if error of selection?	If FV2 opened instead then M might enter the reactor.	Unlikely by trained operator. Operator to dispose of M to dump before correcting see warning above.

TABLE 8.7
Section of Hazop on line A1001, Figure 8.1, page 177

Deviation	Cause and consequences	Action
No flow	Inadequate ethylene supply or FV1 failed closed. No further flow enters reactor and if sequence continued could get runaway.	PAL on ethylene supply. Interlock so FV2 cannot be opened when FV1 open. For an example, see Figure 8.4, page 186.
Reverse flow	Autoclave full, FV2 open and low ethylene pressure could, if NRV fails, give rise to reverse flow	An acceptable risk but warn operator that if level in autoclave starts to fall, this is a possible cause amongst others
	FV2 open when FV1 open, low ethylene pressure and NRV fails	Reverse flow of M into ethylene could promote runaway reaction. Interlock between FV1 and FV2 is needed. Status of valves to be shown on control panel. TAH on ethylene storage. Dump autoclave contents immediately problem is diagnosed.
More level in autoclave	FQC of ethylene fails open	Danger of rapid overpressure on adding M and bursting disc activated. A separate counter and trip on ethylene would prevent the problem.
	FQC of M fails open	Runaway reaction. Install a separate counter and trip on line. Stops leaks.
Less level in autoclave	Inadequate batch of ethylene on failure of FQC or inadequate ethylene supply and operator adds M	Runaway reaction. Separate counter on ethylene — see above. Extra isolation prevents leaks.
	PRC failed open or spurious relief sends material to dump system	Ensure recycle possible. EIV closes on line with warning from dump tank by LAH.
High temperature in autoclave	Excess ratio of M to ethylene	Install trip on high ratio using separate counters of flow. Alternatively trip on high temperature.
	Failure of agitator or agitator off	Low current alarm and interlock between agitator and FV2
	Failure of cooling water	LFA on cooling water with drilled port in TRCV. Back-up supply.
	Runaway reaction for above reasons	Automatic trip with material transfer to dump tank and flow of M stopped

TABLE 8.7 (continued)
Section of Hazop

Deviation	Cause and consequences	Action
High pressure in autoclave	Failure of PRC or excessive reaction	Overpressure. Relief valve feeds contents to dump tank. Bursting disc relief to atmosphere in absolute emergency.
Maintenance (more flow)	FV1 opened when autoclave open for maintenance causing spill and possible explosion	Blanks to be installed or double block and bleed. Project to advise.
More composition	Air present after maintenance causes dangerous explosive mix	Provide a nitrogen purge at start-up venting via the dump tank

ethylene and the autoclave, and reported in Table 8.7. This study is not comprehensive as the aim is to consider features related to valve FV1 for comparison with the study given in Table 8.6.

COMBINED TASK ANALYSIS AND HAZOP

The Hazop study did not identify the types of human error which led to the problem arising. This could have readily been established by introducing a further column termed error mode. The Hazop tackled some of the problems associated with the emergency control instrumentation. It is strongly recommended that a Hazop study is carried out before a full Task Analysis of a new design. Better still is to use a hierarchical step analysis first, as shown in Table 8.8 on page 190, in order to establish the procedure to be used.

SNEAK ANALYSIS

Sneak Analysis originated in the aerospace industries as Sneak Circuit Analysis. The pioneering work was done by Boeing and McDonnell-Douglas in the late 1960s and early 1970s after incidents involving the accidental firing (or not firing) of spacecraft rockets.

Sneak Analysis is a technique which aims to identify hazards associated with the topology of process plant — that is, how the different components are connected together. It can therefore be used as an adjunct to any hazard identification method. The method can be used to examine any system by assuming an initial state of the system and identifying deviations from that state caused by flows of material and energy within the system.

The main value of the method in the process industries appears to be the identification of a sneak path. The term 'sneak' implies an unintended path which might arise from design errors, human actions, control or equipment failure. The best way of carrying out a Sneak Analysis seems to be in marking up sneak paths on the P&ID. Different colours might be used to denote the difference between the three error modes which might cause a sneak path.

TABLE 8.8
Combined Task Analysis/Hazop

Question	Response		
What has to be done? What if omitted? When is it done?	On step 2.2 open valve FV1 and let ethylene enter the autoclave. Autoclave contains no ethylene and sequence of addition could be upset. After reactor emptied and ethylene at adequate pressure.		
Deviation	**Cause**	**Consequence**	**Recommendations**
Reverse flow?	Autoclave full and pressurized when FV1 opened and low ethylene pressure	Contents backflow to ethylene storage	Install non-return valves
Reverse flow?	FV1 open when FV2 opened	If M flows into ethylene could have runaway reaction	Interlock between FV1/FV2. Status of valves on panel. TAH on ethylene storage.
More flow?	FQC1 fails or FV1 fails fully or partially open	Product quality affected	Secondary counter of quantity with further isolation
Less flow?	Inadequate pressure and slow charging of autoclave	Batch could be incorrectly loaded when FV2 open leading to runaway or quality problem	Interlocks highly desirable between valves FV1, FV2, FV3. Secondary counter of quantity with cut-out system.
Error of selection?	Other valve opened in error (any of FV2–4)	Possible contamination of batch, product or feed	Interlocks necessary
More of (composition)?	Air present in autoclave at start of run	Air mixed with ethylene could burn if ignited	Nitrogen purge needed at initial start-up only
Maintenance	Valve opened when reactor open for maintenance	Explosion, spill	Blanks to be installed. Consider spectacles if regular activity.
Fire	A leak to atmosphere can be caused by an external threat or deterioration	Fire or explosion	Local detector for flammables is required

A typical design error might occur when two apparently unconnected tanks are connected by a common drain. This may not even appear on the P&ID. Such an interconnection can result in a change of energy within the line as, for example, when reaction occurs. It is possible, however, for material to pass from a high pressure to a low pressure vessel.

The failure of emergency control valves to open or close when activated can give rise to a sneak path. This is because the operators see that the valve has been activated because the status light in the control room tells them that it has. In fact, the light only indicates whether or not power is applied to the valves; it gives no information whatever about their actual state. In the Three Mile Island incident, the valve was open but the light indicated closed. Sneak indications such as these are very common. This was not the only sneak in the Three Mile Island incident. The chain of events started when water was introduced into the instrument air system via a sneak path.

SNEAK PATHS ON THE ETHYLENE BATCH REACTOR

Another way of identifying problems is to identify all possible sneak paths. It is important not to assume a 100% reliability of non-return-valves (NRVs) as a means of preventing such paths. The main sneak paths considered for the batch reactor are indicated in Table 8.9, where C indicates a closed valve and O an open valve.

The circumstances in which valves are in the wrong status arise from the use of incorrect sequences in procedures and the failure of control systems. Back-up protection comes from NRVs, interlocks and trip systems.

Interlocks and NRVs reduce the likelihood of paths 1, 2 and 3 (Table 8.9). It would be dangerous to stop the dump tank receiving the stream at any time by an interlock. A trip that on any flow through FV4 all other valves close is possible but expensive. An NRV on the dump lines assists in reducing the likelihood of paths 6 and 7.

TABLE 8.9
Status of valves for Sneak Analysis

	FV1 Ethylene	FV2 M	FV3 Product	PCV1 Emission	FV4 Effluent	Path
1	O	O	C	C	C	Ethylene to M
2	O	O	C	C	C	M to ethylene
3	O	O	O	C	C	Ethylene to product
4	C	O	O	C/O	C	M to product
5	C	O	C	O	C	M to dump tank
6	C	O	C	C	O	M to dump tank
7	C	C	O	O	C	Dump to product
8	C	C	O	C	O	Dump to product

O = open valve; C = closed valve

PROCEDURAL SNEAK PATHS

Procedural sneak paths spring from relationships between procedures and possible errors. Examples of errors affecting a component/device/substance Z due to procedures might include the following paths:
- a procedure performed on Z at another time;
- a procedure performed on Y which is physically near Z;
- a procedure performed on Y, similar in colour, form, function or purpose to Z;
- a procedure which can be performed on Z (improvization);
- maintenance or repair at the wrong time;
- a procedure performed on Z without checking preconditions;
- an external threat from painting, cleaning, repairing or building and installing near Y;
- transport or storage near Y;
- mislabelling or misinstruction with respect to Y;
- any common cause failure.

It is unusual for sneak paths to be investigated in this way. The method would need a better structure and the problems can be identified in other ways. However, it is an interesting concept about which the analyst should be aware. For further information see Taylor[12].

FAILURE MODE AND EFFECT ANALYSIS

Failure Mode and Effect Analysis (FMEA) requires a knowledge of each failure mode of the items of plant. The consequences of the failure mode are examined and may be interpreted to give the deviation noted in the Hazop study. In general the method merely examines the physical elements of the plant, such as vessels, pumps and pipes. It is possible to use it to examine actions and procedures. The most critical events can be identified using a criticality index (FMECA).

A Failure Mode and Effect Analysis involves the following steps:
- describe the system;
- list all system items;
- identify all faults for each item;
- determine the effects on other items for each fault and evaluate the resulting impact on overall performance or the integrity of the system;
- estimate the probability and seriousness of each fault: a criticality index may be assigned.

An *item* is any part, component, device, subsystem, functional unit, equipment or system that can be individually considered.

The *capability* of an item represents the ability of an item to meet a specified service demand under given conditions.

A *failure* is the termination of the ability of an item to perform a required function. A failure is an event whereas a fault is a state. After its failure the item has a fault.

A *fault* is the state of an item characterized by inability to perform a required function, excluding the ability during preventative maintenance or other planned actions or lack of external resources.

The *failure mechanism* represents the physical, chemical or other process which has led to a failure. The *failure cause* represents the circumstances which have led to the failure.

A *sudden failure* may not be anticipated by prior examination or monitoring.

A *complete failure* results in the complete inability of an item to perform all required functions; a partial failure or fault allows some but not all required functions to be performed; and a *degradation failure* is both a gradual failure and a partial failure.

An *intermittent fault* is a fault of an item which persists for a limited time duration following which the item recovers the ability to perform a required function without being subjected to any action of corrective maintenance; and a *persistent fault* is a fault on an item that persists until an action of corrective maintenance is performed.

An *error* is the discrepancy between a computed, observed or measured value or condition and the true, specified or theoretically correct value or condition. A *mistake* is a human error which is a human action that produces an unintended result. It is interesting to note that the French term 'erreur' may designate both an error and a mistake.

There is no reason why an item cannot be put in a failed state by a human action such as opening or closing a valve, or disconnecting flanges, interlocks or power supplies.

General information on failure modes is given in *Guidelines for Process Equipment Reliability Data*[13]. They are indicated in Chapter 7.

FAILURE MODE AND EFFECT ANALYSIS OF VALVE FV1
An FMEA has been carried out on valve FV1 and associated control loop (Table 8.10, page 194). The only item not covered here is a small insidious leak of ethylene which would require appropriate analysis. The need for block valves and means of effectively isolating FV1 emerged only in this part of the study. In an FMEA the analysis considers each failure mode associated with the valve. As can be seen from this example, the procedure can be very long although it is thorough.

HUMAN RELIABILITY ASSESSMENT
An assessment of human reliability involves the following steps:
(1) Problem definition involving a review of the problem and a review of the information available.
(2) The qualitative prediction of human errors involving breaking down the individual tasks to a stage where either the actions by the operator cease to affect the safety of the system or are compensated for by the response of system hardware. Performance-influencing factors are identified and analysed to see if they can give rise to significant errors. Human error modes are identified.
(3) The representation of the system in a form suitable for analysis may require the use of fault or event trees and these must encourage further study of the recovery mechanisms involved.
(4) The quantification of the significant human errors can be carried out by many techniques. Some of these adjust values according to many criteria related to experience, familiarity with the system, stress or temporary anxiety, level of alertness, response time available, quality of feedback, information processing skills, judgement and so on. Whether the technique requires art or science is open to question, but the inherent good sense of the analyst will usually stop complete reliance on the values generated. What applies at one site does not necessarily apply elsewhere. Often the response to an error is carried out because others recognize it rather than the actual perpetrator.
(5) A study must be made of all interaction with hardware. Major improvements may sometimes be made with computer control. Flying an aircraft is one system which has clearly benefited from such action.

(6) Human error reduction strategies may prove necessary, should the probability of human error prove significant when the overall risk associated with a particular scenario is evaluated.

(7) Finally documentation of the study is carried out so that the results can be justified and audited. It is important that the study gives recommendations regarding procedures, instructions, training, performance indicators and so on. These may then be implemented.

TABLE 8.10
Failure Mode and Effect Analysis of valve FV1

Equipment	Failure mode	Local effect	Effect on system
Valve FV1 FQC1	Fails open Fails high output	High level in autoclave	Operator alerted by LAH must isolate leak by separate block valve and dispose of some ethylene to dump tank. Repair of valve effected by complete system shutdown.
Valve FV1 FQC1	Fails closed Fails low output	None or insufficient ethylene in autoclave	Separate counter should indicate this problem. Interlock from this prevents further addition of M.
Ethylene supply	Low pressure	Valve FV1 open and limited ethylene supply	Operator may assume autoclave full and start next stage. Interlock prevents this occurrence. Operator alert by PAL.
Valve FV1	Opened at wrong stage in sequence	Sneak path to ethylene storage if NRV fails	M into ethylene prevented by interlock and NRV. Risk acceptable. THA in storage.
Valve FV1	Opened at wrong stage in sequence	Sneak path to M storage in NRV fails	Ethylene into M. Very serious problem avoided by interlock, NRV and design of pump (reciprocating rather than centrifugal). Operator alert by TAH in M storage and must then take preplanned emergency action.
Valve FV1	Opened at wrong stage in sequence	Sneak path to product storage or dump tank	No safety problem but reduce chances by interlocks. See note on recycle of off-specification material.
Interlock FV1/2	Interlock failed or disconnected	FV1 and FV2 can now be opened at the same time	This nullifies the protection, in particular for avoiding a high ratio of M. If interlocks out of action, use the block valves to supplement isolation.
Valve FV1 leaking	Slow leak of ethylene	Loss of quality of product	Nullified by secondary counter. Product testing for quality is further check.

TASK ANALYSIS AND HAZARD IDENTIFICATION

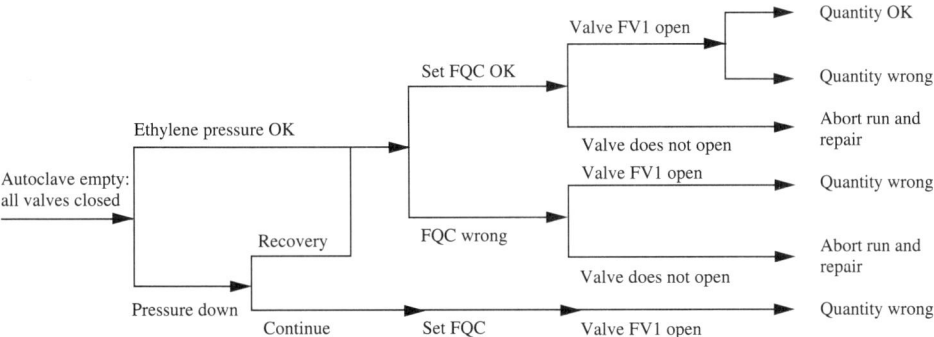

Figure 8.5   Event tree for section of batch plant.

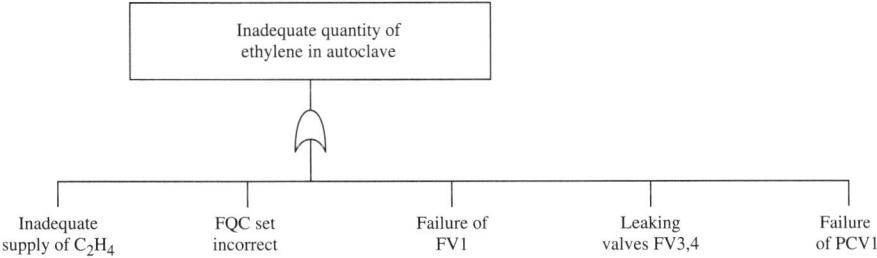

Figure 8.6   Fault tree for section of batch plant.

EVENT TREE AND FAULT TREE ANALYSIS

Event trees and fault trees give a powerful representation of problems. The reduced event tree presented in Figure 8.5 gives a simplified sequence of actions by the operator and the response of the valve FV1. This is the only technique which clearly shows the tendency of the operator, confronted by an infrequent occurrence, to carry on rather than abort an activity.

The fault tree in Figure 8.6 is further simplified and here additional causes have been added. The results of this study support the view that, for design purposes, this method is preferable and quicker than drawing out event trees.

Event trees contain a lot of repetition and for larger systems can be tedious to draw. Only a particularly hazardous system would be studied in the sort of detail reported in Figure 8.5.

HUMAN ACTION ERROR ANALYSIS

A human action error analysis is given in Table 8.11, page 196. This indicates two errors with respect to valve FV1 used in the ethylene batch plant. A number of additional ideas were generated by this study. This is to be expected, as the more ways in which a plant situation is examined, the more likely it is that the completeness of the study will be adequate. But the justification of the cost and time of the study is a different matter.

195

COMPUTER CONTROL OF BATCH PLANT

There is a high degree of reliance on the operator for this plant, leaving a high potential for human error. Two approaches have commonly been used to overcome this problem. The first is to throw instrumentation at the system. The second is to install computer control. The basic process control system of plant is increasingly being based on control and display units built using programmable, computer-based components. Digital controllers and operator workstations based on video displays provide automatic regulation of the system and improve communication to the operators. A supervisory control computer may be added which collects data for manufacturing reporting purposes and may perform advanced control functions. As well as this the computer can act as a data analysis tool to assist the operator when responding to process disturbances. This system lacks reliability and cannot be installed with confidence as the primary risk control system.

TABLE 8.11
Human action error analysis for opening FV1

System state	Normal operation
Step	Open valve FV1
Cause of problem	Valve opened in error
Error mode	Action on wrong object by operator
Error mechanism	Mistaken alternative (spatial disorientation)
Feedback — how observed	High level in ethylene reactor gives an alarm
Recovery	Turn off valve on alarm. This cause must be ascertained by the operator from a small number of options. Failure to close the valve would result in complete filling of the reactor.
Consequence	Product would be off-specification unless adjusted Material recycle necessary
Equipment implications	High level trip considered but rejected
Procedural implications	Check required on procedure for off-specification materials being reprocessed
Training implications	Train and demonstrate necessary actions
System state	**System under repair**
Step	Open valve FV1
Cause of problem	Valve opened in error. Power not disconnected.
Error mode	Wrong choice by operator
Error mechanism	Stereotype take-over
Feedback — how observed	None immediately
Recovery	Isolate vessel and area until problem rectified
Consequence	Ethylene enters reactor when system under repair Probable spill to atmosphere
Equipment implications	Insert blanks and close/lock block valves when reactor opened. Isolate power supply to relevant control valves.
Procedural implications	Note in procedures and on work permit
Training implications	Train operators and maintenance

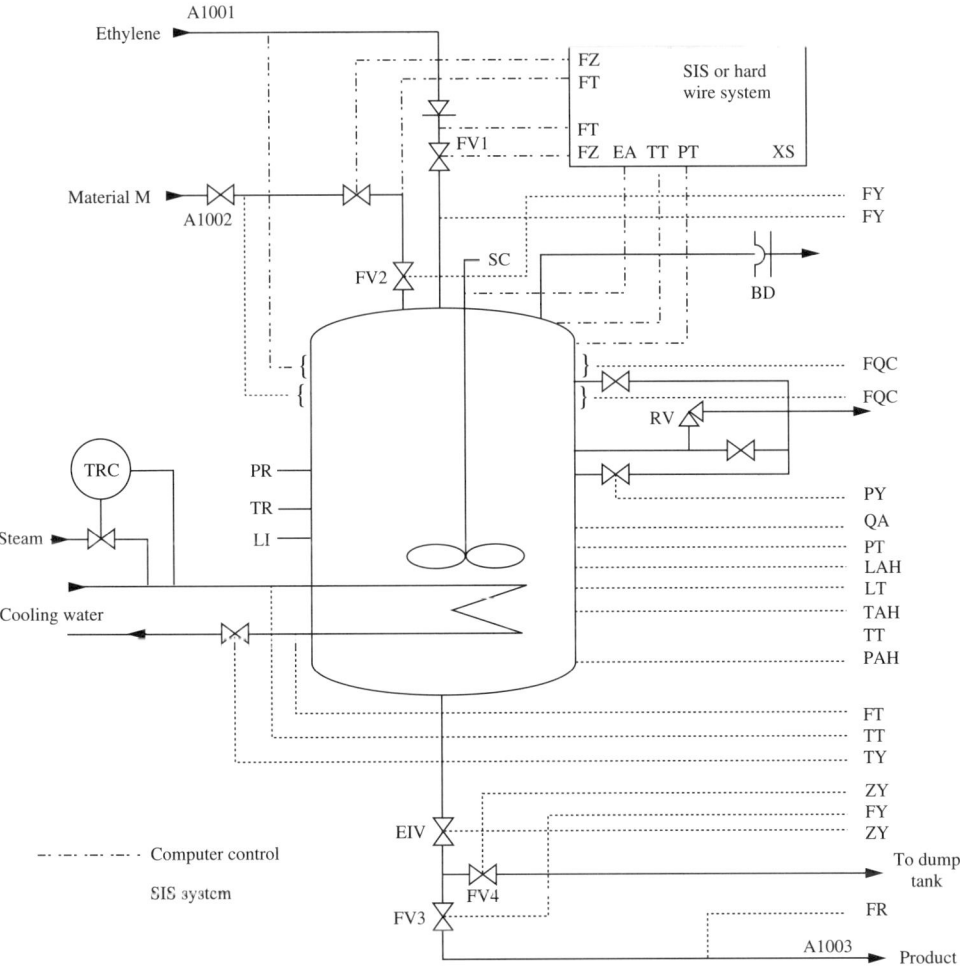

Figure 8.7 Computer control system for batch reactor. EIV — emergency isolation valve. (See page 243 for other instrumentation definitions.)

The AIChE Center for Chemical Process Safety (CCPS) has noted that the basic process control system on such a facility can be the prime cause of a serious accident[14]. The consequences may involve the total loss of information to the operator during normal operation, inadequate and slow access for the operator to plant status during plant upset, disrupted communication among control modules causing unreliable signals to control valves and loss of signals from and to the field instrumentation and valves. Clearly there is a need to carry out effective design of the system to involve appropriate redundant components and fault-tolerant systems.

Figure 8.7 features a computer control system with a hardwire (direct wire) system for protection. This features override control involving actions such as:
- the correct quantity of ethylene is added before M is dosed;
- if the temperature is greater than 40°C the trip system cuts the flow of M;
- M flow is cut at 10 bar pressure;
- no material flow of M when agitator stopped or agitator current is high;

197

- no material flow of M on loss of cooling water;
- FV1/FV2/FV3/FV4 are linked by interlocks;
- emergency push button stops all process material flows and dumps contents of autoclave.

This type of system has largely been replaced by current technology. Any programmable electronic system (PES) is a computer-based system which controls, protects or monitors the operation of plant by sensors and actuators. A PES may range in size from a single processor chip to a large computer system. It can be used as protective control. Furthermore, instead of it simply replacing a simple signal trip device and electromechanical relays, the modern system can carry out more complex monitoring functions. It can, for example, solve process state logic equations, identify acceptable operating envelopes for the plant, warn that abnormal conditions are imminent and interrupt manufacturing operations to prevent the occurrence of a hazardous event. What used to be commonly built as a hardwire system can now be upgraded into a safety interlock system (SIS) and have the continued aims of functioning reliably when an abnormal condition occurs and not giving spurious trips.

Figure 8.7 would be perceived as representing the start of the design process. It is necessary to ensure the integrity of both the basic process control system and the safety interlock system. This will result in improvements in the design as shown. A full Hazop is required which considers not only the process system but also the computer system environment and each input and output signal. On computer systems there is a greater potential for common mode failures. A single failure of a card for an input/output that contains several channels can cause all the channels to fail at the same time. Hence there may be a need to carry out an FMEA of selected items.

It is poor practice to decide to install the computer control system at this stage of the design and in this manner. Such decisions must be taken early in the design so that the system is properly reviewed by Preliminary Process Hazard Analysis and the proposed architecture of the system to be employed for control purposes is properly constructed. Extensive advice on the topic is given by CCPS[14].

For example, Task Analysis has indicated a number of problems which the computer-based system does not enhance. The basic process control system does not include sufficient preventive systems designed to avoid initiating events. The overall system lacks any attention to interlock integrity levels. Basically the system uses single path design throughout. Some redundant sensors are available for diagnostic purposes but basically this system will be prone to failure. Such non-redundant systems have at best an availability of about 0.99. So some upgrade is required, either by using a partially redundant system or a totally redundant system in which a single fault of an SIS component is highly unlikely to result in the loss of process protection.

In the system described the interlock action that automatically takes the process to a safe state is reasonably well acted upon. But the control system completely fails to initiate changes in process conditions to lower operating risks whenever the process is in a failed state. The operator simply takes appropriate action. There has also been little attention to what changes should be initiated when the SIS is in a failed state. Now it may well be the case that such improvements are not necessary for this plant, or that computer control is unnecessary anyway. But it is manifestly absurd that proper consideration has not been given to installing what is the best system to do the task, and this occurred mainly due to a failure to apply the correct control and safety system design philosophy right at the start. The control system was included as an add-on safety system rather than as an essential element of the design and operating philosophy of the plant.

9. RISK CRITERIA

RISK

A *hazard* is a physical situation with a potential for human injury, damage to property, damage to the environment or some combination of these.

The *risk* of an event is the likelihood of a specified undesired event occurring within a given period or in particular circumstances. Risk is usually considered to be a function of the frequency or probability of an event occurring and the consequences of its occurrence, particularly with respect to causing damage and injury. This is the normal definition used in the process industries, although a health hygienist might narrow this definition to consider the hazard of a substance as the way it can harm the individual. The risk of a substance can then be defined as how likely it is to harm you in practice.

The *individual risk* is the frequency at which an individual may be expected to sustain a given level of harm from the realization of a specified hazard.

The *societal risk* reflects the likelihood of accidents involving multiple casualties.

Standards for risk are set considering the views of all related groups. There is, of course, concern from public and employees that the risks to which they will be exposed are tolerable. The public are extremely unlikely to want an LPG unloading station across the road from their home or a nuclear power station within ten minutes' walk. Apart from the risk most people consider these items to be ugly installations anyway. Less clear-cut are environmental objections to a main road through fields at the bottom of the garden, as many people benefit from the detriment to a few. So it has to be accepted that there will never be acceptable criteria to which everyone will agree. If the case seems fair then most social groups accept the majority opinion. Even so there exist groups within the community which tend to be vociferous and committed to a cause and will argue most about the tolerability of risk. So confrontation will always arise.

Risk evaluation is the complex process of determining the significance or value of the identified hazards and risks to those concerned with, or affected by, the decision. It is an emotive topic which will be discussed in the next chapter. But in the first place it is necessary for the process engineer to be able to carry out an initial screening of the risk involved in any activity. This is best done on some numerical basis and hence there is a need for the engineer to be aware of the order of magnitude of the risk which, with luck, will be considered tolerable by the public, workforce, company directors, shareholders, regulators, legislators and planning authorities.

CAUSES OF DEATH

The typical causes of death of 580,000 people in England and Wales are given in Table 9.1, page 200. It is clear that deaths due to work-related accidents are a very small fraction of this number and it might be considered whether the balance of effort put into prevention of such deaths is in the right proportion. Similar observations could be made about the EC and North America.

TABLE 9.1
Typical causes of death in England and Wales

Death from diseases (96.8%)		Death from accidents and violence (3.2%)	
Heart disease	165,000	Suicide	4000
Other circulatory	30,000	Homicide	630
Cerebrovascular	70,000	Accidents in the home	4500
Respiratory	56,400	Road accidents	5500
Cancer	140,400	Other transport	250
Died whilst giving birth	50	Sport and recreation	200
AIDS	1800	Other	2800
Other	90,850	Accidents at work	370
		Struck by lightning	1

TABLE 9.2
Individual risks in Great Britain (causes of death expressed as risk per million per year)

All causes (mainly illnesses from natural causes)	11,900
Cancer	2800
All violent causes (accident, homicide, suicide, etc)	396
Road accidents	100
Accidents in private homes (average for occupants only)*	93
Fire or flame (all types)*	15
Drowning*	6
Gas incident (fire, explosion or carbon monoxide poisoning)	1.8
Excessive cold*	8
Lightning	0.1
Meteorite	0.001
Accident at work — risk to employees	
Deep sea fishing (UK vessels)	880
Coal extraction and manufacture of solid fuels	106
Construction	92
All manufacturing industry	23
Offices, shops, warehouses, etc, inspected by local authorities	4.5
Leisure — risk to participants during active years	
Rock climbing (assumes 200 hours climbing per year)	8000
Canoeing (assumes 200 hours per year)	2000
Hang-gliding (average participant)	1500

Sources include UK Health and Safety Executive document on *The Tolerability of Risk from Nuclear Power Stations* and * from OPCS Monitor Series *DH4 No 11*, 1985, and Registrar-General for Scotland, *Annual Report*, 1985.

The balance of work fatalities even in the process industries do not occur as a result of major incidents involving the release of process material. Thus in a typical oil exploration and production company most fatalities will occur from transport by road, air and sea, other drowning incidents and other normal accidents such as falls and crushing. Some 10% of fatalities might be caused by fire, explosion and asphyxiation.

INDIVIDUAL RISK

The individual risk (IR) is the frequency at which a given individual may be expected to sustain a given level of harm from the realization of specified hazard.

Table 9.2 lists IR estimates from various causes taken from UK Government sources. These figures are given as the chance in a million that a person will die from that cause in any one year, averaged over a whole lifetime (except where otherwise stated).

An estimate of individual risk generally includes consideration of both the probability of being killed and being present — that is:

$$IR = F P_c P_p$$

where F is the frequency of an undesired event, P_c is the probability of the nearest person being killed and P_p is the probability of that person being present.

Thus the person most at risk in a group of houses would probably be old or infirm. Also such people tend to be always present.

OCCUPATIONAL RISK

Occupational risk has frequently been measured as the fatal accident rate (FAR), which is defined as the number of fatalities expected per 10^8 exposed hours and roughly corresponds to the number of deaths among a group of 1000 workers. Some values of FAR from data in the past are given in Table 9.3. These occurred some time ago and subsequent improvements in these industries show

TABLE 9.3
Estimated fatal accident rates (FARs) due to disease attributed to types of chemical or physical exposure

Occupation	Disease	FAR
Shoe industry	Nasal cancer	6.5
Printing trade workers	Cancer of lung	10
Uranium mining	Cancer of lung	70
Coal carbonizers	Bronchitis, cancer	140
Asbestos workers	Cancer of lung	160
Amosite asbestos factory	Asbestos, cancer of lung/pleura	460
Cadmium workers	Cancer of prostate	700
Nickel workers	Cancer of nasal sinus	330
Naphthylamine	Cancer of bladder	1200

HAZARD IDENTIFICATION AND RISK ASSESSMENT

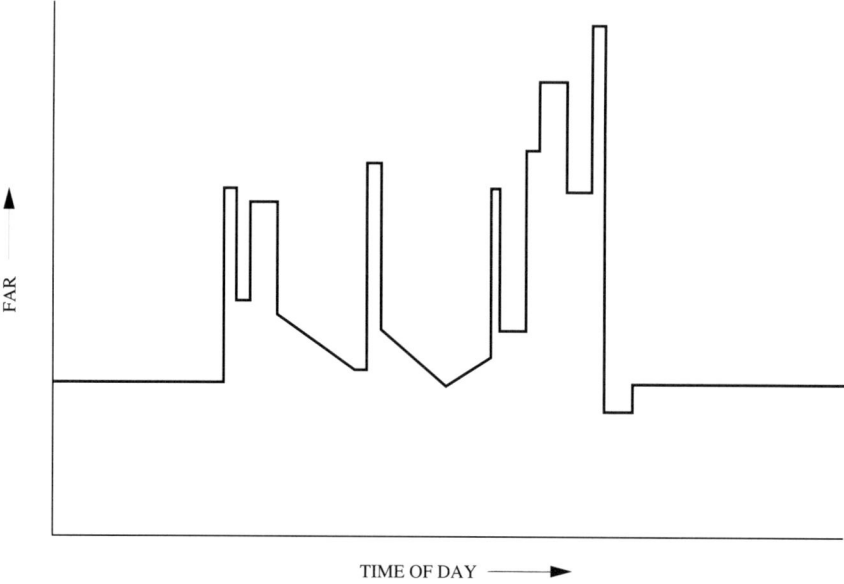

Figure 9.1   A day in the life of ...

the value of logging data in this manner. The advantage of such a study undertaken from Government records is that dangerous trends can be identified and appropriate steps taken to reduce, eliminate or protect against a given health hazard.

An approximate relationship between individual risk and fatal accident rate is:

$$IR = 2.2 \times 10^{-5} \text{ (FAR)}$$

Figure 9.1 shows how amusing it can be to plot FAR throughout the day for an individual. You can interpret your own events in the diagram: fun activities and eating in dubious canteens might play their part. The graph should not be taken too seriously.

SOCIETAL RISK, SR

The societal risk (SR) reflects the likelihood of accidents involving multiple casualties (three or greater). The results are normally plotted on so-called F-N curves which show the frequency of N or more people being killed against the number of people at risk: see Figure 9.2. The values here are based on four categories of incident over two decades. The results are only approximate and refer to a particular data bank. But they give an indication of the relative frequencies of some major accidents.

Note that F refers to the frequency of events greater than or equal to the number of fatalities, N, in any incident. This means that the value at the curve for the frequency of 10 or more deaths includes the frequency of all events giving rise to 10 or more deaths.

Note also that a one in 100,000 annual chance of killing 100,000 people should not be equated to one death per year; the response by society is different. Also society is more willing to accept voluntary risk than involuntary risk.

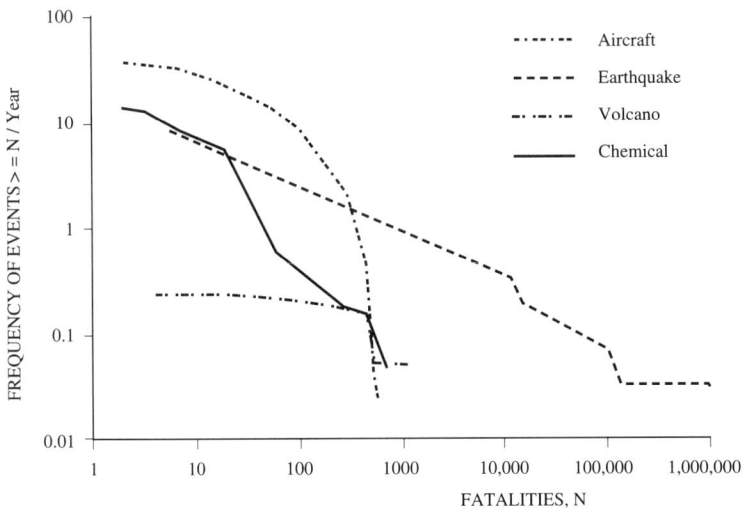

Figure 9.2   F-N graph showing societal risk. Fatalities for given categories worldwide over two decades.

In modern times no instance has been recorded of a member of the public being killed as a result of an accident on any chemical plant in the UK. This excludes possible health damage from incidents in the past; for many years that type of incident was not reported and standards of environmental monitoring have since been gradually improved. Thus the risk is very low and, unlike the case of risks to employees, cannot be derived from the historical record.

CRITERIA FOR ACCEPTABLE RISK

The earliest studies of risk used the term 'acceptable risk'. The adoption of quantitative techniques enabled hazards to be ranked numerically in order or priority. The question immediately arose whether any or all of these hazards were 'acceptable', or of sufficient magnitude to require expenditure on potentially expensive safety measures. This required some form of guidance on criteria against which the acceptability of estimated risk levels may be judged.

The search for acceptable levels of risk for industrial activities generated a considerable volume of literature. Various criteria were put forward for specific applications in a number of countries, but no clear consensus emerged. This reflects the contentious nature of the essentially political value judgements involved. It was slowly appreciated that the relative importance of different risk parameters for making decisions on safety varied considerably from case to case, depending on the numerical results of the specific analysis to be evaluated; for example, individual versus societal risk, or public versus occupational risk.

The simplest and most common basis for arguing that some potentially hazardous activity should be accepted was that of risk comparisons. The predicted risks (individual, societal or occupational) from the activity were shown to be lower than 'background' levels of risk already present and apparently acceptable to society. Following early applications of Quantified Risk Assessment or QRA (see Chapter 10) in the 1970s, extensive tables of such comparative risk data were compiled for this purpose relating to natural hazards such as lightning and flooding, everyday activities such as car driving and use of domestic appliances, and occupational hazards in different industries. Whilst the simplicity of its approach was attractive, it was open to severe criticisms for failure to distinguish between acceptable risks and those which were merely tolerated. It also fell into disrepute when it emerged that deaths were occurring from radiation long after the Windscale accident of 1957 occurred in the nuclear industry (in which radioactive material was released following a fire in the air-cooled nuclear reactor). To this day the episode is seen by many as a deliberate cover-up by the UK Government.

It is clear that in the past many industrial hazards were accepted in ignorance by public and employees. This was a function of the technological, social and economic conditions of the time. It is now argued that the risk comparisons failed to address the 'benefits' from hazardous activities. Where 'benefits' are not recognized by an affected individual or community, no finite level of risk — no matter how small — may be judged as tolerable by them. So while comparative risk studies provide a useful perspective on numerical risk results, they offer little to justify such risks in 'absolute' terms of acceptability.

In contrast to this simple risk comparison approach, extensive research undertaken by psychologists and other social scientists over recent years has revealed a complex series of factors, values and beliefs which in practice underlie the public's perception of the risk of industrial activities. In particular, it has been shown that perceived risks cannot be readily correlated with those 'objective' risk estimates which may be statistically derived. Rather they are influenced by qualitative characteristics, such as the level of free will associated with the risks involved, their catastrophic potential and their relative unfamiliarity. The perceived importance or benefits of industrial activities have also been demonstrated to play a significant role in conditioning public attitudes.

The reconciliation of the public view as based on perceived risks with estimates of objective risk is now recognized to be an issue of considerable political importance — as, for example, reflected in many national debates concerning the acceptability of nuclear power. However, as acceptability criteria are intended solely for the evaluation of QRA results, they should be based primarily on objective risk estimates. Risk perception by public or trade unions, for example, may then influence decisions on the control of major hazards as separate political factors, considered in a subsequent step.

A FRAMEWORK OF RISK CRITERIA

An approach is being widely promulgated in the UK that the general form or framework for acceptability criteria should be represented as the three-tier system illustrated in Figure 9.3.

It involves the definition of the following elements:
- an upper-bound on individual (and possibly, societal) risk levels, beyond which risks are deemed unacceptable;
- a lower-bound on individual (and possibly, societal) risk levels, below which risks are deemed not to warrant regulatory concern;
- an intermediate region between the upper and lower bounds, where further individual and societal risk reduction are required to achieve a level deemed 'as low as reasonably practicable' (the ALARP principle).

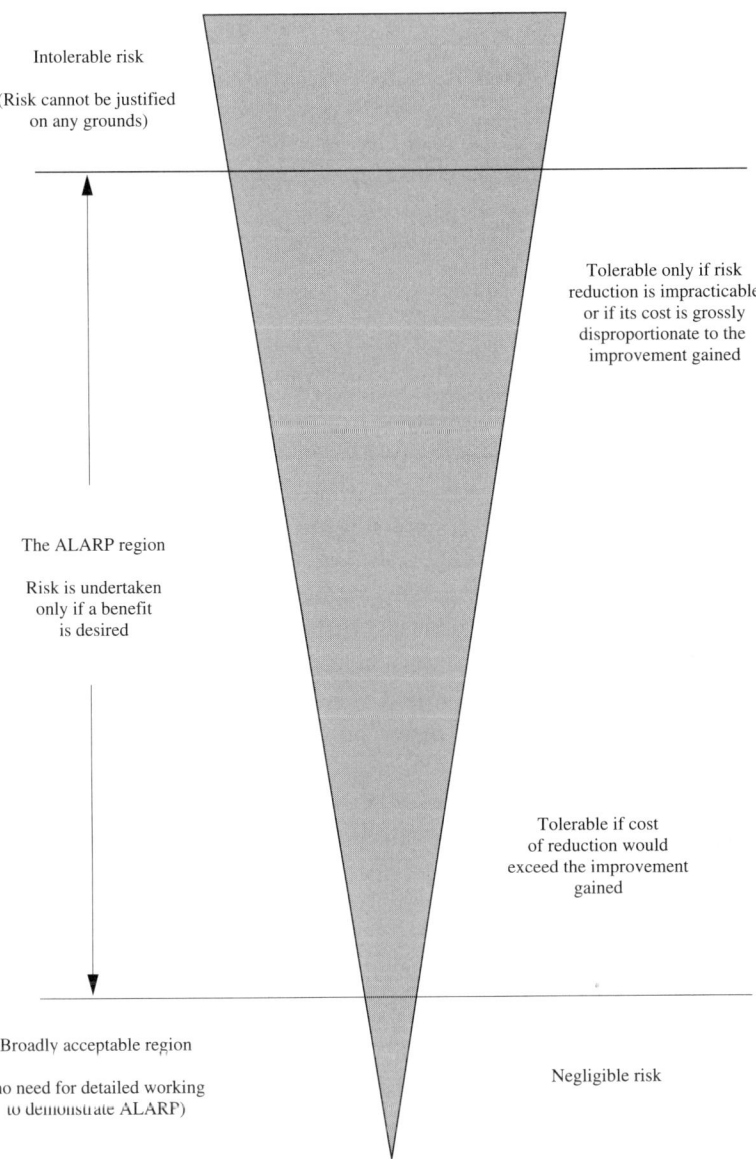

Figure 9.3 Levels of risk and ALARP. (Source: Health and Safety Executive, 1992, *The Tolerability of Risk from Nuclear Power Stations*. Crown copyright is reproduced with the permission of the Controller of HMSO.)

TABLE 9.4
Published acceptable values of risk

Canvey Island report, UK (1978)		35×10^{-6}/y
Netherlands	N = 10	1×10^{-5}/y unacceptable
		1×10^{-7}/y acceptable
	N = 100	1×10^{-7}/y unacceptable
		1×10^{-9}/y acceptable
Health and Safety Executive, Royal Society, UK (1983)		1×10^{-5}/y upper level
		1×10^{-6}/y acceptable
		1×10^{-7}/y sensitive population
Du Pont		0.3×10^{-6}/y
BNFL (Thorp reprocessing plant, UK)		0.01×10^{-6}/y
UK Central Electricity Generating Board nuclear power		1×10^{-6}/y upper level
Sizewell B, UK (1987)		0.04×10^{-6}/y design
		0.3×10^{-6}/y normal

The risk levels of the boundaries will vary according to the number of people at risk. Table 9.4 shows some early published values of acceptable risk.

The numerical values which have been suggested for upper and lower bounds show significant variation, not only in the values themselves but also in the extent to which these concepts are applied to societal risk. The values also vary between industries, reflecting considerable differences in philosophy and value judgements as expressed by individual authorities. From a practical point of view they also have a profound effect on the use of QRA as a means of demonstrating compliance. For instance, where the difference between upper and lower levels is smaller than the uncertainty band typically associated with a risk assessment, the practicality of applying the criteria is called into question. Also if the numerical values of the levels are very low, the technical uncertainties inherent in all risk assessments may make it questionable whether compliance can be demonstrated.

TOLERABLE RISK

Following the public inquiry into the Sizewell B nuclear plant, it was decided that the term 'acceptable risk' failed to reflect the importance of the problem, and in particular the reluctance that people commonly show towards undertaking certain dangerous activities. Subsequently the UK Health and Safety Executive[15] defined the phrase 'to tolerate a risk' as follows:

> 'Tolerability' does not mean acceptability. It refers to the willingness to live with a risk to secure certain benefits and in the confidence that it is being properly controlled. To tolerate a risk means that we do not regard it as negligible or something we might ignore, but rather as something we need to keep under review, and reduce still further if and as we can.

This definition of tolerability is taken to imply that risks should be monitored, balanced against possible benefits, and wherever possible reduced to 'as low as is reasonably practical' (the ALARP principle discussed earlier). But it is not the norm at present to carry out a cost-benefit study to see whether this is correct. When it comes to public harm there is a tendency to use absolute risk values supplied by the regulators for land-use planning.

Furthermore, the definition of tolerability does not satisfy all parties. It accepts that judgement on what is tolerable is a political matter[16]. However, it probably does not relate benefits clearly enough to tolerability or address the critical issue of how public input to tolerability decisions might be achieved.

The target of any company is to have zero accidents. It is inevitable, however, that an unlikely major event will occur at some location because of the large number of companies worldwide. So arguments have been accepted by regulators that when the staff of an operating company carry out a design they should use a company standard which sets target values for the maximum risk which might be tolerated from their activities. The current values for land-based operation which companies in the UK appear to be find acceptable are similar to those given in Table 9.5.

These are compromise best estimates obtained from canvassing opinions in industry and the values given in Table 9.5 quoted by HSE. The highest targets might be exceeded by an order of magnitude ($\times 10$) in circumstances deemed important by the company, although the value of 10^{-4}/y should not be exceeded. This would move the values into a zone corresponding to the intermediate ALARP region. Conversely if any vulnerable groups were in the vicinity threatened, then more restrictive values would apply.

Such target values of maximum tolerable risk would probably be considered acceptable by many social categories of people and those working at the location. But they certainly would not be accepted by groups including a majority of egalitarians or sectarianists. Any member of the public would be aggrieved to find a chemical development in 'their backyard'. So the debate will

TABLE 9.5
Target values of maximum risk not to be exceeded

Employee individual risk	
• all process causes	10^{-4} per year
• specific process cause	10^{-5} per year
Public individual risk	
• all process causes	10^{-5} per year
• specific process cause	10^{-6} per year
Risk of major incidents (that is, societal risk)	
• near miss from all process causes	10^{-4} per year
• accident from all process causes	10^{-5} per year
• catastrophic accident from all process causes	10^{-6} per year
• accident from specific process causes	10^{-6} per year
• catastrophic accident, specific process causes	10^{-7} per year

continue and increasingly people have the right to know. This has increased the extent of disputes over risk values and marked the end of pronouncements from on high of acceptable risk ... and quite right too, given the uncertainties in the data indicated in the next chapter.

HSE DANGEROUS DOSE

The UK Health and Safety Executive (HSE) frequently uses the term 'a dangerous dose'. A dangerous dose has the potential to cause death but it will not necessarily do this. It represents a dose of toxic gas or heat or explosion overpressure which causes severe distress to everyone. A substantial fraction of people may require medical attention, some may be seriously injured and require prolonged treatment, and highly susceptible people might be killed.

Thus for land-use planning, the lower bound figure of individual risk suggested by the HSE[17] is one in a million per year receiving a dangerous dose or worse for a typical pattern of user behaviour in a development near an existing major hazard installation. This figure should be evaluated as a cautious best-estimate. For developments with higher proportions of highly susceptible people, an additional lower bound of 1/3 in a million per year is suggested. An upper bound figure of 10 in a million per year is suggested.

The HSE has demonstrated the use of the bounds around a 50 te LPG tank (Figure 9.4). When residence time in a development (for example, housing) is high, IR criteria are used to reach a decision on the appropriate advice. The HSE is likely to advise against housing developments

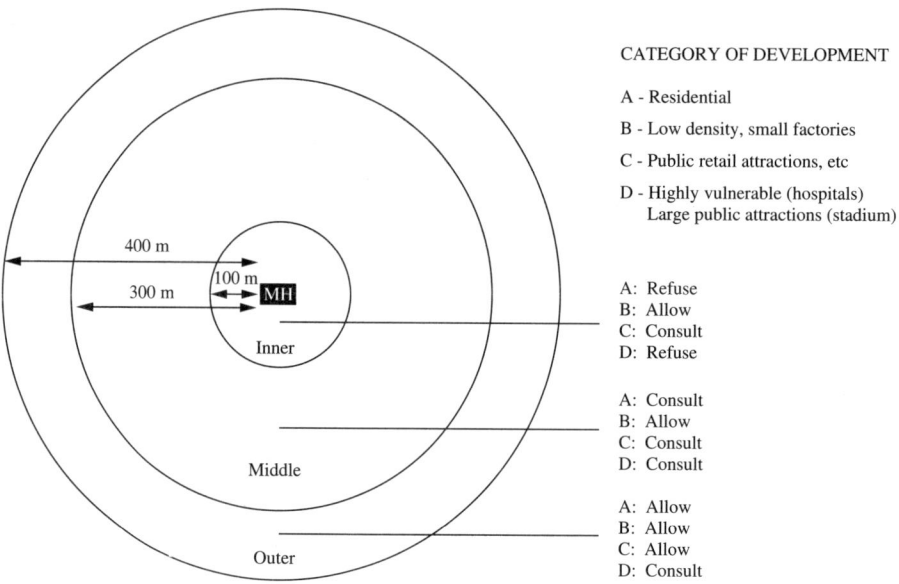

Figure 9.4   Three-zone, four-category development control policy, 50 te LPG tank[17].

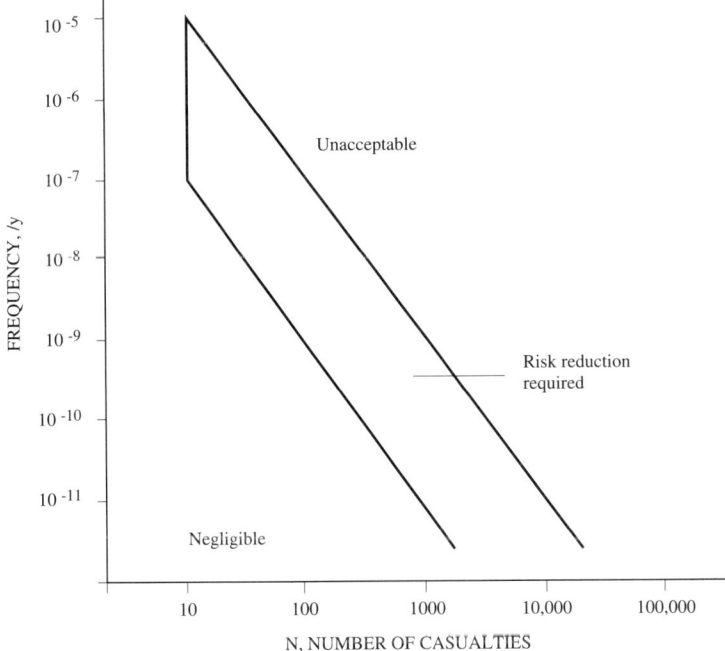

Figure 9.5 Societal risk criteria. (Source: Health and Safety Executive, 1989, *Risk Criteria for Land Use Planning in the Vicinity of Major Industrial Hazards*. This document is in the course of a fundamental revision to be published in 1996. Crown copyright is reproduced with the permission of the Controller of HMSO.)

for 25 and 75 people at an IR of 10^{-5} and 10^{-6} per year respectively. Larger occupancy developments where residence is very short and hence where societal risk dominates are equated to hypothetical housing developments and similar advice is given. For a very large development (for example, hospitals) the criterion may be extended below 10^{-6} per year to 0.33×10^{-6} per year. Similar arguments for a public retail outlet would relate to the maximum number of people at risk during peak hours. Note that all values are likely to change, so the recent literature must be consulted for current values used by regulators.

The impact of societal risk is well illustrated by the chart developed by regulators in Holland (Figure 9.5). This gives the group risk per activity (for death all at one time of N or more people). An increase in the number of deaths by a factor N in a given situation is only acceptable if the probability of this event occurring is a factor N^2 lower for both types of level. The maximum permissible risk levels for disasters are defined as 10^{-5} per year for N = 10 or more deaths and 10^{-7} per year for N = 100 or more deaths, etc. The corresponding negligible levels are defined as 10^{-7} per year for N = 10 or more deaths and 10^{-9} per year for N = 100 or more deaths, etc. This should not be interpreted as it being possible to evaluate figures to this order of accuracy, except from the historical record where it relates to floods, earthquakes and so on.

10. RISK ASSESSMENT

INTRODUCTION

Quantified Risk Assessment (QRA) is widely used as a technique to aid management decision and is defined by the CONCAWE organization as 'The identification of causes of possible accidents followed by a technical analysis to determine the likelihood of occurrence and potential consequences of those accidents leading to a numerical estimate of an appropriate measure of risk, together with the value judgements made with regard to the significance of the estimated level of risks'[18].

Five specific elements are involved:

(1) *Hazard identification* to determine the incident scenarios, hazards and hazardous events, their causes and mechanisms.
(2) *Frequency estimation* to determine the frequency of occurrence of identified hazardous events.
(3) *Consequence analysis* to determine the extent and probability of the consequences of identified hazardous events.
(4) *Risk evaluation* to determine the risk levels.
(5) *Sensitivity analysis* to prioritize further studies of risk, evaluate the significance of risk levels and set a schedule for implementation.

The elements of the procedure are used both to generate information and as an aid to decision-making. For decision-making, the procedure is only taken as far as is necessary to generate the information required or to make a decision. The extent of application of the various elements and degree of quantification employed therefore varies significantly from one situation to another. Some general features of the elements are outlined in Table 10.1.

Elements (1), (2) and (3) are normally developed using hazard analysis and semi-quantitative values of elements (2) and (3) may be all that is required. A Short-Cut Risk Assessment method (SCRAM, as recommended when carrying out a Preliminary Hazard Analysis) can be used to prioritize and recommend where it is appropriate to carry out a full QRA. The latter involves the use of a full decision-aiding framework as indicated in Figure 10.1, page 212.

A QRA involves several alternative cycles to generate values. The analysis proceeds by initially deriving order-of-magnitude estimates for frequency causes, consequences, costs and the like, and examining the sensitivity of these quantities to initial assumptions. When evaluating results against criteria in this way the degree of sophistication and analytical effort can be progressively tuned to reflect the aim of the study, the overall scale of the problem, the relative magnitude of the individual failure mechanisms or incident scenarios and the uncertainties involved in the evaluation.

QRA is considered to be inherently better suited for decision problems associated with failures of technical equipment rather than of management/organizational systems. This is in part a function of inadequate models for the latter but also the absence of data. Active errors can readily be evaluated but latent errors present great assessment difficulties. Increasingly shortcomings

TABLE 10.1
Elements of Quantified Risk Assessment

Hazard identification:	• check-list • Preliminary Process Hazard Analysis • Hazop
Frequency estimation:	• historical data • Fault Tree Analysis • reliability and availability studies
Consequence analysis:	• Event Tree Analysis • incident categorization/definition
Risk evaluation:	• combine frequency and consequence
Sensitivity analysis:	• compare relative risk estimates • prioritize further studies • evaluate absolute uncertainty

in management are being seen as 'aggravators' which can be assessed as factors which are used to adjust the value of the frequencies and probabilities representing the likelihood of an event[19].

WHY ASSESS RISK?

LOADS
Risk assessment has developed over the last twenty years into an essential part of process safety. It is still usual to design to an accepted standard, regulation or code of practice but it is appreciated that although such practice should give a safe reliable design, this must be further assessed allowing for features such as the following:

Loads in excess of design
Codes of practice normally only take into account representative or typical potential loadings which are outside the design basis. An external event such as an aeroplane crash onto a plant is one such excessive load.

Loads not recognized by the design
Loads and physical processes may not be recognized in the standard at the time the plant or item of equipment was designed. Numerous incidents occurred before brittle fracture and fatigue were understood and incorporated into standards.

Human or computer error
A mistake or a deliberate violation may cause failures in design, manufacture and construction, operation or maintenance.

HAZARD IDENTIFICATION AND RISK ASSESSMENT

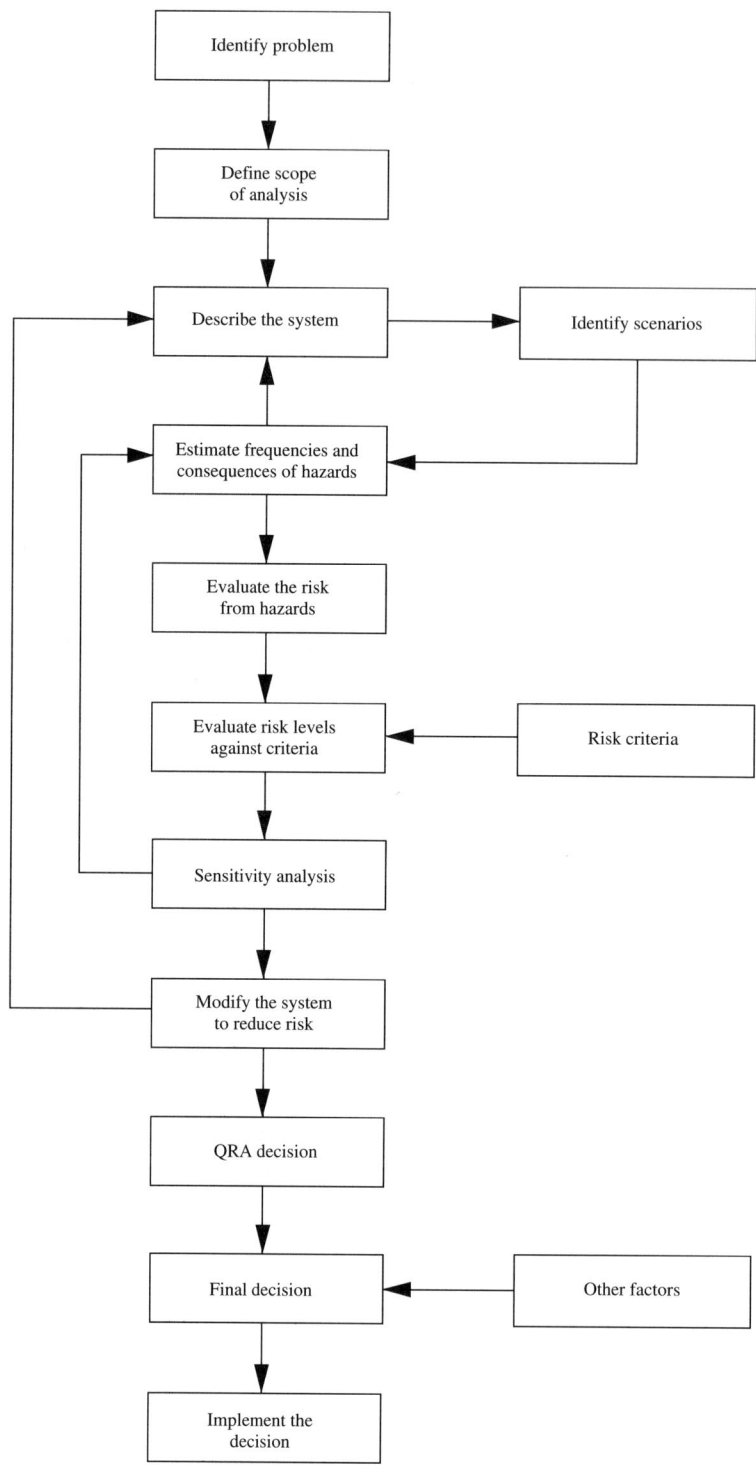

Figure 10.1 The role and use of Quantified Risk Assessment.

Furthermore, it is appreciated that when the consequences of failure are serious — involving damage to people, property and the environment — then studies in addition to normal design are essential. This requirement has arisen following the disasters at Windscale, Flixborough, Seveso, Bhopal, Mexico City and Chernobyl which were followed by great media attention. People have become deeply aware of the hazardous potential of process and nuclear installations.

This has placed an onus of proof on these industries to indicate that a particular type of failure is extremely remote and preferably has not happened before to specific types of plant.

COMPANY HEALTH AND SAFETY POLICY

A proper company health and safety policy should allow all activities to be managed without causing any unnecessary or unacceptable risk to the health and safety of all employees, customers and members of the public who may be affected by its operations. Similar aims affect property and the environment, with particularly tight constraints on risk for listed buildings and sites of special scientific interest. Such objectives require not only compliance with all legal requirements and appropriate codes of practice. Additional measures must be taken to eliminate hazards whenever reasonably practical, and reduce the average accidental risk for employees at work to no more than their average risk at home. This target is itself lowered thanks to the employees' training and awareness of good safety practices.

In the process industries such a policy requires that hazard studies are commenced prior to detailed design in order to identify associated significant hazards. Then, by appropriate design and selection of process materials, equipment and operations, the residual risks are reduced to acceptable levels. Subsequently vigilance during production ensures that risks are maintained at or below an acceptable level.

A statement of policy similar to that given in Table 10.2 should be signed by the head of the company and promulgated throughout the company. It should be backed by action for enforcement.

TABLE 10.2
Statement of principle

The company health and safety policy is that all activities will be managed so as to avoid causing any unnecessary or unacceptable risk to the health and safety of all employees, customers and members of the public who may be affected by its operations.
- The company will conform with all legal requirements and appropriate Codes of Practice and will take any additional measures it considers necessary.
- Hazard studies will be commenced prior to detailed design to identify associated significant hazards and where reasonably practicable those hazards will be eliminated.
- The risks from hazards remaining will be reduced to acceptable levels by appropriate design of equipment and operations which will be based on the relevant most up-to-date Codes of Practice available supplemented by safe practices developed from operational experience. The proposed activity will not be operated if the risks cannot be reduced to an acceptable level.

RESOLVING DIVERGENT VIEWS

The general standards of safety in the process industries are high due to the implementation of rigorous management systems, the safety awareness of the workforce, and compliance with regulations, codes of practice and standards. Such traditional approaches may usefully be complemented by full or partial QRA, particularly for the assessment of major hazards where there is a remote chance of major accidents occurring.

In such instances a significant divergence of views may arise between those who, irrespective of consequences, tend to dismiss the risk by, for example, referring to the soundness of construction and those who tend to concentrate on the possible consequences, irrespective of their likelihood of occurrence, and may thereby promote allocation of resources to safety expenditure of limited real value.

These divergent views are often difficult to resolve qualitatively, even where professional safety advisers and engineers are involved. Furthermore, the problem becomes increasingly significant for regulatory decisions in the public domain, since public pressures and the understandably cautious inclination of regulators may tend to focus on 'worst-case' consequences of accidents to the exclusion of considerations of their likelihood.

Against this background, a primary advantage of QRA is that it requires an explicit, numeral consideration of both frequencies and consequences, promoting a more balanced perspective in the assessment and control of major hazards. It can provide direct inputs to resource allocation decisions. This enables costs and risk reductions associated with differing safety provisions to be combined in a single measure with which to assess the cost-effectiveness of alternative options. As such, QRA reduces the effect of 'consequence-driven' decisions, which otherwise might place an excessively onerous burden on industrial siting, layout and costs.

Risk assessment is quantitative in nature. Therefore, although this does not stop people arguing about a potential hazard or its acceptability, it does require these arguments to be related to quantitative values and subjected to rigorous scrutiny. It leads to a better understanding of the system and its potential weaknesses and this enables the identification of improvements and significant reduction of the risk.

QRA requires that arguments be developed to a much greater extent, withstand close scrutiny, and highlight outstanding points of contention. The framework of QRA can be used to promote an active dialogue between regulators and industry and thereby, on occasions, may provide the fastest and easiest route to resolving particular issues of regulatory concern.

LEGISLATION

QRA helps to resolve any divergent views on the allocation of resources to safety expenditure, and subsequently on the need to cease production whilst repairs are carried out. Of course there will still be disagreement about the potential hazard or its acceptability, but the arguments are then related to quantitative values which are capable of rigorous scrutiny. Hence QRA leads to a better understanding of the system and its potential weaknesses, enabling a significant reduction of the risk to be achieved.

The questioning attitude of public opinion has affected legislation. Now regulatory authorities in the UK like the Health and Safety Executive (HSE) and the Nuclear Installations Inspectorate (NII) require some form of risk assessment in looking at situations where the consequences of failure might be severe in terms of the effect on the general public. In 1984 the Seveso

Directive came into force for process industries in the European Community. In the UK it was enacted as the Control of Industrial Major Accident Hazard (CIMAH) Regulations. In America similar requirements are imposed by API 750 and OSHA 29 CFR 1910. This means that if more than a threshold limit of a hazardous substance is stored, a detailed safety report must be presented demonstrating that the likelihood and consequences of a serious accident involving that material can be shown to be tolerably low.

Risk assessment is something the public do all the time; crossing the road, travelling by bus, even smoking. On the basis of experience, the risk of suffering harms is acceptably low. It even includes value judgements; for instance, jay-walking in order to get to a meeting or lecture in time. Sometimes a higher risk than usual is acceptable for purposes of enjoyment; for instance, playing cricket or riding a horse.

USE OF ABSOLUTE AND RELATIVE RISK ESTIMATES

Risk assessment involves no preconceptions about the credibility of any type of accident. Any hazard or incident chain that can be identified is included. Scenarios are discounted if it can be shown that either:
- the consequences of their occurrence in terms of their effect on adjacent population is negligible;
- the likelihood of their occurrence is negligible when compared to the likelihood of occurrence of other accident scenarios that have been identified.

Risk is a function of the likelihood of occurrence of an undesired event and the consequences resulting from its occurrence. To reduce the risks from a particular event, it is necessary to either seek to reduce the frequency of occurrence of the undesired event or ameliorate the consequences of its occurrence.

An *absolute estimate of risk* can be compared with a value of estimated risk normally expressed as the likelihood of a specific consequence or set of consequences. It is therefore sensitive to uncertainty resulting from errors in the evaluation due to incompleteness, inaccurate manipulation of data, inappropriate or unavailable data.

Typically QRA estimates have an absolute uncertainty of one or more orders of magnitude (that is, a difference in likelihood, say, of 10^{-4} to 10^{-3}). Such an order of magnitude uncertainty in individual risk may, however, correspond to a much smaller uncertainty in physical location of the contour of equal risk as many physical effects diminish rapidly with distance.

The *relative use of risk estimates* is less sensitive to error because the resulting risk estimates due to variation are subject to similar uncertainties, many of which will cancel out when evaluating the change in risk. It is therefore possible to estimate the reduction in risk achieved through the modification of a system with considerable accuracy. Only cases falling near or into an intolerable risk zone need to be prioritized for detailed study.

THE SHORT-CUT RISK ASSESSMENT METHOD

The need for a Short-Cut Risk Assessment Method (SCRAM) at the various stages of design, and at the time of Hazop, has led to an increased emphasis on Concept and Preliminary Process Hazard Analysis. These methods are designed to evaluate the incident scenarios within the area of identification of emergency control, 'dangerous disturbances' and consequence analysis, but the analyst accepts that the identification of all immediate causes of incidents will not be achieved. But is this method sufficiently accurate to justify its use to prioritize further studies of risk? It is too early to say, but initial studies are promising.

The SCRAM procedure generates a list of possible incidents as top events together with an indication of their incident scenario record on a risk evaluation sheet. Order of magnitude values are generated for average duties for both plant and human actions. The method probably reduces the frequency of improper selection of incidents for analysis but clearly this is at the expense of complete generation of individual incident scenarios.

The evaluation of a Risk Rating using SCRAM proceeds as follows. The Risk Rating is here defined as the likelihood, L, of a specific undesired event occurring within a given period or in particular circumstances. The likelihood measure is in units of frequency (per year). It is a negative value. The severity, S, is a measure of the expected consequence of an incident outcome as given in Table 10.3.

The Risk Rating is represented by the equation:

$$\text{Risk Rating} = \log^{10} 10^L + \log^{10} 10^S$$

$$= L + S$$

where

L is the exponent of likelihood as measured by frequency (a negative value),
S is the severity category as given in Table 10.3.

A Risk Rating is only acceptable when its value is equal to, or less than, zero.

Given a likelihood of −2 and a severity of 3 means that the risk rating is −1 and further reduction of risk is required.

There are several other versions of this method. Some methods classify severity and likelihood between the numbers 1 to 10 and multiply the values together. Others simply add appropriate ranges. The advantage of these methods is that the higher the risk ranking number, the greater the need for further study. The best advice is to use a method with which all in the company are familiar.

SEVERITY CATEGORIES

The severity categories basically are considering certain types of harm and damage. This can take the form of fatalities and injuries to people and damage and harm to the environment, damage to plant and property, damage to business and public relations.

These consequences are examined to determine whether they affect only the specific plant or extend to other process and storage areas. The works facilities such as offices, canteens and similar areas are usually considered, along with damage outside the works boundary as part of the community or public areas.

The severity of the harm and damage and whether it is of a short-term or long-term duration must also be evaluated. This must be treated as a near-miss incident even if no immediately visible harm is apparently done by a release.

TABLE 10.3
Severity categories

Catastrophic consequences: Severity 5
Catastrophic damage and severe clean-up costs
On-site: loss of normal occupancy of three months
Off-site: loss of normal occupancy of one month
Severe national pressure to shut down
Three or more fatalities of plant personnel
Fatality of member of public or at least five injuries
Catastrophic damage and severe clean-up costs
Damage to site of special scientific interest or historic building
Severe permanent or long-term environmental damage in a significant area of land
Acceptable frequency 0.00001 per year

Severe consequences: Severity 4
Severe damage and major clean-up
Major effect on business with loss of occupancy up to three months
Possible damage to public property
Single fatality or injuries to more than five plant personnel
A one in ten chance of a public fatality
Short-term environmental damage over a significant area of land
Severe media reaction
Acceptable frequency 0.0001 per year

Major consequences: Severity 3
Major damage and minor clean-up
Minor effect on business but no loss of building occupancy
Injuries to less than five plant personnel with one in ten chance of fatality
Some hospitalization of public
Short-term environmental damage to water, land, flora or fauna
Considerable media reaction
Acceptable frequency 0.001 per year

Appreciable consequences: Severity 2
Appreciable damage to plant
No effect on business
Reportable near-miss incident under CIMAH Regulations
Injury to plant personnel
Minor annoyance to public
Acceptable frequency 0.01 per year

Minor consequences/near misses: Severity 1
Near-miss incident with significant quality released
Minor damage to plant
No effect on business
Possible injury to plant personnel
No effect on public, possible smell
Acceptable frequency 0.1 per year

The harm done to people, flora and fauna should be a standard category which is set regardless of company size. For example, if a member of the public is killed this is a severe incident. But £100,000-worth of damage to plant will be major to one company and severe to another, according to the size of the installation or company. Hence such categories must be defined by the company.

Similarly business damage and bad publicity can have such knock-on effects that any major accident can affect company activities elsewhere. This may mean upgrading the incident to a severe category.

The severity appropriate to a given incident is always assigned according to the highest rating identified.

At present five categories are defined as assigned against likelihood varying in frequency from one to 10^{-5} per year. It is up to companies to decide what likelihood should be applied to every case. It is also likely that the appropriate values will change with time. Already a catastrophic incident for land-based activities in the UK is probably more appropriate as 10^{-6} per year due to pressure by regulators and public.

PRIORITIZATION AND SENSITIVITY STUDIES

It is convenient to link prioritization for further study of risk with sensitivity analysis as one is simply a crude form of the other. Should the risk values appear sensible when tested against experience and should the accuracy required be perceived as appropriate, then a prioritization can be carried out for further study as indicated in Table 10.4.

A sensitivity analysis can demonstrate the significance of effects, although the absolute degree of uncertainty of ultimate effects cannot be demonstrated by short-cut methods. The sensitivity analysis should determine whether the range of uncertainty prevents making a decision on the basis of Table 10.4.

TABLE 10.4
Prioritization table for risk

Severity category	Risk rating			
	−2	−1	0	1
1	None	None	None	C
2	None	None	C	B
3	None	C	B	A/B
4	C	B/C	B	A
5	B	B/C	A	A

A — immediate attention needed; B — further study probably required;
C — further study may be necessary

RISK ASSESSMENT

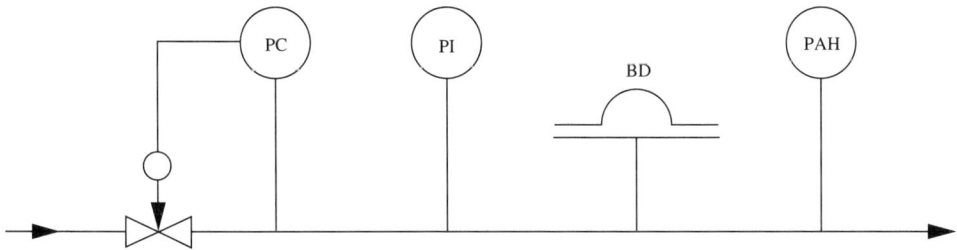

Figure 10.2 Process selection for risk assessment.
BD — bursting disc; PAH — pressure alarm high; PI — pressure indicating; PC — pressure controlling.

A RISK ASSESSMENT STUDY USING SCRAM

The section of plant shown in Figure 10.2 contains a toxic gas. The plant has been examined by a Hazop study as reported in Chapter 5, page 91.

As a result the following three cases are to be studied using risk assessment:
Case A — the bursting disc discharges to atmosphere;
Case B — the disc discharges to a vent scrubber kept in constant readiness;
Case C — the disc fails to discharge and the pipe fails catastrophically.

The main causes of failure as indicated by the Hazop study have been assigned probabilities and frequencies as follows:
- failure of control loop, 0.1/y;
- spurious failure of disc, 0.05/y;
- operator fails to act on gauge, 1;
- alarm failure or operator inactive, 0.1;
- blockage of scrubber, 0.1;
- disc fails to discharge on demand, 0.01;
- vent scrubber ineffective, 0.1;
- escalation of toxic event, 0.1.

More causes exist in reality but this list serves for the purposes of the planned analysis.

EVALUATION OF LIKELIHOOD

No other events need to be considered in this study. The fault trees given in Figures 10.3, 10.4 and 10.5 (pages 220 and 221) were developed for the three cases.

Figure 10.3 considers the discharge of the pipeline assuming the failure of the process control loop as the initiating event. The operator fails to stop the pressure rise. The inadequate action on alarm includes a variety of faults including the failure of the alarm, the failure of the procedure and the failure of the operator to take appropriate action. The discharge can also occur on spurious failure of the disc at normal operating pressure.

Figure 10.4 allows for the above eventualities but also allows for the failure of the vent scrubber. For example, this might be switched off, be unable to cope with the demand or fail in service.

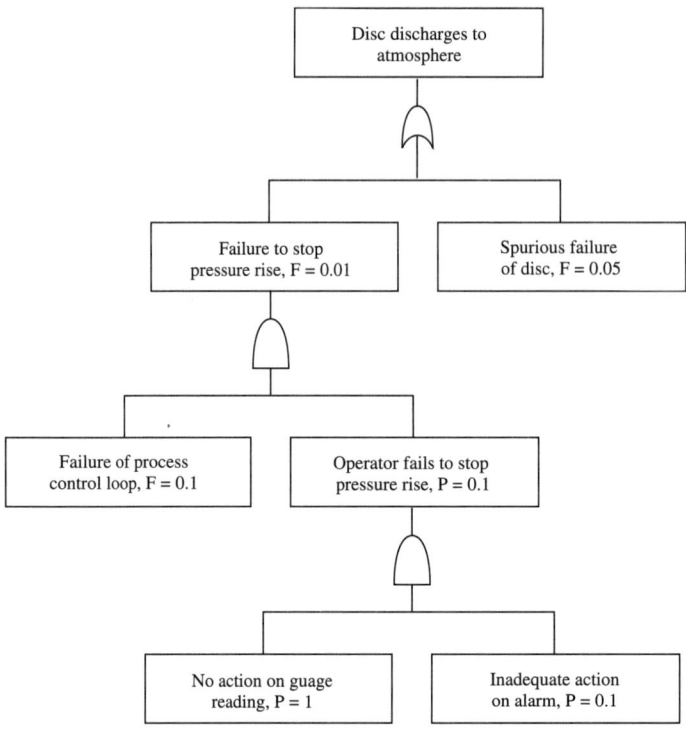

Figure 10.3 Fault tree for Case A.

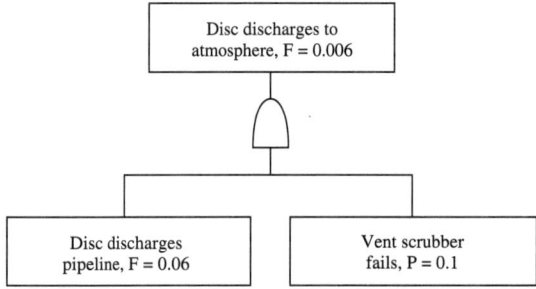

Figure 10.4 Fault tree for Case B.

Figure 10.5 considers the effect of the failure of the disc to discharge or cope with the flow due to back-pressure effects.

Figure 10.6 presents an event tree for the cases of failure to stop the pressure rise and spurious failure of the disc. This also summarizes the risk information which is given in the next section.

RISK ASSESSMENT

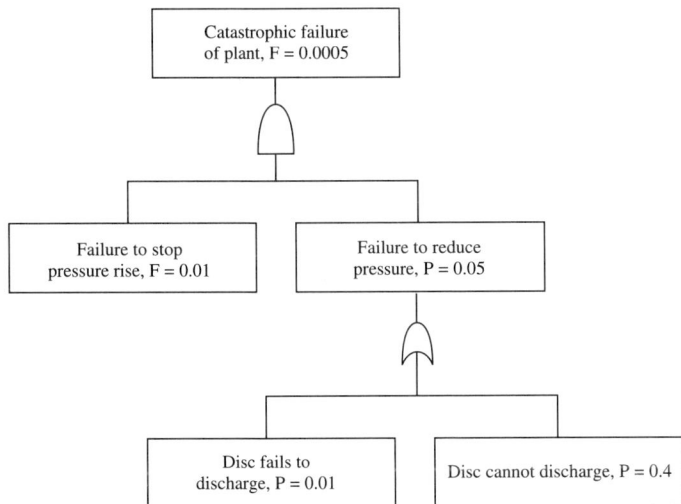

Figure 10.5 Fault tree for Case C.

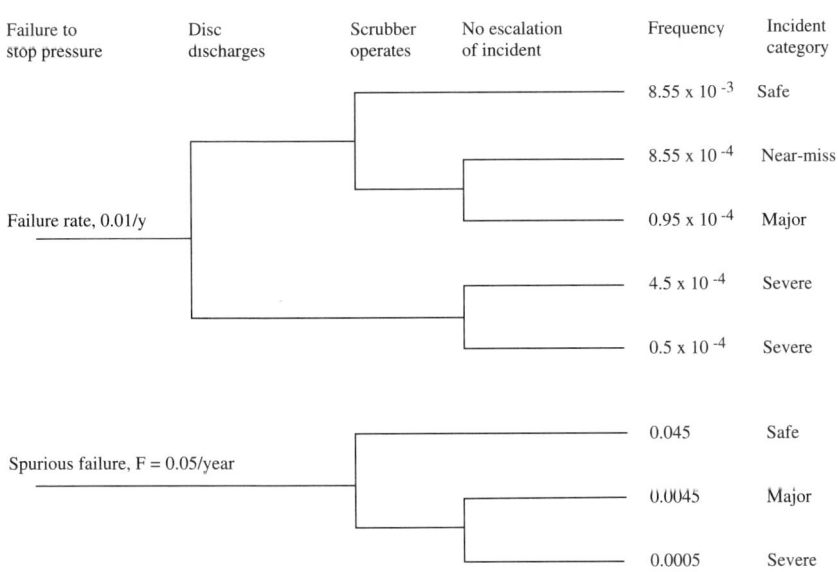

Figure 10.6 Event tree for Case B.

RISK EVALUATION OF THE CASE STUDY

Risk assessments of the likely severity of incidents suggested the following minimum likelihoods as being acceptable:
- likelihood of an appreciable event, 0.01;
- likelihood of a major event, 0.001;
- likelihood of a severe event, 0.0001.

The system has been evaluated for risk as shown in Table 10.5 and justifies the installation of a vent scrubber.

REQUIREMENTS FOR A QUANTIFIED RISK ASSESSMENT

A full QRA can then be carried out as appropriate, making:
- a 'precise' prediction of the realization of the hazards;
- an improved and realistic estimate of the level of damage from the hazards;
- a comparison with guidelines for risk and acceptability criteria;
- modification of the system to reduce the risk;
- resolution of the problem with provision made for feedback, feed-forward and monitoring of the system.

A complete and accurate QRA involves the following actions:

TABLE 10.5
Risk evaluation of the plant section

Case study events	A	B	C
A. Failures of control system per year	0.1	0.1	0.1
B. Operator fails to stop pressure rise	0.1	0.1	0.1
C. Failures to stop pressure rise per year (AB)	0.01	0.01	0.01
D. Spurious failures of disc per year	0.05	0.05	
E. Disc discharges per year (C + D)	0.06	0.06	
F. Disc fails to discharge per year			0.01
G. Ineffective discharge per year			0.04
H. Failures to reduce pressure per year (F + G)			0.05
I. Vent scrubber ineffective per year		0.1	
J. Toxic discharges to atmosphere per year	0.06	0.006	0.0005
K. Escalation given discharge	0.1	0.1	0.1
L. Appreciable toxic event (Severity 2)	**0.06**	**0.006**	**0.0005**
M. Major toxic event (Severity 3)	**0.006**	**0.0006**	**0.00005**
N. Severe toxic event (Severity 4)			**0.00005**

RISK ASSESSMENT

Definition of scope
- The definition of the goals, objectives and depth of system.
- Identification of economic criteria for justifiable expenditure and consequential loss.
- Determining the acceptability and consultation requirements of outside bodies.

Description of the system
- Information on organization, management, procedures, communications, etc.
- The availability of local expertise, skills and training.
- Information on the environs, adjacency to industry, population centres, geography.
- Data on the weather patterns affecting a release.
- The identification of external threats to the plant.
- Information on the plant, process chemistry, process diagrams, procedures, etc.
- Data on the input and output requirements and operating parameters.
- Information on the planned emergency response.

The precise prediction of the realization of hazards
- All hazards, pollutants and wastes are identified and their likely contributions are assessed.
- The frequencies or probabilities of hazards are reliably estimated.
- The consequences of hazards are accurately enumerated.
- The modification of data due to management factors.

A realistic estimate of the level of risk from hazards
- The exposure of people is sensibly defined.
- The level of harm is accurately correlated and the hazardous effect measured.
- The financial, environmental and business damage is estimated and recovery costs evaluated.

Risk level evaluated against criteria
- An evaluation is made against criteria for individual risk, societal risk, occupational risk and dangerous dose.
- Comparison with acceptance criteria and consents for effluents, emissions, wastes and noise.
- Impact of the development on the notifiable status of the site.
- A comparison with economic and business criteria.

Sensitivity analysis
- The significance of effects should be demonstrated by sensitivity analysis.
- An indication should be given of whether the range of uncertainty prevents making a clear-cut decision.

The system is modified to reduce the risk
- Consider the elimination, alteration or reduction of hazards.
- Consider the reduction of the frequency of events.
- Consider the reduction of the consequences of significant events.
- Consider the reduction of exposure and repeated exposure.

Resolution of the problem
- Generate a QRA and other data.
- Allow for economic, business, social and political factors.
- Ensure the systems can meet legal and regulatory requirements.
- Safety investments should be prioritized.
- Ensure the system can be monitored and audited and that all safety 'barriers' are in place.
- Information must be provided for on-site and off-site emergency planning.

TRANSPORT EXAMPLES

HELIPADS IN THE JUNGLE

This example shows how every eventuality must be allowed for in a QRA study. Helipads (23 × 170 m) are required every 4–6 km along a narrow track through the jungle. With the size of helicopter used there is a serious accident if one engine of two fails during part of the flight path shortly after take-off. This problem can be eliminated by clearing larger helipads of 23 × 300 m, a size which is recommended by the Civil Airways Authority for commercial and air operations. However, there is an increased potential for accidents in clearing a larger area of jungle, due to the incidents inherent in such a hazardous activity in individual clearings. This adds up to a greater risk.

Similar factors affect modifications. It may seem extremely sensible to change the location of a key valve on an operational plant to make it easier to operate and protect it from damage in the event of an incident. But in so doing it is necessary to isolate the line, make the line safe, effect the change and start the plant up again safely. All such activities generate opportunities for matters to go wrong. The risk in carrying out the modification may outweigh the improvement in residual risk for the ongoing system.

SHIP COLLISION WITH AN OIL RIG

The collision of vessels with oil rigs is of great concern in the North Sea and similar locations. Four types of collision mode might be considered:
- a vessel under power in passage;
- a vessel drifting without power;
- visiting vessels on approaching the rig;
- vessels when alongside.

Clearly the damage caused depends on the impact energy of the collision. A collision having an energy greater than 14 MJ would cause catastrophic damage to the rig due either to structural collapse or loss of buoyancy. An impact strength between 4 and 14 MJ would cause severe damage to one compartment on the rig but be less likely to damage two compartments. A collision of less than 4 MJ should be insignificant on a rig with adequate physical fendering arrangements, provided that no individual was in the immediate vicinity.

The worst case situation is likely to arise for a vessel under power in passage. A collision would result from failure to keep watch or inadequate visibility and inoperable radar. But a large vessel or a semi-submersible drifting without power at low speed can also be dangerous. For instance, on Christmas Eve the Shell FSU, a converted tanker without any propulsion units, broke away from its mooring mast in severe weather. It drifted past the Clyde platform at a distance of

500 m. The collision energy would have been of the order of 200 MJ. Severe weather is a major problem: it makes measures to direct the vessel away more difficult, and hinders complete or partial evacuation, assuming that a warning has been given of a possible impact.

The collision energy of a visiting vessel might be at service speed and of the order of 4 MJ. Warnings may be given on the ship and by the field protection vessel. The reliability of the dynamic positioning system might be improved. Simple rules also help, such as ensuring that vessels visiting should operate downwind and that any activities placing people in relevant danger areas on the rig should be avoided when vessels are visiting. For example, diving operations might be stopped.

With appropriate action it is considered that the frequency of collisions below 4 MJ is unimportant as a cause of death, the frequency of collisions between 4 and 14 MJ is between 10^{-3} and 10^{-4} per year, and the frequency of collisions of 14 MJ or more, without warning given, is between 10^{-4} and 10^{-5} per year. No such incident causing major platform damage or fatality had occurred by 1993 in the North Sea in 2000 platform operating years.

OBJECTIVES AND SCOPE OF QUANTIFIED RISK ASSESSMENT

In defining the objectives and scope of QRA, it is essential to acknowledge explicitly the boundaries of limitations of the study, since these condition the extent to which definitive conclusions can be drawn. Within the established boundaries it is important to ensure that all potentially relevant failure mechanisms and incident scenarios are identified for consideration. Compliance with this requirement for completeness may be difficult to demonstrate, as it is based on the experience and judgement of technical analysts.

It may be necessary to widen the boundaries. The measures and guidance values of the risk to be considered must be clearly specified in defining the scope of a QRA. These may include individual and societal risk to members of the public, or occupational risk to workforce who may be affected.

It is recommended that the analysis should initially take the form of deriving order-of-magnitude estimates of frequencies, consequences, costs, etc, and the sensitivity of these quantities to initial assumptions when evaluating results against criteria. In this way, the degree of sophistication and analytical effort employed can be progressively tuned to reflect the aim of the study, the overall scale of the problem, the relative magnitude of individual failure mechanisms or accident scenarios, and the uncertainties inherently involved.

SOURCES OF UNCERTAINTIES IN QUANTIFIED RISK ASSESSMENT

Some of the sources of uncertainty in QRA are noted in Table 10.6, page 226.

These uncertainties can be classified into three principal sets:
- *model uncertainty* — the model is inadequate or not used within its range of validity;
- *data uncertainty* — data are not complete and relevant to the situation;
- *quality uncertainty* — the analysis is not at a sufficient depth, all initiating events are not considered and all undesired events and sequences not included.

HAZARD IDENTIFICATION AND INCIDENT SCENARIOS
Uncertainties which arise during the hazard identification stages include the following.

TABLE 10.6
Sources of uncertainty in Quantified Risk Assessment

System description:
- process description, drawings or procedures do not represent actual operation
- site area maps and population data may be incorrect or out of date
- available weather data may be inappropriate
- the emergency response may be inadequately assessed

Hazard identification:
- identification of major hazards and their causes may be incomplete
- hazard screening techniques may omit important cases
- incident scenarios may not incorporate all control measures

Frequency techniques:
- exploration of historical data may overlook hazards introduced by scale-up
- limitation of fault tree theory requires system simplification
- incompleteness in Fault and Event Tree Analysis
- data may be inaccurate, incomplete or inappropriate
- there may be inherent problems in ascertaining human factors
- failure to modify frequencies due to different management and maintenance factors

Consequence techniques:
- inappropriate model selection
- incorrect physical basis for model
- inadequate validation of model
- uncertainties in physical model data
- source terms selected incorrectly
- uncertainties in damage effects
- mitigating effects incorrectly applied
- uncertainties affecting emergency response

Risk estimation:
- assumptions to reduce the depth of treatment
- restricted conditions of wind speed and stability

Failure to identify all the significant failure events:
- lack of knowledge;
- failure to carry out a formal hazard identification technique;
- choice of unsuitable techniques for hazard identification;
- inadequate identification of external threats to the plant;
- poor application of the hazard identification techniques;
- poor modelling of the incident scenario and emergency response.

Failure to include all significant events which were identified:
- events are deliberately omitted (poor judgement);
- important events are inadvertently omitted.

Only a small proportion of the incident scenarios which can occur can be identified in the time available, so the emphasis is on successfully identifying the significant events. Where important causes are omitted the subsequent analysis underestimates the risk levels. The change in risk due to failure to identify all causes is often small, however, and some of the emphasis on achieving completeness in terms of identifying all immediate causes of incidents is misplaced as many are comparatively rare events. Also the main system defences may well act to reduce the propagation of the event chain from unknown causes.

Serious errors arise from failure to evaluate the incident scenario correctly. The model of the scenario to study has been described in Chapter 1. Conventionally it is rare to study the propagation of an incident chain in key system safety reviews, particularly during Hazop studies. Important human actions on failure of emergency control systems are rarely checked and yet they are often vital in reducing the consequences of an undesired event.

FREQUENCY ANALYSIS

Frequency analysis is used to estimate the likelihood of each of the undesired events or accident chains identified at the hazard identification stage. Two basic approaches are used:
- the use of historical data;
- the synthesis of event chain frequencies using techniques such as Fault Tree Analysis and Event Tree Analysis.

The approaches are complimentary and wherever possible both should be followed.

FREQUENCY ANALYSIS USING HISTORICAL DATA

Relevant historical data can be used to determine the frequency with which incidents have occurred in the past and hence make judgements about the likelihood of their occurrence in the future. A typical source is the MHIDAS data bank of AEA Technology.

If the type of installation under study has suffered no major design changes then past experience should be a reasonable guide to the likelihood of future events. Thus the frequency of crashes on unmanned railway crossings gives a good estimate of event frequencies, but the study of accidents on steam boiler installations from their conception is less valid as design methods and construction have improved.

Such frequencies are best used for simple systems where there are few variables which affect the outcome. If one factor changes then the average historical data must be subjected to critical review. The variables might be related to hardware or software, including features such as improved training of the workforce. Technology changes and improved protective measures may be applied. Anyone studying the frequency of BLEVEs from the historical records would generate the wrong conclusions because techniques for protection and firefighting have radically changed and are widely applied. Any adjustments can be based either on detailed analysis or on judgements made by people familiar with the system under consideration.

In theory certain hazards are common to similar facilities or types of equipment throughout the country. Thus the average of pooled experience from similar facilities should give a reasonable approximation to the likelihood of an accident at any one of them. Thus, if 2500 chlorine storage and distribution facilities in the country resulted in one serious death in 40 years, this corresponds to a serious accident rate of one per 100,000 facility years or a frequency of 10^{-5} per facility year for continuous and semi-continuous exposure mechanisms. Some exposure mechanisms are extremely infrequent. It could be that the return period of a significant earthquake which could seriously damage a chlorine facility might be around 200 years. Thus, the pooled experience of 2500 facilities over a 40 year interval is not a reliable guide to the earthquake hazard on chlorine plants, since none of them may have been subjected to this type of loading over the period in question. This type of hazard must be assessed separately and its possible significance and likely frequency estimated separately. This is then added to the average frequency derived from pooled experience.

Historical data enables the analyst to focus directly on the overall undesired event, but cannot indicate the relative likelihood of different routes leading to the overall undesired event. It cannot be used as a basis for determining weaknesses in the design and hence for making recommendations about where these can be eradicated and overall safety improved.

In many cases there is an insufficient base on which to make historical judgements. The process may be comparatively recent, few installations exist and no incidents have occurred. Thus it is only possible on this basis to estimate the risk as, say, less than one in 100 years, but it does not clearly indicate whether this might also be less than one in 1000 years. This may be the information which is required. For example, at Oppau, Germany, in 1925 the separation of ammonium nitrate and ammonium sulphate using explosives had been carried out thousands of times without incident until, one day, a major explosion occurred.

SYNTHESIZED FREQUENCIES

A failure rate is not an intrinsic and immutable property of a piece of equipment. Values vary as a result of factors such as:
- change in the defined equipment boundary;
- the wide range of sizes in the sample;
- the process severity of the processing medium;
- the environment of operation;
- the suitability for service (quality);
- maintenance strategies adopted;
- the variations in data selection.

Failures are recorded as time-related or demand-related. In reality it is often difficult to distinguish the two. This creates problems in correlation of data and also complicates the evaluation of the fault tree.

The quality of data for use in risk assessment is generally poor, with much published data stemming from sources pre-1980. These have achieved a certain status because they are perceived as giving historically correct answers. But much of the data does not distinguish, for example, between 'low recovery, fast and dangerous failures' and 'high recovery, slow and safe failures' in control systems, or data on or changes in the historical record. In addition some data can only be very approximate due to the large number of assumptions and uncertainties in the model.

THE SHORT-CUT RISK ASSESSMENT KNOWLEDGE BASE

This section deals with values for failure rates to use in a Short-Cut Risk Assessment Method (page 215). *On no account should this data be used for industrial purposes.* The values are based on general information given in the literature and in many cases average values are used.

There is no record of all the original sources and in many cases the data is out of date following technological improvements. For example, the design of bursting discs has been greatly improved together with better methods for their design to fit in the system.

Rule of thumb values for different events are given in Tables 10.7 and 10.8, pages 230 and 231. These values merely show how failure rates appear to follow a certain logic in their frequencies; progressively reducing in a manner as the user might expect. Similarly for pipe failures. In general, a guillotine break of a 3 inch pipe is a magnitude of 10 times more likely to occur than that of an 8 inch pipe; a hole which weeps is a factor of 10 times more likely to occur than a split.

A range of failure probabilities might apply to a control system depending on the reliability target of the design.

Single channel system	2×10^{-1}	to	10^{-2}
Redundant system (M from N)	5×10^{-2}	to	10^{-3}
Partly diverse system	10^{-2}	to	10^{-4}
Fully diverse system	10^{-3}	to	10^{-5}
Two separate diverse systems	10^{-4}	to	10^{-6}

If anything the general values given in Tables 10.7 and 10.8 are believed to be low, as it is assumed that the system is being operated by well-trained staff in a good environment. Conditions are probably free from vibration in a controlled environment with regular testing and appropriate condition monitoring. Clearly conditions for moving equipment and stressful conditions are much worse. Also process conditions can be particularly harsh. Undoubtedly, maintenance and management factors have a major effect on the failure rates of equipment.

Similar criteria affect the failure rates of personnel. Further basic data on human reliability is given in Table 10.9 from Williams[20] (page 232). These generic values are then multiplied by a range of error-producing conditions given appropriate factors according to features such as inexperience, use of opposite technique, risk misconception, conflict of objectives and low morale.

The other techniques used in human reliability estimation rely heavily on the experience of the user and the specific application.

KEEPING FAILURE RATES IN CONTEXT

It is important to appreciate that systems can fail quickly or slowly and give rise to fail-danger or fail-safe states. In some cases only those corresponding to 'fail-danger fast' need be evaluated.

An analysis of control loop failures shows that failures can be classified as follows:

Fail-safe mode	1 in 2 failures
Fail-danger mode	1 in 2 failures
Fail-sudden mode	1 in 4 failures
Fail-slow mode	3 in 4 failures

Hence the hazard rate is F/8 per year for loops to fail dangerous and fast. Historical experience justifies this as a safe value; recommendations as low as F/20 have been found to be acceptable.

TABLE 10.7
Generalized data on frequency per year for failures in average duties

Complex equipment prone to breakdown:
- continuous condition — 10
- conveyors — 10
- rotary and large centrifugal compressors — 10
- process analysers — 10

Complicated equipment or loops:
- pumps, turbines, small compressors — 1
- mechanical equipment — 1
- pressure circuits — 1
- control loops — 1
- process upsets — 1
- power/utility failure — 1
- gas analyser — 1

On-line computers:
- mainframe — 10
- micro systems — 1

Passive or simple active equipment:
- exchangers — 0.1
- fans, motors — 0.1
- valves — 0.1
- process alarms, sensors, transmitters — 0.1

Small items:
- switches, relays — 0.01
- gauge — 0.01
- annunciator — 0.01

Leaks:
- sliding seal — 0.01
- valve passing — 0.01
- ducts — 0.01
- hose failure — 0.01
- compressors — 0.01
- plastic pipe — 0.001
- pumps — 0.001
- metal pipe — 0.0001

Major rupture or disintegration:
- pressure vessels (gas, liquid) — 0.0001
- disintegration of rotor — 0.0001
- pressure circuit, pump — 0.0001
- pressure vessels (liquid and vapour), compressor, filter — 0.00001
- valve — 0.00001
- metal pipe (small, diameter < 50 mm) — 0.00001
- metal pipe (medium, 50 mm < diameter < 150 mm) — 0.000001
- metal pipe (large, diameter > 150 mm) — 0.000001
- act of nature, impossibility — 0.000001
- aircraft impact — 0.000001

FAILURE OF SYSTEMS ON DEMAND

Some systems contain components which are normally dormant, but are required to respond in the event of a demand. For example, a trip system is required to guard the system in the event of a hazard. Its failure could therefore be considered a fail-danger as opposed to a fail-safe fault.

The probability that the system will not operate on demand is a conditional probability. It is an event A which occurs given that event B has occurred. It is possible to define this conditional probability by the expression:

Hazard rate = demand rate × conditional probability of system failed-danger.

This special conditional probability is often termed the fractional dead time of the system (FDT).

TABLE 10.8
Typical ineffectiveness probabilities of protection

	Ineffectiveness
Human:	
• no action taken	1
• response to gauge when in automatic operation	1
• required action outside control room, automatic operation	0.9
• response to gauge when in manual operation	0.1
• failure to confirm shutdown	0.15
• response to multiple alarms	0.15
• critical familiar action under high stress	0.3
• critical action, written procedures available	0.1
• response to single alarms	0.02
• procedure on check-list	0.02
Protection:	
• automatic control system, simple protective system	0.1
• relief valve, tested, all failures	0.1
• pressure rise exceeds relief system capacity	0.04
• excess flow valve, shut-off valve	0.02
• relief valve does not open	0.01
• relief system under capacity	0.01
• protective system, with redundancy	0.01
• mechanical interlocks	0.01
• alarm malfunction	0.01
• interlock system tested frequently	0.001
• motorized valves	0.001
• overpressure of relief device	0.001
• check valves	0.001
• pressure switch, relay, limit switch	0.0001

TABLE 10.9
Some data on human reliability, selected from Williams[20]

Generic task	Nominal human unreliability	
	Range	Average
Totally unfamiliar, performed at speed, with no real idea of consequences	0.35–0.97	0.55
Shift or restore system to a new or original state on a single attempt without supervision or procedures	0.14–0.42	0.26
Complex task requiring a high level of comprehension or skill	0.12–0.28	0.16
Fairly simple task performed rapidly or given scant attention	0.06–0.13	0.09
Routine, highly practised, rapid task involving relatively low level of skill	0.007–0.045	0.02
Restore or shift a system to original or new state following procedures, with some checking	0.0008–0.007	0.003
Completely familiar highly practised routine task performed by well trained individual with time to effect recovery	0.00008–0.009	0.0004
Miscellaneous task for which no description can be found	0.008–0.11	0.03

In order to limit the possible time that the protective system could be out of action, it should be proof tested at regular intervals appreciably less than the mean time to failure of the system.

For a component with a failure rate per year of λ and proof tested with test interval T, the FDT between tests is given by

$$FDT = 0.5\, \lambda\, T \text{ when } \lambda T \ll 1$$

For example, if a relief valve is proof tested once per year and it shows fail-danger faults at a rate of $\lambda = 0.1$/year then the fractional dead time is 0.05 which is a realistic value.

For a multiple protective system which is a series system of n units, each of failure rate λ_i, which can fail to danger, the system FDT, FDT_s, is given by:

$$FDT_s = \sum_{i=1}^{n} 0.5\, \lambda_i\, T$$

Always check that the answer is sensible, as the expression is an approximation.

VOTING SYSTEMS

A system which operates when at least r channels are triggered (an r-out-of-n system) has a larger FDT than a 1-out-of-n system but has a smaller spurious trip rate as the system is activated when r channels fail-safe. In order to reduce the FDT of voting systems it is possible to arrange for the tests of individual components to be staggered.

TABLE 10.10
Rules of thumb for ineffectiveness probability of mitigation

Delayed ignition of LPG leak from vessel	0.5
Immediate ignition of flammables giving pool or torch fire	0.3
Immediate ignition of large leak of flammable vapour	0.1
Contamination of environment given that fire has occurred	0.1
Flammable release catching fire (all releases)	0.05
Explosion given delayed ignition and some confinement	0.2
Missiles given explosion	0.1
Major escalation given missiles	0.1
Toxic event given release	0.1
Toxic event with multiple public casualties	0.01
Escape from toxic exposure (depends on chemical)	0.95
Escape indoors from toxic vapour	0.8
Escape indoors from toxic vapour	0.2
Escape from pool or jet fire	0.95
Escape from missiles	0.05
Escape from flash fire, BLEVE, explosion (no warning)	0.01

If the testing time is small compared with the test interval then the FDT is reduced as the testing interval is reduced. However when the system is tested then the system is inoperative. Also each test can introduce safety problems and the maintenance may leave the system with a fault, and in a failed-danger state.

INEFFECTIVENESS PROBABILITY OF MITIGATION
Some values have been assembled in Table 10.10 for ineffectiveness probability of mitigation. These are very difficult to determine and are largely based on historical records.

GENERAL FACTORS AFFECTING RISK
It is accepted that if the standard of various factors is improved then the values of frequencies and probabilities change. The key factors appear to be:
- the quality of engineering, design and construction;
- the quality of procedures, information and communication systems;
- the capabilities, experience and safety awareness of management and organization;
- the control of change and external threats;

- plant specific factors:
— the internal process environment;
— the external process environment;
— the location and accessibility of plant;
— safety awareness and supervision of personnel;
— the quality and experience of personnel;
— the quality of maintenance and its system of operation;
- the effectiveness of the emergency plan;
- the business climate, safety culture, external pressures, resources.

Levels of risk are affected by the root causes of incidents but it is impracticable to adjust all data according to these factors, and not feasible at an early stage of design. Fortunately for many processes the factors affecting consequences are not critical because the extent of possible damage is restricted and well-defined. Such plants would normally only pose severe or catastrophic consequences following extremely rare events. It is also not difficult to distinguish between a works which is soundly run with operating technology well-understood by the workforce and one which is badly maintained, having new or novel technology outside the skills of the staff. A process unit having novel technology at a site close to a populated area, but remote from technological support, is obviously going to be more at risk.

Basically the approach recommended is to look for major variances in one or more factors and change values by up to an order of magnitude. If this change affects key variables, such as the likelihood of immediate cause and the probability of inadequate emergency control, then the risk is almost certainly going to appear unacceptable. This will then direct appropriate remedial action to either eliminate the deficiency or reduce its effects.

The following approach is recommended:

- Determine the case represented by plant conditions:
(a) Average case using appropriate baseline failure rate data.
(b) Excellent case conditions in which the internal duty is clean and maintenance performance is good on a well-established plant.
(c) Worst case conditions in which the duty is severe or maintenance performance poor or the plant employs novel technology.

- Adjust the likelihood of failure in average duties as follows:
(a) For the worst case multiply by 10.
(b) For an average case multiply by 1.
(c) For an excellent case multiply by 0.5.

- Modify the ineffectiveness probabilities of control and mitigation:
(a) If a protective system is in the failed state, $P = 1$.
(b) In the worst case *increase* calculated values of P as follows:
   0.01 to 0.1; 0.1 to 0.5; 0.5 to 0.9.
(c) In an average case do not adjust probabilities.
(d) In an excellent case *reduce* calculated values of P as follows:
   0.001 — no change; 0.01 to 0.005; 0.1 to 0.05; 0.5 to 0.1.

Predictions are only as accurate as the model and the base data. Most data is very general in nature and items vary considerably in performance. The analyst may subconsciously be working towards a conceptual target and be satisfied when this is achieved. Also large fault trees are virtually impossible to check. Nevertheless, the answers given are a useful relative measure and the reductions in risk conceived from the exercise are indeed meaningful.

RISK REDUCTION

The process risks in the ongoing operation are identified, evaluated and compared with targets. Risk reduction measures may then be considered. Such action is readily possible as the variations in the process are relatively minor and appropriate data are usually available.

More radical methods involve examination of the design; these should have been carried out at an earlier stage of hazard review. There is no reason why major changes should not be considered again, particularly when they involve a major reduction in risk — for instance, by substituting less hazardous materials, reducing inventories, avoiding extremes of temperature and pressure, increasing separation distances between plant and people, strengthening or otherwise improving materials of construction and any other means of improving the inherent safety of the system. The problem is that such changes are costly to implement at the later stages of the design. Other measures are related to the specific chain of events and relate to increasing or improving add-on safety features, and improving control, monitoring and testing.

Clearly the likelihood of the identified immediate cause of incidents has a considerable effect on risk, and reduction might stem from changes in specific root causes. Improved training of operators with enhanced assistance to help them recover in a given situation is often invaluable. Similarly mitigation action related to response during emergencies both on and off the works can be improved.

The organization of a risk reduction programme involves reviewing the results of the risk analyses and then deciding upon appropriate courses of action such as:
- requesting additional information;
- examining options in more detail;
- seeking further quantification or greater accuracy;
- implementing the risk-reduction measures;
- accepting and managing the residual risk;
- stopping existing operations;
- not going ahead with the project.

Semi-quantitative guidelines are given by AIChE[13] in Table 10.11, page 236. These can readily be combined with the values used in SCRAM.

AIChE[13] also emphasizes the need for effective implementation of risk-reduction measures and commitment of resources to this activity. The implementation may require measures like quality assurance, further training, supervision and change of procedures. Appropriate communication of ideas from staff at all levels on the measures being adopted is essential. Implementation must include the adoption of appropriate control measures and the monitoring of performance indicators reflecting the standards adopted.

TABLE 10.11
Semi-quantitative ranking of risks, based on AIChE[13]

Class I Risk

Risks (related either to specific processes or general management systems) significant enough to require immediate shutdown of an operating unit until the hazard is mitigated. A unit may not be operated while risk-mitigation measures are being implemented. Stop-gap measures may be used only if a detailed analysis shows that such measures can effectively mitigate the hazards; such measures may be used only on an interim basis (following established variance procedures). In addition, there must be an immediate initiation of a program to provide a permanent solution.
Example: Bypass of an emergency shutdown system.

Class II Risk

Other serious process risks or deficiencies in risk management systems which require immediate action to mitigate the risk. A program to provide a permanent solution should be initiated immediately.
Example: No test program to verify operability of remotely operated block valves.

Class III Risk

Process risks or deficiencies in risk management systems of a less serious nature than those in Class II, and that have less immediacy than those in Class II. These situations should be corrected as soon as reasonably possible to bring the facility's program up to good industry practice.
Example: Process flowsheets need updating.

Class IV Risk

Other areas of possible risk reduction or improvement in risk management systems (advisory in nature).
Example: Failure to follow up near-miss incident reports.

UNCERTAINTIES IN QUANTIFIED RISK ASSESSMENT AND DECISION-MAKING

Performing a QRA requires the introduction of data for which precise information is lacking. Experienced judgement is needed, particularly when considering the frequency of unlikely failures. A rigorously performed QRA would require that confidence limits are set on the figures but, because of the difficulty in making best assumptions in the first place, setting confidence limits would be even less certain and is seldom attempted.

The extent to which the uncertainty of a numerical value presents a problem depends on a number of factors. The uncertainty should be tested by sensitivity analysis to see whether the event being considered is much influenced by changes to the figure; for instance, a particular uncertainty is less important if it relates to one of the less likely causes of an event.

Another consideration is whether the range of uncertainty in the final result would prevent clear-cut conclusions being drawn. In this regard the uncertainty of the QRA figures may pose particular difficulties when seeking to judge compliance with acceptability criteria involving very low risk levels.

Special attention must be given to present analytical results to decision-makers in a manner which is not open to misinterpretation and which reflects the degree of detail in the analysis. The scenarios that underlie the significant risks should be described in a way that provides realistic insight into the cases for which the probabilities and consequences have been estimated.

To aid decisions, it is necessary to combine the numerical estimates and express the results in terms of those risk parameters considered in the course of the study (for example, individual or societal risk) for comparison with appropriate criteria. The uncertainty of the results should be clearly stated, whether expressed numerically or graphically. Great care should be taken to avoid giving the impression of accuracy, such as quoting computer-generated results to many significant figures.

In the case of individual risk, it is essential to specify whether the calculated frequencies of death or serious injury relate to the person or the 'critical' group of people most at risk from a given activity (as a result of their location, habits or time periods for which they remain vulnerable), or whether they relate to an 'average' value, representative of all individuals potentially affected. It is usual to estimate individual risk levels as a function of distance from a hazardous installation. The estimated values can be expressed as a series of iso-risk contours overlaid on a map, to reflect visually the spatial distribution of assessed risk. Note that the presentation belies the accuracy of the data.

Similar considerations apply in the case of societal risk, which is typically adopted to supplement individual risk results, by reflecting the likelihood of accidents involving multiple casualties. Here, the relationship between frequency and the number of people suffering a specified level of harm may be expressed graphically in an F-N curve, with the frequency of exceeding given numbers of casualties plotted cumulatively. Alternatively, where such detail is considered unwarranted, the statistical expectation value of this frequency distribution may be calculated, with the analytical results simply expressed in terms of 'expected' casualties per year.

The findings of QRA may be made public in the cases involving land planning in relation to proposed emergency procedures or following an accident in a court of law. It must be recognized that low probability risks are conceptually difficult, even without the extra complication of uncertainty. The public presentation of QRA results must avoid arousing undue anxiety; the right balance must be found to convey the findings accurately while making them as understandable as possible. It is by no means easy.

DECISION-MAKING WITH QUANTIFIED RISK ASSESSMENT

Risk evaluation has been discussed earlier and it was noted that the development of cost-benefit approaches and measures to determine whether or not risks are 'as low as reasonably practicable' (ALARP) remain at a very early stage. No European consensus on acceptability criteria appears likely within the foreseeable future. The conclusions which emerge from QRA (together with appropriate warnings concerning limitations of scope, uncertainties, etc) are typically forwarded to those making decisions on applications as a formalized safety input to the decision. But this input should not be the sole consideration of regulators or Government; they generally need to account for a diverse range of other factors.

The conclusions from QRA usually rank the proposed measures into one of the following categories:
- Category 1 — clearly necessary or worthwhile;
- Category 2 — no strong indication of whether measure worthwhile or not;
- Category 3 — clearly unnecessary or not worthwhile.

Although such considerations are mainly concerned with safety, there will be issues to address outside the scope of QRA. They may be management or organizational systems of safety control, or concerns expressed by groups such as the media, local residents or trade unions. Indeed, in the case of a Category 2 conclusion suggesting the proposal is marginal on QRA considerations, such 'other' factors will be overriding in the ultimate decision.

Against this background it is neither surprising nor unreasonable to expect the decision-making process in some instances to outweigh the conclusions of a QRA on other legitimate grounds. Such flexibility is an essential and inevitable response to the complexities of present-day risk management decisions concerning major hazards. This is preferable to a rigid imposition of low risk limits which would give a disproportionate weight to quantified risk considerations at the expense of all other factors. This necessary degree of flexibility does not in itself detract from the potential value of QRA in clarifying the main relevant factors and the implications of the value judgements underlying the final decision.

QRA is not and should not be represented as a precise scientific exercise. Uncertainties remain and the QRA conclusion forms only one of several factors to be taken into account in the final decision. Whilst this in itself does not detract from the potential value of QRA in clarifying for the decision-maker the relevant factors and the implications of the ultimate choice, it confirms that QRA should not be regarded as a universal panacea for easy decision-making. When correctly applied in appropriate circumstances, QRA can be a valuable tool to aid safety decision-making. However, if imposed indiscriminately it may not only involve commitments to unwarranted analysis but also yield misleading results, giving false sense of security or leading to a waste of resources. Consequently, the development and use of QRA in regulatory decision-making must be highly selective.

RESIDUAL RISK AND THE MANAGEMENT OF PERFORMANCE
QRA is valuable at all stages in the life of a plant. All plants contain residual risk. Action to evaluate and reduce residual risk should continue throughout the life of a plant by seeking to eliminate the root causes of incidents, particularly in maintenance, external threats, procedures, information, information transfer and information processing, the abilities of personnel in the task, and the capabilities of management and organization.

Changes in process requirements, plant modifications and external threats must be analysed and monitored carefully. This can be effected by an overall system safety *culture* that encourages the interchange of ideas between operators and management and the right attitude to process safety and loss prevention. This can partially be achieved by the principles of Total Quality Management, backed by regular process safety audits. Efforts should not only be aimed at the organization, its structure and safety awareness, but also towards the task- and performance-influencing factors which derive from it.

Despite awareness of the danger there is much evidence that degradation of performance and increase in residual risk arises regularly. The design intent represents the mode of operation of the system as designed or intended. Consider a plant which has overcome initial problems from any design functional deficiencies. It is unlikely to be running exactly at the design intent and further drift away from this standard will occur, particularly if demand for product is high or low.

Normal practice represents the normal operating mode of the system. The features of such operation which differ from the design intent will have accumulated over many hours of operation, and provide the setting for a subsequent incident. The changes may well help to amplify the root causes of any eventual incident or may indicate a reason for the escalation of the incident.

Actual practice represents the actual operating mode immediately prior to an incident. In normal operation the accident did not happen. On this particular occasion it did. Any difference between normal and actual operation will usually identify the immediate cause of the incident or the reason for its escalation.

The study of serious accidents often shows that an immediate reduction of safety follows the removal or degradation of some clearly identified defence against incidents. This normally causes at least an order of magnitude increase in the probability of the protection failing. The effect of failing to maintain the effectiveness of an emergency shutdown system, coupled with an inadequate assessment of the effectiveness of mitigation and a resulting increase in the likelihood of escalation of the incident, is illustrated in Table 10.12. What seems from the Short-Cut Risk Assessment to be, arguably, an adequate design has suddenly been changed to a design with extremely high risk. At Bhopal, for instance, the protective systems were inoperable and the mitigation systems proved woefully ineffective.

Similar comments also apply to features external to the plant, such as change of external threats or deliberate action to change the operation of plant to conditions outside those of the design. Also the plant environs may change, making the consequences of an accident greater. An example is when dwellings are built close to the site, as explosions killing many people in Mexico and Pakistan have shown.

Full motivational safety awareness must be maintained. People choose to behave safely if they realize the consequences. Economic strictures might suggest continuing operation when precautions are faulty but all concerned should be aware of the possible consequences of this decision and its effect on residual risk.

TABLE 10.12
Design study and state before incident

	Original	Actual
Immediate cause, F/y	1	1
Ineffective control, P	0.1	0.1
Ineffective protection, P	0.1	1
Ineffective mitigation, P	0.1	1
Escalation, P	0.01	0.1
Frequency of top event	10^{-5}	10^{-2}
Consequences of top event	5	5
Risk rating	0	3

It has been indicated that risk assessment contains many assumptions and uncertainties. A continuing programme to manage residual risk requires the upgrading of studies as better information becomes available, as for example from the improved modelling of consequences.

Instigation of investigation is also important and a study of operational performance indicators may show that the residual risk is greater than anticipated and further studies may be required. Periodic auditing of the system is also desirable.

Over the last few years the toxicity of certain chemicals and the criteria promulgated for evaluated residual risk have been getting more stringent. So further reduction of risk or stopping of activities may be necessary on even the best run plants.

COMPANY EXPANSION

The tragic event at Bhopal had a major effect on companies who obviously did not wish an event to occur which caused such loss of life. At a less important level, the effect of the incident on the financial standing of Union Carbide has been noted, and there has been an appreciation that serious accidents affect the viability of the company concerned and the industry in general. There might be extenuating circumstances which increase risk, as when oil exploration is carried out at a remote country location (for example, in the jungles of Vietnam), where company staff are not as readily backed by experienced operatives. However, this would not be an acceptable reason for going ahead once media hype had distorted the cause of the incident.

Such effects mean that:
- new projects in obscure locations may not be worthwhile;
- acquisitions of an existing business may not be desirable;
- the historical risk associated with a certain process under consideration may cause such a development to be perceived by the public as high risk. The use of a specific chemical or a chemical new to the company on that site may not be feasible.

Any new business must be compatible with existing risk guidelines. If it is not, the costs associated with bringing it into internal compliance must be ascertained. The company should encourage any client or supplies company or associated transport company to use similar risk management practices in order to avoid knock-on effects of poor risk management.

THE APPLICATION OF RISK ASSESSMENT

At the engineering interface there is now a good understanding of the principles and techniques of QRA. The requirements for risk assessment are identified in codes of practice and the need for specialist help is recognized where the consequences are great. It is desirable to establish procedures and standards for the application of QRA methods and make hazard identification and risk assessment an interactive part of engineering at all stages of development and design. In particular all personnel involved in operation of the plant should receive training in the development of incident scenarios and acquire some understanding of risk assessment methods. This also applies to senior management throughout the company. It is particularly important for key personnel to be able to interpret risk assessment results and incorporate them in decision-making.

The growth of legislation on health and safety issues indicates the way the application of risk assessment is growing:
- Health And Safety At Work Act, 1974, 'Ensure as far as is reasonably practicable ... '.

- EEC Council Directive, the Seveso Directive, 1982, On the major-accident hazards of certain industrial activities.
- Control of Industrial Major Accident Hazards Regulations, CIMAH, 1984.
- Federal Law SARA Title III, 1986. Information spread of emergency plans.
- OSHA Rule 29 CFR 1910.119, 1992, Process Safety Management of Highly Hazardous Plants.
- Offshore Safety Case Regulations, 1992. Full QRA required.

The setting of regulatory targets is favoured by the 'homeostatic' approach to set determinate goals as represented by ALARP, ALARA and BATNEEC. Those arguing against identify the lack of homogeneity of social attitudes to risk. There are too many uncertainties in the estimates so there is a need for balancing arguments. This has been termed the collaborationist approach, in which the opposing views remain in a balance of opposed tension, and risk values used change according to the current force of argument.

The use of anticipation in risk management is favoured by the use of more audits and precautionary measures. There is a strong emphasis on learning from hindsight. Some argue that this can reduce the response to the unexpected (resiliencism). It also means that excessive complexity is brought into the system.

There seem to be two views on the need to increase the blame and liability in the event of incidents. One view argues to increase the financial and legal liability on those best able to take action to minimize risk. The legal blame can be targeted on designers and managers. The other view favours a no-blame approach which gives free flow of post-hit information (absolutionism). But it also causes people to work 'by the book' instead of modifying their actions as the situation demands. It also reduces motivation to undertake activities.

More use of QRA certainly suits the legal and bureaucratic requirements. It avoids anomalies and special pleading. But there remain many limitations in practice and it is claimed that many assumptions can be value-laden and implicit.

There are many advocates of improvement in institutional management. This emphasizes the commitment to safety at the highest level, the adoption of low and zero accident targets, the provision of training and resources, the use of management and safety management audits with safety assurance and an improved safety culture. Whether the slim management of today can continue to deliver is open to question. Supervision, experience and reduced pressure of work have always been high priorities for effective safety management. It is also not at all certain that contradictory approaches yield similar outcomes.

There is concern about the high cost of risk reduction. Fortunately, this can often be offset by meeting other goals such as improved quality and higher production gains. Such an emphasis on positive benefits may be a better approach than the trade-off against other basic goals, such as by adjusting BATNEEC.

Today there is a broader participation in decision-making and increased accountability of decision-makers. People in general should be trained to understand risk at every level in society. This broadens the base of knowledge, consensus and responsibility. It would be sensible to give more training in risk awareness in schools. Also, operators should learn the basis of Short-Cut Risk Assessment. After all, the man in the street who is interested in betting knows far more than the average non-betting reader ever will about probability. In-company short courses can readily be created for this purpose.

APPENDIX 1 — THE DEVELOPMENT OF P&I DIAGRAMS

GENERATING P&I DIAGRAMS
Piping and instrumentation (P&I) diagrams are produced by marking up a process flow diagram and ensuring that all operating modes of the plant are taken into account, including:
- initial start-up;
- normal start-up and shutdown;
- start-up after emergency;
- abnormal operation;
- normal operation;
- partial shutdown;
- emergency shutdown.

Do not forget the need for maintenance and inspection.

P&I diagrams show the engineering details of equipment, instruments, piping, valves, fittings and their arrangement. Some of the letter codes from British Standards are given in Tables A1.1 and A1.2 (page 244). Upper case letters are used. To denote HIGH or LOW use H or L in association with the symbols. To avoid excessive detail in the presentation, details of piping are not shown.

HAZARD CONTROLS
Hazard controls are introduced to limit undesirable consequences induced by residual hazards which remain despite the best efforts of designers. Features should be designed to minimize the probability of occurrence of initiating events and their propagation to undesirable events. Capabilities for the implementation of corrective functions or contingency functions, or both, should be built into the design in order to prevent the propagation of initiator events to undesirable events. Special procedures should be developed to counter undesired events which cannot be satisfactorily controlled by design or by safety/warning devices.

SYNTHESIS OF HAZARD CONTROL SYSTEMS
The synthesis of the control systems is normally carried out in the following sequence:
- Identify the key features of the process which are to be controlled: main and secondary flows, recycles, products, turndown requirements, location of surge tanks.
- Synthesize the material balance controls to ensure the proper material flow through the process.
- Develop the product quality controls.
- Determine the controls for secondary flows, pressures, temperatures and level.
- Determine the controls for abnormal operations.

TABLE A1.1
Letter codes for instrument functions selected and adapted from British Standards

	First letter	Succeeding letters
A	Alarm	
B		State — for example, motor running
C		Controlling
D	Density	Difference
E	All electric variables	Sensing element
F	Flow rate	Modifier or ratio
G	Gauging, position, length	
H	Hand operated	
I		Indicating
J		Scan
K	Time	
L	Level	
M	Moisture/humidity	
N	User's choice	User's choice
O	User's choice	
P	Pressure or vacuum	Test-point connection
Q	Quality	Integration or summating
R	Nuclear radiation, relief	Recording
S	Speed, frequency	Switching
T	Temperature	Transmitting
U	Multivariable	Multifunction unit
V	Vibration	Valve
W	Weight of force	
X	Unclassified variable	Unclassified function
Y	User's choice	Computing relay, relay
Z		Emergency or safety acting

H or L after a succeeding letter denotes high or low
HH or LL after a succeeding letter denotes a subsequent alarm to the above

TABLE A1.2
General symbols for valves, actuating elements and instruments

An automatic actuating element (a thin circle of 5 mm diameter) and a correcting element of unspecified type.		An automatic actuating element with integral manual actuating element and a valve.	
A manual actuating element acting on a valve.		A control valve which opens on failure of actuating energy.	
A control valve which closes on failure of actuating energy.		A control valve which retains position on failure of actuating energy.	
A flow-rate recording controller adjusting a valve. The instrument is mounted on a local control panel.		A locally mounted instrument to record and control flow rate with summation of volume. The valve can be isolated manually on failure.	

- Add constraint/override controls. Consider both active and passive control systems.
- Add start-up and shutdown controls.
    Improvements to the basic control system are generally affected by actions such as:
- Considering the availability of measurements and manipulations.
- Considering the use of electronic programmable systems, feed-forward and feedback.
- Examining control system response. A plant is easier to control with a flat or slow response. Damp rather than magnify trends and reduce the interaction of loops.
- Examining monitoring requirements. Note in particular adverse influences, incipient problems, evidence of deterioration and the defined point of failure.
- Incorporating user-friendly features:
— status clear on valves, instruments, on-off signals, etc;
— design tolerant of maloperation or poor maintenance;
— operator response easy and consequences of failure low.
    The current trend is to divide these systems up into the basic control system and the safety interlock systems. This permits a thorough consideration of all the activities with the added

bonus of considering improved logic in making control decisions. If computer control is widely used this must be built into the system from the start. Monitoring is an important feature of any control system. All plant outputs and inputs should be measured. This includes efforts directed towards studying the impact of the release of chemicals and radiation on humans, fauna and flora. Techniques used include environmental monitoring, biological monitoring, biological effect monitoring and health surveillance. Environmental monitoring in the workplace can be used to assess compliance, test ventilation and identify hazardous areas. But it is the uptake of a substance by a worker and not its individual concentration in the workplace which is of greatest concern. This can only be achieved by biological monitoring of the individual worker which accounts for exposure of the body by all routes.

SECOND-CHANCE DESIGN

Second-chance design implies that two measures must be provided against the occurrence of any event deemed to be significant. These measures include controls, protections and emergency responses which should be provided against the occurrence of undesired events. More layers of protection with enhanced availability and reliability are required against high risk events.

A primary measure might be part of a normal control system backed up by an attention-seeking device on deviation demanding action by manual intervention. Similarly, a simple interlock designed to prevent an immediate action such as opening a valve would be regarded as a primary measure. A secondary measure might be a passive safety device (for example, a vent, drain or overflow) or an active safety device (automatic shutdown systems or relief valves). Multiple actions by such systems can be built into a comprehensive safety interlock scheme.

Passive protection systems are also important as measures for limiting the consequences of an accident. Methods to note include:
- improved protection against fire, non-ionizing radiation, etc;
- segregation of people, plant and external cause;
- open construction, fire-breaks, fire-walls;
- improved drainage and spill containment;
- improved personal protection, havens, etc.

PRESSURE RELIEF

The purpose of pressure relief is to prevent the mechanical failure of vessels and equipment due to excessive internal pressure or vacuum. A vessel may still fail at design pressure given operating temperatures above the design limits, high loads or deterioration of materials of construction. The maximum allowable pressure or vacuum within the vessel needs to be identified; then a pressure relief system is provided to allow sufficient flow into and out of the vessel to ensure that the maximum allowable pressure or vacuum is not exceeded.

SAFETY VALVES

Safety valves (safety relief valves) and fire relief valves are either direct spring-loaded valves, supplementary loaded valves or pilot operated valves. A safety valve starts to open at its set pressure, it becomes fully open at some overpressure above set pressure (usually 10%) and it closes when the pressure falls. It has a slow response and thus is inapplicable for explosion relief. There is a risk of blockage and of trace leakage if the operating pressure is close to the set pressure.

A bursting disc is designed to burst open or burst at a specific pressure somewhere in the range between the minimum and maximum specified bursting pressures. There is usually a 10% difference over the minimum pressure but discs with a smaller range are available at a price. Bursting discs cannot close after opening so they may allow a large discharge. They are less likely to block than a relief valve and have a very fast response. They can be manufactured in a wide range of materials and there is no leakage until failure. Given correct selection and specification they are cheap to install and maintain. However, incorrect assembly is more likely than for relief valves.

Other systems used for relief include combined pressure/vacuum relief valves on low pressure storage tanks, relief hatches, surge relievers, fusible plugs, vents, vents via seal-pot or flame arrestors, restriction orifices and vacuum relief.

VENTS

An open vent is the relief valve of the vessel and must be tested and correctly maintained. Its size should not be changed without proper evaluation. More than one vent may be required. If a vent discharges into a closed system then the pressure drops in the connecting lines and the head in the vessels should be checked.

Flame arrestors or traps prevent the unrestricted propagation of flame through gas-air mixtures. They are unsuitable for flammable dust or other aerosol suspensions. They contain an assembly of narrow apertures through which gas can flow but which subdivide a flame into flamelets which are quenched before they reach the outlet. They are used, for example, on vents from storage tanks, on inflammable gas pipelines and at the base of flare stacks. Normally they should be positioned near the ends of lines and, if possible, vertically. Pressure drop must be maintained and monitored as apertures are easily blocked, causing overpressure or vacuum in tanks. The use of common vent-lines should be avoided. Typically arrestors comprise wire gauze, crimped ribbon, perforated or parallel plates, sintered metal, packed towers, metal foam, hydraulic systems and water systems. Correct design is essential and the arrestor will normally prevent an external source of ignition igniting a flammable air/vapour mixture in the tank.

GENERAL COMMENTS

The flow sheet or line diagram needs to be split into a number of systems for pressure relief purposes. Vessels which are connected together by piping of adequate capacity, and without valves which can isolate any vessel when connected to its source of pressure, may be considered as a system of vessels for any application of process relief. It is important that blockages are not possible. If such a system is totally blocked in then energy inputs will cause overpressure; if all the outlets from a system are blocked or inadequate, then flow from a higher pressure will cause overpressure. If all the inlets from a system are blocked or inadequate, then cooling or flowing out can cause a vacuum. Chemical reaction may be initiated or exacerbated by the pressure-initiating event.

All relief devices must be correctly sized, chosen, installed and maintained. Isolation devices on relief valves should not be left closed longer than necessary for testing and should not be fitted below relief devices unless there are two devices with interlocked valves. The relief system protects equipment and piping from pressure outside the operating limits of the plant. A relief system does not necessarily protect the materials of construction from failure. An increase in temperature or deterioration of the materials of construction can result in loss of containment by rupture at normal pressure. All relief systems should vent to a safe and remote area — for example, a flare stack, a scrubbing system, a receiver, back into the process or to an incinerator. In many cases a condensing system and evaporators may be considered. Any venting to air may be undertaken only if all the risks involved and the environmental restrictions have been taken into account.

Always consider carefully the phase of the flow, the position of the device, isolation, seating difficulties, testing, activating pressure, required flow rate, corrosion, blockage and incorrect installation. Design the relief system to relieve pressure resulting from the vaporization of a liquid due to heat input from a nearby fire. Fire relief valves if fitted should be set as little as practicable above the operating pressure.

DEPRESSURING SYSTEMS

Depressuring systems include the use of pressure control valves, dump valves and other valves actuated manually. Complete depressuring of a system is occasionally necessary. This can take a long time for a large container or process system which cannot be isolated into sections. Emergency depressuring by the operator or an automatic system must take the integrity of a vessel in a fire into consideration. If material from emergency depressuring passes into the relief header then this must be listed as part of the load on the line.

INTERLOCKS

These days there is little distinction between interlocks and emergency shutdown systems as the latter become known as 'safety interlock' systems. The change is logical but the older terminology was helpful, as an interlock was thought of as essentially something which prevented an action. It was directed towards preventing the initiating of an unsafe act.

Interlocks control operations which must take place in a given sequence. They also stop a response to a preconceived action which is incorrect at a given time or when another item of plant has failed. Items of equipment must have specified relationships between their states.

Examples of different kinds of interlocks include a padlock and chain on a hand wheel, mechanical interlocks, a switch to prevent access on a centrifuge whilst it is rotating. The start-up of machinery should be prevented when it fails to meet all preconditions, correct sequences and conditions for transmission. Interlocks include the use of keys, time-delay switches and means of preventing the disarming of protective systems or their isolation. Interlocks can be applied when a vehicle is loading to prevent it being driven away. They might prevent the opening of a valve when others are open, or stop a robot arm from moving when the operator is too close.

A good interlock controls operations positively, is incapable of defeat, is simple, robust and inexpensive, is readily and securely attached and is regularly tested and maintained.

All tasks can be examined to see what should be the immediate response to error. For example, when using a word processor the keyboard command may be queried or ignored. If a file is deleted it will ask if the user is certain about this action. Even major mistakes may be recovered under certain circumstances. Clearly computer control can be geared the same way. The general philosophy is of value, as it shows the range of possibilities of effecting early recovery of an incorrect action.

EMERGENCY CONTROL SYSTEMS

Emergency control systems are usually active protection systems activated manually or automatically. They include the following:
- emergency relief and relief treatment systems;
- emergency shutdown and process abort systems;
- emergency isolation and interlocks;
- emergency depressurizing and transfer of material;
- emergency input of material and inerting.

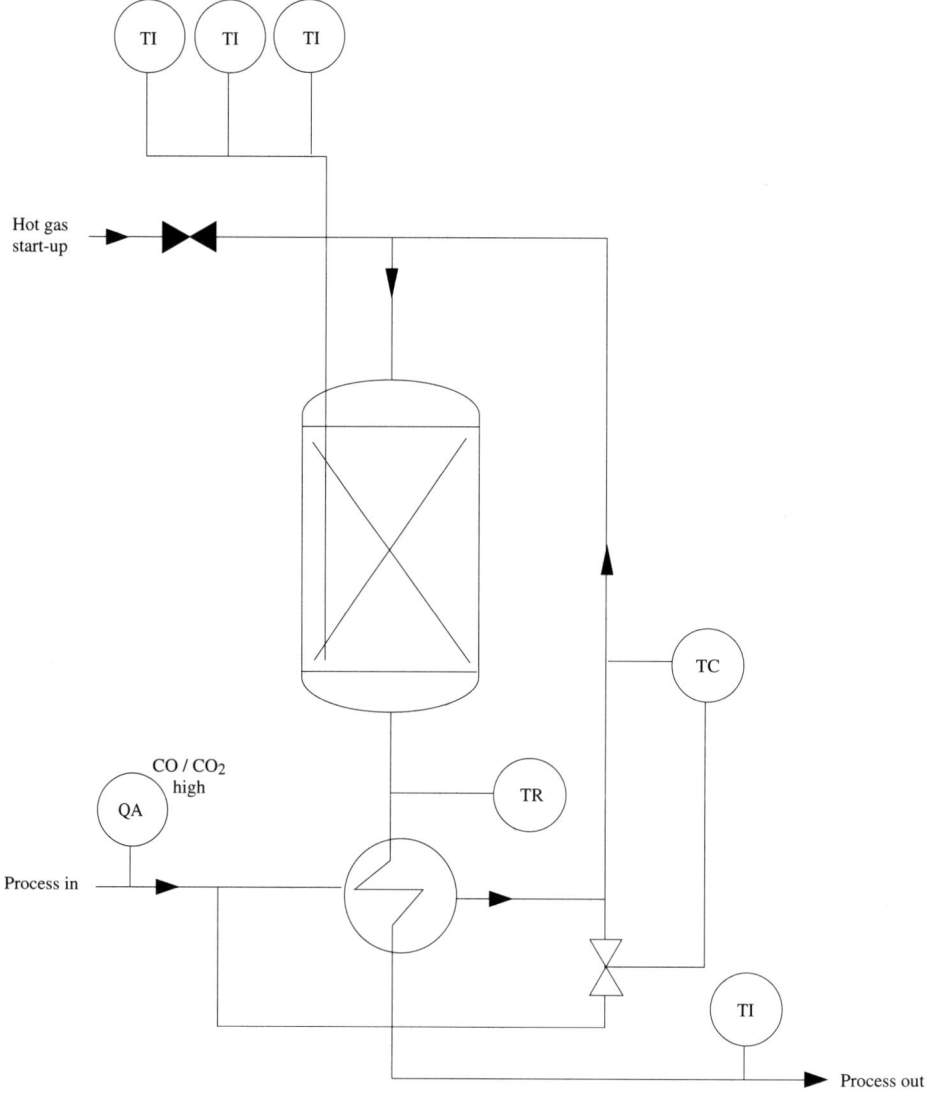

Figure A1.1 Exothermic reactor system. (See page 243 for instrumentation definitions.)

An emergency shutdown system carries out appropriate activities when a set point is passed, preferably after an earlier alarm. It should always give an alarm when activated, either automatically or by the operator. What happens when the trip occurs, or when it is removed, tested or disarmed, should be studied. It is important to appreciate that activation of an emergency shutdown system does not mean that the appropriate action has been taken by the system. Many accidents, including the famous discharge at Three Mile Island, have occurred because the operators did not appreciate that valves had not opened or relief valves reset. Indicating that a trip has been activated does not mean it has responded. Systems can also be used to limit flows.

CASE STUDY — EMERGENCY SHUTDOWN SYSTEM

Consider the design of an emergency shutdown system for the exothermic reactor system shown in Figure A1.1. Information on what typically might be necessary and what can go wrong with the process is given in Tables A1.3 and A1.4 (pages 250 and 251). Such information is helpful and forms a small equipment knowledge base. The actual trip system for this plant is shown in Figure A1.2 (page 252). Do you think that it could be improved? In order to help with this task, here is a check-list describing problems with valves:

What happens if the control valve fails in a way which:
- reduces heat transfer, flow, pressure, level?
- increases heat transfer, flow, pressure, level?
- assures adequate flows to prime movers, etc?
- blocks in the equipment item?
- results in a vent to atmosphere, etc?

What happens if:
- the valve sticks in the opposite position to that set for design failure?
- the valve sticks in any other undesirable position?
- the valve fails open, etc?
- several valves fail?

Check for:
- overpressure when control valve closes;
- overpressure of downstream equipment/piping when control valve fails fully open;
- effects on equipment particularly sensitive to control valve failure;
- reactor temperature runaway;
- full-path opening not available for pressure relief;
- uncontrolled loss of material to atmosphere;
- off-specification products and sneak paths;
- spurious failure and its effects.

Examine this list to check for failure or malfunction of valves on the trip system. It should be possible to discover two major problems in the trip system as designed. For example, on the actual plant there was a high level of oxides of carbon in the feed to the methanator. The trip system was activated but the temperature continued to rise. What was the cause and what action should be taken by the operator to prevent the rupture of the reactor? An improved system is given in the chapter on Preliminary Process Hazard Analysis on page 46.

TABLE A1.3
Equipment knowledge base — reactor

Reactor: Gas-phase reactor	
Design intent: Convert reactants a, b, c ... to products z, y, x, for a specified range of temperature, pressure, flow rate, flow ratio and in specific conversions for a particular catalyst	**Undesired events and their causes:** Extra reaction: • impurities in feed, common contaminants • charge of wrong material • incorrect composition or flow ratios • loss of feed
Type of unit: • tubular • catalytic bed or beds • fluidized or fixed bed	• incorrect residence time or distribution • hotspots • high inlet temperature • active catalyst
Ancillary equipment: • heat exchangers, start-up heater • catalyst regeneration system • distributors, packing	• delayed onset of reaction • incorrect equilibrium temperature • autoignition of reactor contents • reactor produces unwanted by-products
Normal control: • temperature control (inlet and outlet) alarm • flow ratio control/alarm • flow control/alarm • pressure control (normally downstream) • pressure high/low alarm • composition analysis (inlet and outlet) • intermediate quench control/alarm	Incorrect temperature: • high inlet temperature • extra reaction takes place • loss or change in heating or cooling High or low pressure: • failure of pressure control • imbalance of flow • lack of demand for product • downstream blockage • off-specification product • relief or vacuum relief failed closed
Emergency control: Emergency shutdown or process abort on high temperature/pressure, incorrect flow ratio, or incorrect outlet composition possibly involving: • cooling, quench, deluge • bypass, depressuring • purging, inert gas, steam • dilution by reactants, products or other • shut-off feed, increase off-take • suppression of reaction, add inhibitor Emergency relief and relief treatment Emergency depressuring and material transfer	Off-specification product: • impurities in feed • less reaction, low inlet temperature • unwanted reaction • inadequate catalyst • incorrect or no catalyst, inactive catalyst • catalyst spent or reactor contaminated Other: • reactor lining fails • coolant or heating medium leaks into reactants

TABLE A1.4
Section from Hazcheck

Change leading to increase in exothermic reaction

i. Set parameters wrong or uncontrolled.
ii. Change in flow, quantities, throughput, batch ratios or sequence.
iii. Change in conversion — selectivity, residence time, local effect, composition or phase, suppressants or initiation, change of mixing/agitation, contamination, internal leak. Excess or deficiency of one reactant. Back flow of reactant.
iv. Reaction outside the reactor.
v. Increase in temperature, preheat, or failure in cooling.
vi. Delayed onset of batch reaction whilst adding reactant.

Operators fail to correct dangerous trend

i. No action taken as incapacitated, work overload, insufficient time, lack of knowledge.
ii. No action taken as unaware, inadequate interface, failure of alarm.
iii. Action taken but ineffective as procedure inadequate, insufficient time, design or equipment failure.
iv. Escalation due to wrong action taken.

Control system fails to correct dangerous trend

i. Reading or indication is invalid.
ii. Malfunction or failure to control system.
iii. System mis-set.
iv. Control system design error.
v. Maximum response to change is limited.
vi. Correction not possible due to external interference.

Emergency control systems fail to correct trend

i. Malfunction or failure.
ii. Spurious failure or activation which increases the danger.
iii. System mis-set or isolated.
iv. Emergency control system design error.
v. Maximum response to change is limited.
vi. Correction not possible due to external interference.

The process operator is a key factor in all features of control. In particular, the operator can note deviation from normal control and effect correction before action is taken by emergency control systems. The operator can activate the emergency control system ahead of sensing elements and can check if this system has operated correctly or averted release or discharge to atmosphere. The objectives set for personnel must be achievable and it may be necessary to improve instructions or information about the system and improve the stimuli and interface for action and

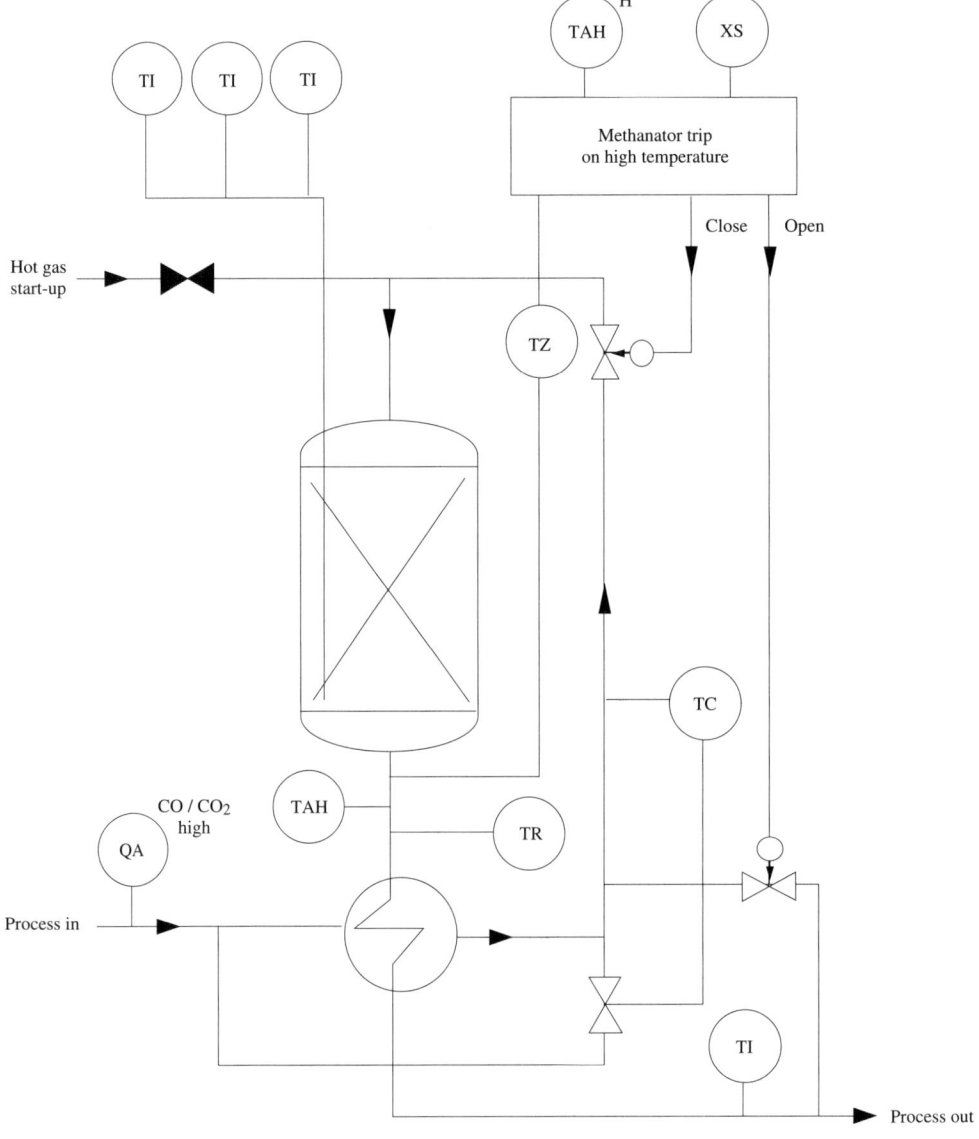

Figure A1.2 Exothermic reactor system with trip system. (See page 243 for instrumentation definitions.)

response. The interface for communication of requirements to the decision-taker can be enhanced. The system should be compatible with the sophistication of the operator. This may mean modification of the design according to where the plant is located. The tasks of the personnel should be studied, together with the means of recovery from human error and equipment failure.

CASE STUDY — A P&I DIAGRAM OF A COMPRESSOR SYSTEM

Develop a P&I diagram for the knock-out drum and positive displacement compressor system shown in Figure A1.3. A separator and a compressor knowledge base are given in Tables A1.5 and A1.6 (pages 254 and 255).

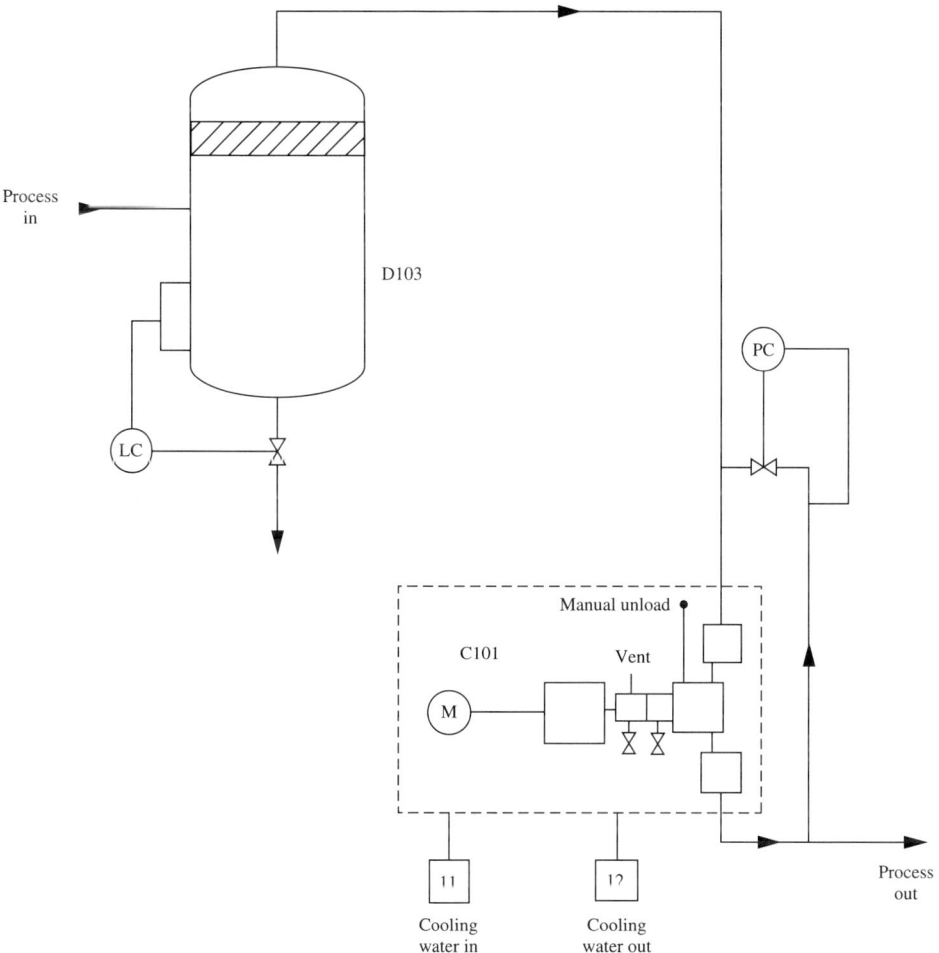

Figure A1.3 An abbreviated diagram of a compressor system.
LC — level control; PC — pressure control; M — motor.

TABLE A1.5
Equipment knowledge base — separator

Separator: Gas-liquid gravity separator	
Design intent: To remove liquid from a two-phase gas-liquid stream for a specified range of temperature, pressure, flow rate and composition of each phase	**Undesired events and their causes:** Extra phase Gas blowby: • loss of level control Liquid blowby:
Type of unit: Knockout pot	• loss of level control • inadequate separating device • overload of liquid output line
Ancillary equipment: Dephlegmator	Inadequate separation Incorrect pressure
Operating mode: Continuous	Incorrect temperature Impurities in feed Failure to remove aerosols
Basic process control system: Level control High/low level alarm Temperature control/alarm on inlet Pressure control/alarm (normally downstream)	
Safety interlock system: High level trip/alarm Low level trip/alarm Pressure relief on system	For leaks, rupture, failures of parts or instruments, etc, use Hazcheck — see Appendix 2

The use of knowledge bases such as those given in Tables A1.3, A1.5 and A1.6 is strongly recommended. They can be used in conjunction with the Hazcheck knowledge base — see Appendix 2. Initially they are of value during Concept Hazard Analysis and Preliminary Process Hazard Analysis in identifying equipment problems. They give an indication of the basic process control system and the safety interlock system which is commonly applied on such a process unit.

The tables also indicate the basic process knowledge required. This can serve as a guide during process design. It can be augmented by appropriate information from a process design manual. The knowledge base can be extended as the plant safety schedule is developed to include equipment specifications and drawings, maintenance requirements and so on.

The knowledge base has been extended in some standards such as API750 to provide an invaluable guide for the safety professional. They remain, however, only guides and the eventual plant must be examined at some time using Hazop.

The knowledge base can be extended to specific items of equipment as these are selected. This should include plant diagrams. Data on the requirements of maintenance and inspection should be added prior to Hazop so that it can be checked.

APPENDIX 1

TABLE A1.6
Equipment knowledge base — compressor

Compressor	
Design intent: Increase pressure of a gas from a specified inlet pressure to a specified outlet pressure for specified ranges of flow, temperature and composition	**Undesired events and their causes:** Excess temperature: • loss of cooling or lubrication • excess recycle flow • compressor valve failure • excess compression ratio • insufficient flow • increase in feed temperature or fire
Type of unit: Centrifugal, reciprocating, various stages Provide spare units on-line Dual units of 60% capacity	Mechanical damage, possibly with explosion: • liquid carry-over • air entry due to vacuum • excessive speed, reverse rotation • loss of lubrication
Ancillary equipment: Intercoolers, knockout pots Alternative motive power	Overpressure: • suction valve fails open • excess recycle flow • blocked discharge
Basic process control systems: High temperature alarm (inlet and outlet) Flow control/alarm Pressure control/alarm Speed control/alarm Manual unloading Pressure alarm (lubrication) Vibration alarm Leak detectors in buildings	• increase in feed pressure • overspeed • failure of pressure control Underpressure: • suction valve restricted or closed • low flow/low feed pressure • underspeed Mechanical failure/problems • vibration giving loosening • stopping
Safety interlock systems: Emergency shutdown or process abort: • high level in knockout pot • low lube oil pressure • low pressure due to leak • high temperature • shut-off feed, increase off-take Safety relief: • discharge • suction	• deterioration of construction or seals • isolation and start-up after maintenance • freezing ambient conditions • solid impurities

TABLE A1.7
Safety analysis table — compressors

Undesirable event	Cause	Detectable condition	Warning	Protection
Overpressure of suction	Failure of upstream pressure control	High pressure	High pressure alarm	Process safety valve
Leak (suction)	Deterioration and rupture	Low pressure	Low pressure alarm	Emergency shutdown system
Leak into building		High gas concentration in building	Composition detector with alarm	
Overpressure on discharge	Blocked discharge or excess back pressure	High pressure	Pressure alarm high on discharge	Process safety valve on discharge unless compressor is incapable of overpressure
Leak on discharge	Deterioration/ rupture	Low pressure and backflow	Pressure alarm low on discharge	Process safety valve and emergency shutdown system
Excess temperature	Compressor valve failure, cooler failure, excess compression ratio, insufficient flow	High temperature	Temperature alarm high	Emergency shutdown system

CASE STUDY ANSWERS

An appropriate answer to the second problem can be found in Chapter 4 within the case study on the methanator. Standards such as API14C give helpful advice on the selection of plant items. Compressor units transfer hydrocarbon gases within the production process into pipelines leaving the platform. Recommended safety devices for a typical compressor unit are shown in Table A1.7. Further information is given in API14C, the standard on the provision and location of pressure safety devices, flow safety valves (check valves), gas detecting devices, temperature safety devices and shutdown devices.

APPENDIX 2 — HAZCHECK LISTING

MODELS OF INCIDENT SCENARIOS

A study of the methodology generally applied in the process industries shows that the key methods of hazard identification and risk assessment are all used to model part of the incident scenario leading to a possible accident. An incident scenario represents an event path or chain leading from the initiating event to consequences, via undesired events in a specific propagation time.

An *undesired event* is defined as an event which is either itself inherently unsafe if somebody or something is exposed directly to it, or part of a chain of events which lead to an inherently unsafe event.

It is convenient to have a list of typical undesired events and their causes. Hazcheck provides this type of information. It is emphasized that the suggestions in Hazcheck merely represent ideas to be used to generate actual events which occur on a particular plant. Hazcheck is intended solely as a general knowledge base and is not intended for general use in hazard review meetings.

The *immediate causes* are perceived as the initiating events of an incident scenario. They relate to the circumstances that immediately precede an undesired event. These can normally be seen or sensed as the direct cause of an incident and can be recognized as taking the form of a fault, error or an unsafe act. Thus the action by plant personnel or equipment may be inadequate or the plant construction may become defective or deteriorated in service, or the plant may be subjected to environmental or external threats.

If the normal control systems do not immediately adjust the change in process conditions effected by any immediate causes, the overall effect is a *process deviation*. Often at this time a process alarm signals that such a deviation has taken place and the operator may be able to correct the situation immediately. If not then the process conditions correspond to those representing a *hazardous deviation*. At this time automatic emergency control systems may act on the system and if there is a combined result of *failure to control the situation* by normal and emergency control a *dangerous disturbance of plant* is created. *Failure to recover the situation* will result in a release of process material by rupture or discharge.

In Hazcheck the process and hazardous deviations are grouped together, as is all failure to recover the situation. This avoids considerable repetition in the text.

A significant release of material is one that might at the least do harm to people, business, plant and the environment. What is significant is left to the good sense of the safety analysts.

The *failure of countermeasures for a release* and the *escalation by energetic and toxic events* will usually give rise to serious *impact on plant, people and the environment*. The extent of this event will be influenced by the nature and quantity of the release and the extent of any escalation by fire and explosion. It will also be affected by the emergency response, both on- and off-site, and the quality of the post-accident response. When such a release has no effects it should be reported as a near miss. The process deviation represents the starting point in the search for a Hazop study. The dangerous deviations are the starting points for a Preliminary Hazard Analysis.

Hazcheck release scenario

IMMEDIATE CAUSES

Keywords	Preconditions for failure
Action by plant personnel inadequate	Error of omission Error of commission Failure of information processing or check Deviation in the flow of information Action based on incorrect/inadequate information Task not completed correctly Inadequate immediate reversal of error Improper and inadvertent actions
Defects directly cause loss of integrity	Incorrect materials of construction Other defect in plant or support
Technical deviation in the equipment, feed streams or services	Deviation in the flow of material Sudden failure causing functional defect Gradual or partial failure Degradation or catastrophic failure Equipment cannot be used when required Design functional deficiencies
Technical deviation in the control system of the plant	Sudden failure causing functional defect Gradual or partial failure Degradation or catastrophic failure Equipment cannot be used when required Incorrect use or abuse by operators Removal of protective systems
Change from design intent	Design functional deficiencies Use for purposes outside design specification Incorrect modification Change in environment or public domain
Environment and external causes	External energetic or toxic events General accidental impact damage Falls and slips External interference causing loosening Adverse ambient conditions Act of God, natural causes Force majeure, sabotage Theft and hooliganism

PROCESS OR HAZARDOUS DEVIATIONS

Keywords	Preconditions for failure
Construction deteriorating in service	Accelerated corrosion or erosion
	Loosening or vibration
	Impingement and impact blows
	Overstress of moving machinery
	Out of tolerance faults
	Excessive thermal expansion or contraction
	External effects as noted above
	Failure of supports or foundations
Abnormal opening in equipment	Device fails causing loss through opening
	Incorrect status of valves, etc
	Disconnecting and maintenance activities
	Planned operating procedure done incorrectly
	Deliberate discharge
Defect left in construction	Defect left on initial construction
	Defect left after maintenance
	Incorrect design
Adverse change in planned product or other release	Change in periodic or fugitive emission
	Change in emergency discharge
	Change in product
	Adverse change after leaving plant
Deviations in flow of material	No flow
	Less flow
	More flow
	Reverse flow
	Less level or load
	More level or load
Change in mode of operation	Planned change in normal operation
	Abnormal operations or maintenance

PROCESS OR HAZARDOUS DEVIATIONS

Keywords	Preconditions for failure
High pressure	Explosion or reaction Connected high pressure source Vaporization of liquid in closed system Pressure surge Blockage or incorrect isolation Internal leak
Low pressure	Connected source of vacuum or draft Reduction in quantity inside closed stem Imbalance in input and output Blockage or incorrect isolation
High temperature	Connected high temperature source Mismatch in heat exchange or cooling Change in mixing conditioner delayed mixing Input of energy from machines Planned internal explosion or exothermic reaction Unexpected explosion or exothermic reaction External fire or explosion Other external heating source Internal leak
Low temperature	Connected low temperature source Vaporization of low-boiling liquids Expansion of gas External source of cooling Internal leak
Overload or stress	Change in external or internal loading Machine overload or malfunction Incorrect or extraneous phase
Change in operating parameters	Change in composition, phase or size Change in time or operation Change in stress Change in physical properties Accumulations, contaminants or blockages

APPENDIX 2

FAILURE TO CONTROL THE SITUATION

Keywords	Preconditions for failure
Emergency control systems fail to correct the situation	Emergency release can cause hazard Emergency control promotes dangerous or hazardous disturbance Misuse of emergency control system Emergency control inadequate in duty Emergency control systems fail on demand Emergency control system failed or isolated Emergency control not provided or installed
Operators fail to correct the situation	Inadequate monitoring or alertness Failure of alarm Actions of operators cause or increase hazard Operator cannot recover the situation Operator action inadequate Corrective action not taken by operator No action possible by operator
Normal control fails to correct the situation	Hazard on change in planned release or product Normal control promotes dangerous or hazardous disturbance Misuse of normal control system Normal control inadequate in duty Normal control system failed or isolated Reading or indication invalid Normal control not provided or installed
Maintenance fail to correct the situation	Maintenance causes hazardous disturbance Maintenance response or action inadequate Maintenance action not taken Inadequate monitoring of situation Maintenance unable to take action

DANGEROUS DISTURBANCES

Keywords	Preconditions for failure
Rupture on exceeding mechanical design limitations	Overpressure Underpressure, usually vacuum Overtemperature Undertemperature, usually below 0°C Overload or stress
Rupture within normal operating conditions	Critical defect left after construction Critical deterioration
Flow through abnormal opening	Opening left in plant Opening made in plant
Adverse change in planned product or other release	Change in planned product Change in planned emission or effluent Change in emergency discharge
Entry into vessel	Planned entry when hazard present Planned entry when hazard added Unplanned entry such as fall

SIGNIFICANT RELEASE OF PROCESS MATERIAL

Keywords	Preconditions for failure
Release of process material	Rupture on mechanical design limits exceeded
	Rupture due to defective or deteriorated construction
	Discharge through available opening
	Adverse change in a planned release
Failure to recover situation	Failure of operator to attenuate release
	Failure of automatic controls to attenuate release
	Significant inventory and dispersive capability within system

FAILURE OF COUNTERMEASURES TO RELEASE

Keywords	Preconditions for failure
Immediate response inadequate	Inadequate detection and warning
	Inadequate response of people caught in release, fire, etc
	Inadequate activation of response
Response by countermeasures inadequate	Countermeasures disabled, isolated or not installed
	Countermeasures inadequate or failed
	Primary explosion or fire
	Maloperation of countermeasures
	Countermeasures cause more danger
Release fails to disperse	Planned discharge fails to disperse
	Change on ignition
	Spread by liquid run-off, waterways, etc
	Accumulation or deposition after release
	Entry into food chain
Failure to avoid ignition of flammables	Release of flammables
	Significant flammable mix
	Flammables ignited before release or self-ignite
	Source of ignition
Ignition of flammable mixture	Ignition after delay causing vapour cloud explosion or flash fire
	Ignition giving torch or pool fire
Failure to avoid toxic release	Release of toxic material
	Emergency discharge treatment fails

ESCALATION OF RELEASE

Keywords	Preconditions for failure
Escalation by fire	Release ignites
	Flash fire
	Jet fire
	Pool fire
	Boilover
	Electrical discharge
	Further release of material following fire
	Fire-fighting ineffective
	Further spread of fire
Escalation by explosion	Physical explosion
	Chemical explosion
	BLEVE
	Dust explosion
	Electrical explosion
	Vapour cloud explosion
	Escalation by explosion and missiles
	Secondary explosion, dust explosion
Escalation by toxic event	Release is in toxic concentration
	Accumulation after release
	Accumulation in biota
	Change in material after release
	Secondary loss of toxic material
	Toxic effects from materials used to fight fire
Emergency response inadequate	Inadequate on-site response
	Inadequate off-site emergency response
	Inadequate protection of personnel
	Inadequate secondary protection of process plant
	Inadequate protection of environment
	Contamination due to fighting emergency
Post-incident response inadequate	Inadequate health control
	Failure to determine immediate response
	Failure to remove continued hazard
	Failure to isolate affected area
	Inadequate general clean-up of affected area
	Inadequate restoration of business activities
	Inadequate counselling of victims
	Inadequate collection of data
	Inadequate reporting of incidents
	Inadequate action on incidents
	Inadequate response to insurance enquiries

IMPACT OF RELEASE	
Keywords	**Preconditions for failure**
Catastrophic consequences	Catastrophic damage to plant and severe clean-up
	Loss of normal occupancy on and off site
	Three or more fatalities on plant
	Fatality of member of public or five public injuries
	Damage to site of special scientific interest or historic building
	Severe long-term damage to significant area of environment
	Catastrophic effect on business with national pressure to shut down
Severe consequences	Severe damage to plant and major clean-up
	Loss of normal occupancy on site
	Single fatality on plant
	One in ten chance of fatality of member of public
	Five plant injuries
	Short-term damage to significant area of environment
	Severe effect on business with media reaction
Major consequences	Major damage and minor clean-up
	No loss of building occupancy
	Some hospitalization of public
	One in ten chance of fatality on plant
	Five plant personnel injured
	Short-term damage in section of environment
	Minor effect on business with considerable media reaction
Appreciable consequences	Appreciable damage to plant
	Minor annoyance of public
	Injury to plant personnel
	No damage to environment
	Some loss of production or quality
	No effect on business
	Reportable as near miss under CIMAH Regulations
Minor consequences	Near-miss incident
	Minor annoyance to plant personnel
	Minor production or quality lapse or shutdown

APPENDIX 3 — A HAZOP STUDY

This Hazop study was carried out by David Rochford as part of his dissertation for the degree of MSc (Eng) in Process Safety and Loss Prevention at the University of Sheffield. It was deliberately made broader than a conventional study and includes some unusual keywords. It demonstrates the power of the method and contains a number of innovations.

The system under study (Figure A3.1, pages 268–269) is part of an inland oil-producing site and requires a disposal route for produced water and a source of injected water for enhanced oil recovery. Both requirements are satisfied by the re-injection of produced water after suitable treatment. Oily water from miscellaneous sources is dosed with oxygen scavenger and commingled with produced water which has been degassed after separation in a gathering centre. This stream is then passed to the storage tanks, T710 and T720, which are nominally at atmospheric pressure. This material is purged with sweet fuel gas to exclude oxygen. The tanks act as surge and header vessels to the two existing positive displacement pumps, P715A/B. These pumps are fed via a common manifold. They supply injection water to various well sites. The section reported here was concerned with the review of the storage tank T710 at the main gathering centre and a suction line to the associated water injection pumps P715A/B and P716.

PROPOSED SYSTEM

At the gathering centre, the existing 4 inch suction pipework from the two surge tanks is to be replaced with larger 8 inch piping in order to remove the tendency to cavitation which is a feature of the existing system. The suction and discharge manifolds will be replaced and extended to serve the new reciprocating pump, P716, which will be in parallel with the existing pumps. The existing pumps are to be upgraded.

The overall operating philosophy appears to be to keep the injection pumps running. If a shutdown or mechanical problem occurs, the operators should review the reason for the shutdown and restart if appropriate. Note the limited use of protection devices. There are, for example, no low or high pressure alarms/switches in the injection pump suction or discharge lines. In the case of a high discharge pressure upset, reliance is placed entirely on the discharge pressure safety valves which return high pressure water to the produced water storage tanks, T710 and T720. The return manifold could suffer from various line-up and valve closure errors, and the scope for human error is considerable.

Documentation from the study appears on the following pages. Figures A3.2 and A3.3 (pages 270–273) show P&I diagrams both before and after Hazop. Not all the issues were resolved because the study was undertaken by a single team member. In a normal study the other team members contribute their experience and help eliminate specific issues which are not significant. More causes would also be raised with a larger group. The response to any action in a Hazop study should reflect the concern for safety, design and operability raised by that action. However the purpose of the Hazop is to raise concerns rather than to determine the best means of resolving them, or decide which of them should be considered and adopted.

HAZARD IDENTIFICATION AND RISK ASSESSMENT

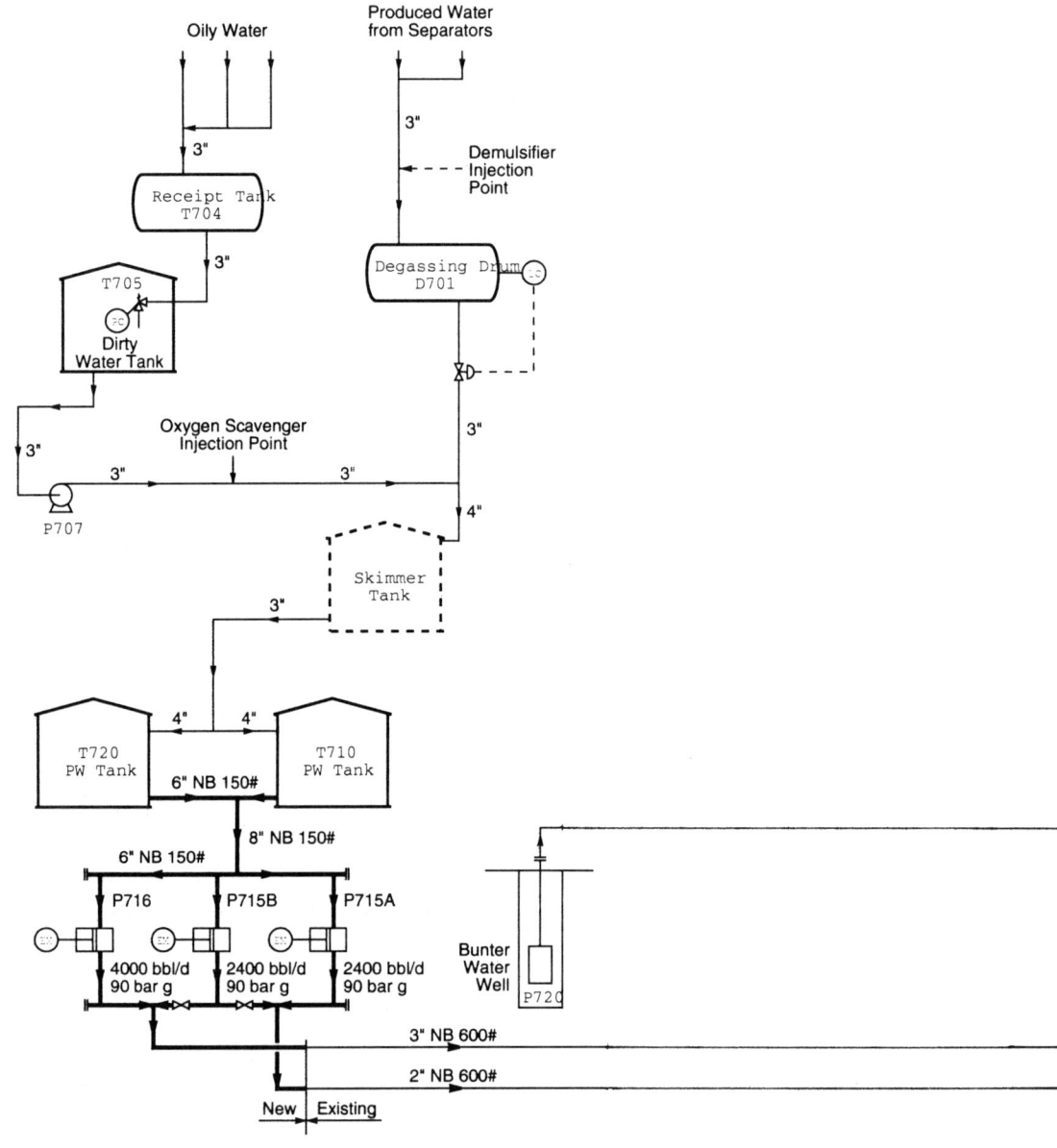

Figure A3.1 Water injection upgrade — the overall picture.

APPENDIX 3

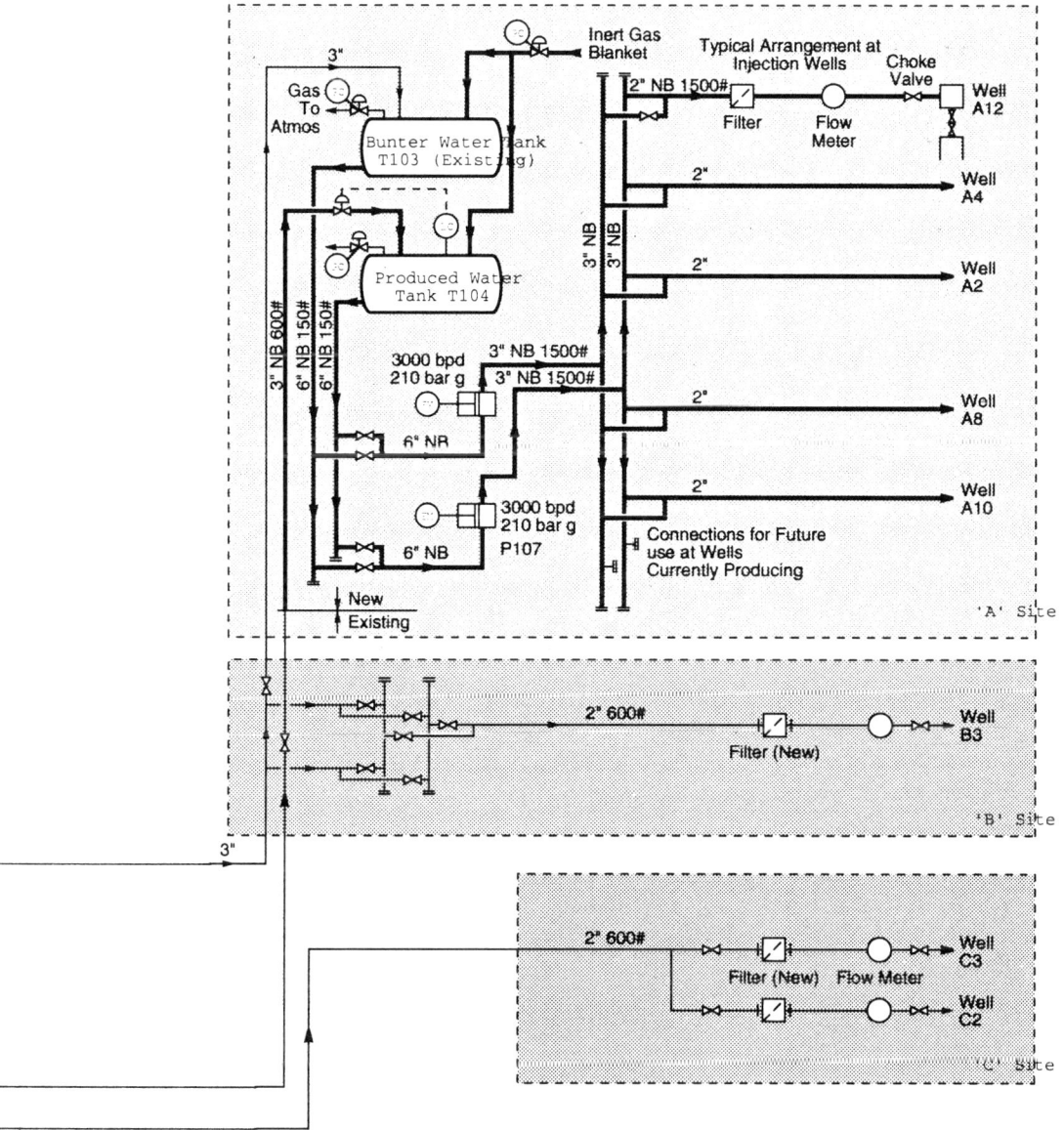

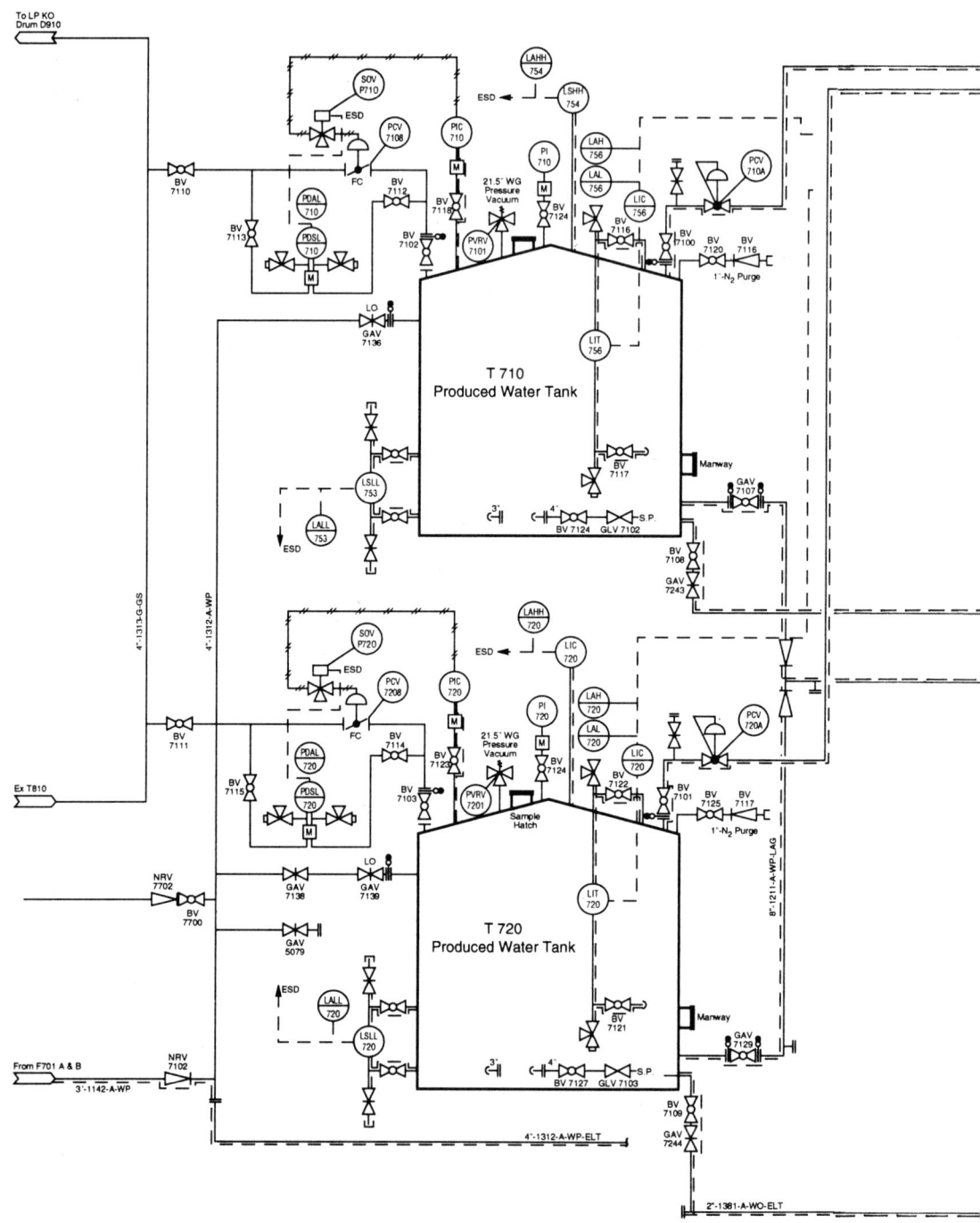

Figure A3.2 Water injection upgrade P&I diagram — produced water storage and injection before Hazop.

APPENDIX 3

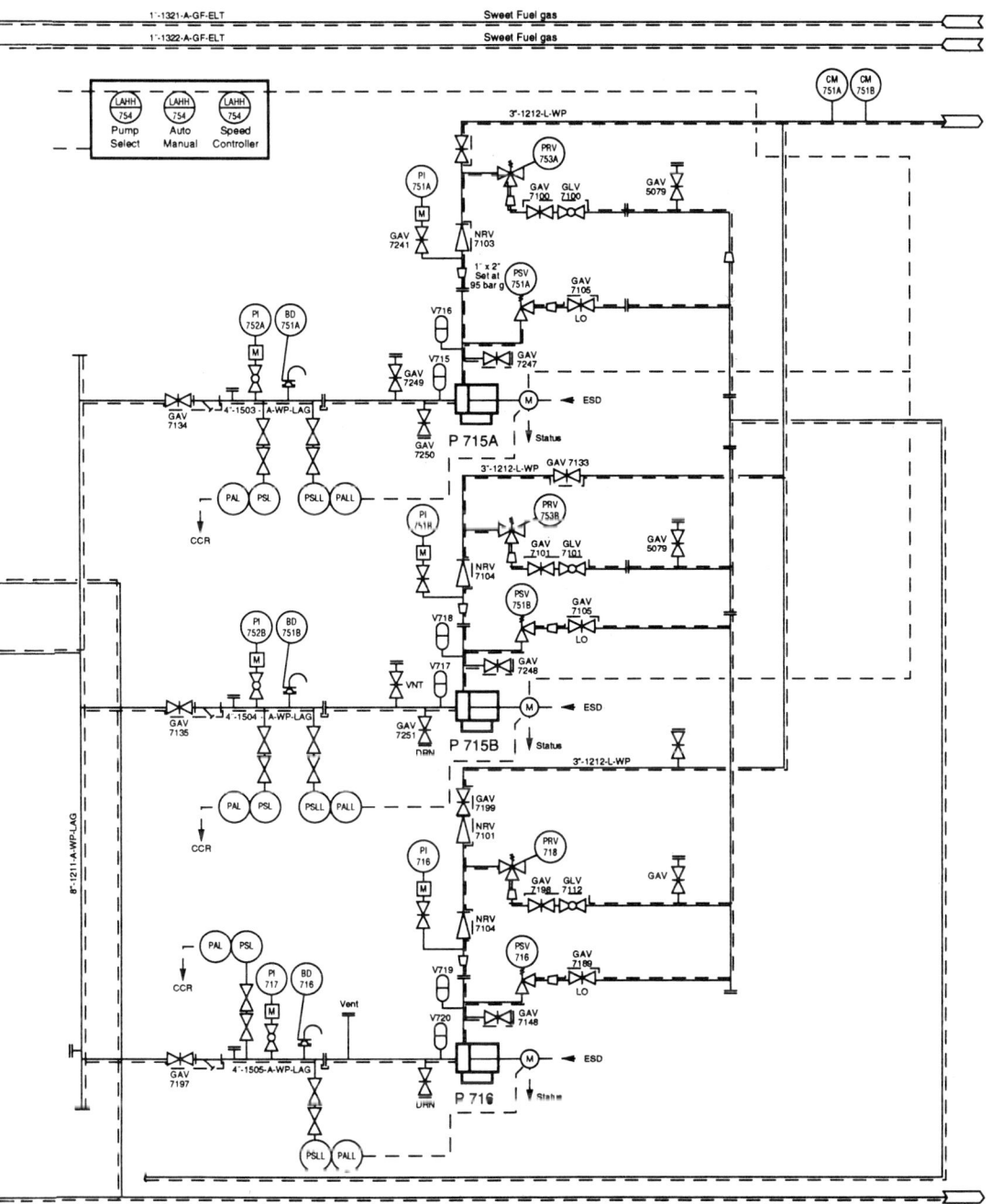

HAZARD IDENTIFICATION AND RISK ASSESSMENT

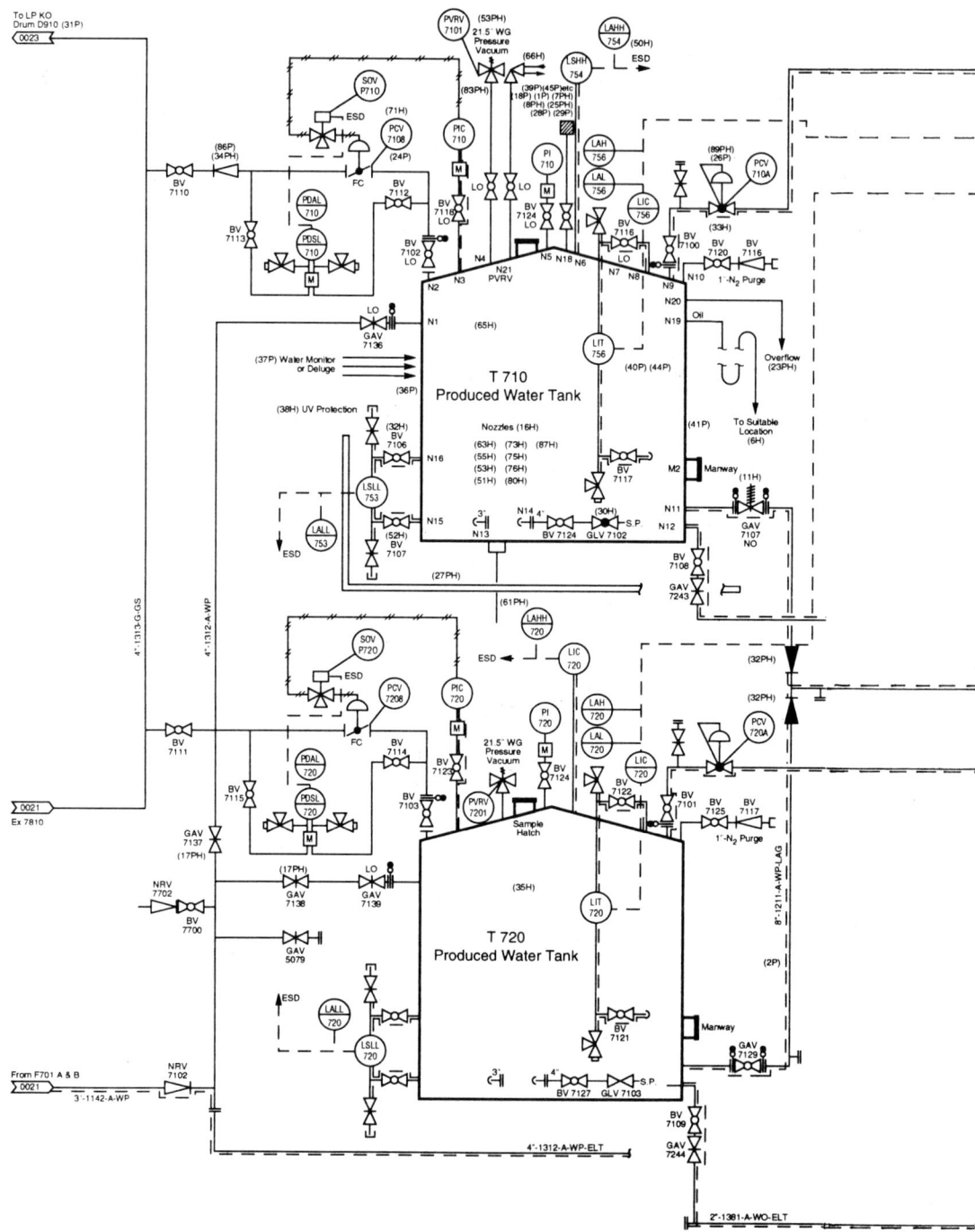

Figure A3.3 Water injection upgrade P&I diagram — produced water storage and injection after Hazop.

APPENDIX 3

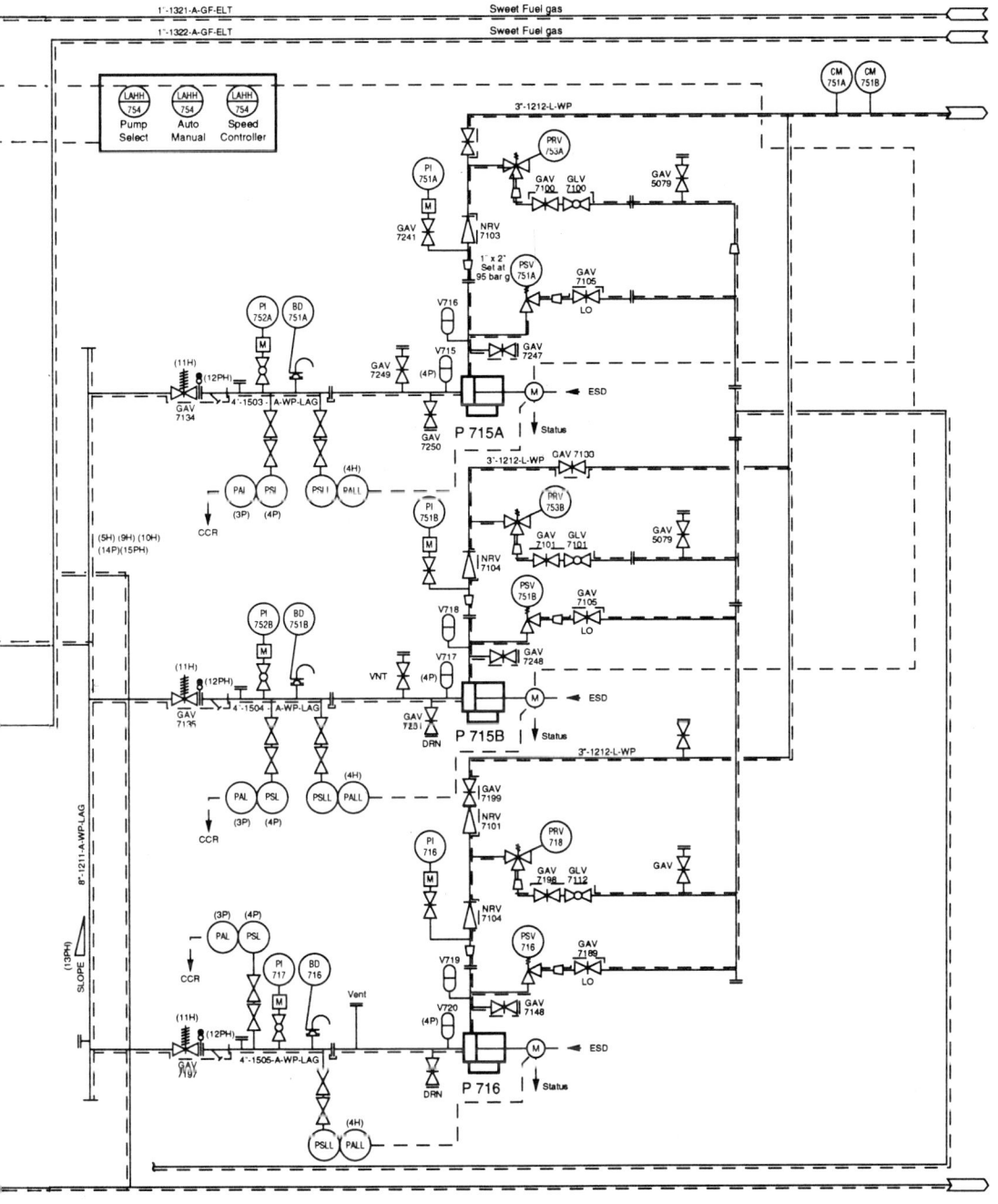

273

HAZARD IDENTIFICATION AND RISK ASSESSMENT

Hazop study action report form: water injection upgrade, Sheet no. 1

Line section: Produced water tank T710 connection and outlet line 8"–1211–A–WP–LAG to suction valves of pumps P715A/B, P716. Tank T720 considered to be isolated at valve GAV7129 and out of service.

Design intention: Delivery of produced water to pump P715A/B, P716 from tanks T710 and T720.

PARAMETER	DEVIATION	POSSIBLE CAUSES	
Flow	None	Line rupture	
Temperature	Lower	Dead leg at T720 suffers from cold winter conditions	
Level	Lower	Original design feature	
Composition	Changes	Reservoir need may change requiring the addition of additives to produced water (PW)	
Composition	Loss of circulation	Oil separates from water as residence time in T710 is long	
Cracking	Brittle fracture	Cold water conditions and vibration from pumps	
Impact	Dropped objects	Maintenance with heavy plant and limited access	
Reliability	Availability	Lack of inspection through, for example, corrosion	
Routine	Slips	Boring repetitive inspection methods	
Checking	Line of sight	Either all isolation valves closed or partially closed	
Shutdown	Normal	Lack of isolation of all pumps P715A/B, P716 and no spectacle blinds downstream of suction isolation valves	
Shutdown	Orientation	1" blinds potentially suitable for draining purposes if located at base of pipe: also manifold end blinds suitable	

APPENDIX 3

CONSEQUENCES	ACTION REQUIRED	PROJ	OPS
Potential to pull vacuum on T710	1. To confirm that vacuum relief capacity is adequate to compensate for maximum outflow rate.	*	
Possible freezing of pipe adjacent to T720	2. Suggest consideration of steam tracing.	*	
LALL/LSLL 753 set too low to provide adequate net positive suction head (NPSH) especially for P716	3. Ensure NPSH requirements are met. 4. Include accelerative NPSH requirements in review. Consider installation of pressure switch (PS) or pressure switch level indicator (PSLI) in suction.	*	
Potential corrosion of tank (base, sidewalk) and piping	5. Suggest a chemical compatibility test be conducted and discussion with reservoir engineers.	*	*
Possible oil slug into pump and more importantly to P715A/B, P716 resulting in injectivity loss at well bore	6. To ensure PW treatment functions adequately. Consider means of removing oil layer from T710.		*
Flange leaks and/or pipe failure	7. See action 1.	*	*
Pipe damage/rupture	8. See action 1.	*	*
Water injection shutdown required and loss reservoir pressure support	9. Ensure maintenance routines include pipe inspection.		*
Trends in pipe thickness go unnoticed	10. Consider multi-tasking of workforce and further checks.		*
Lack of flow, loss of suction, two phase flow, etc	11. Consider the use of rising stem gate valves	*	
Dangerous working practice unless complete shutdown acceptable	12. Consider installation of spectacle blind downstream of the three suction valves.	*	*
No means of draining down pipe section for maintenance	13. Review need for draining. Does a slope exist between tank connection to pumps? Consider valve and end cap at manifold blinds.	*	*

HAZARD IDENTIFICATION AND RISK ASSESSMENT

Hazop study action report form: water injection upgrade, Sheet no. 2

Line section: Produced water storage tank T710.
Design intention: Receives PW from storage surge facility and provides suction to P715A/B and P716.

PARAMETER	DEVIATION	POSSIBLE CAUSES	
Simultaneous operations	Too many	All three pumps on line and maldistribution	
Emergency recovery	Reaction time	Varying size of pipe hole would affect the rate of water leakage from the pipe	
'Start study'	Cannot	No nozzle references on T710	
Flow	No	Isolation valve GAV1737 closed in error	
	No	No nozzle available to purge out vapour space to exclude oxygen	
	No	Ball valves associated with N2 and N3 closed in error	
	No	Ball valves associated with N5, N7 and N8 closed in error	
	No	Valve GAV7107 closed in error	
	No	Valves BV7108, GAV7243 open in error (and BV7124 on nozzle N14)	
Flow	More	Inflow above outflow when P715A/B, P716 out of service	
	More	Pressure letdown valve PCV710B oversized	
	More	Pressure vacuum valve on V4 leaks in service. Occurs on pumpout or rise in barometric pressure	

APPENDIX 3

CONSEQUENCES	ACTION REQUIRED	PROJ	OPS
Individual pumps starved of suction pressure, vibration, etc	14. Review hydraulic calculation to establish flow capacity in worst operating scenario — see action point 11 for consequences.	*	
Varying rates of response needed	15. Review alarms, training of operators and available time to react to major damage.	*	*
Cannot reference action items to nozzles only lines	16. Nozzle references added to allow Hazop to proceed.	**	
No flow to T710 via nozzle N1	17. Consider interlocking GAV7137 and GAV7138 to provide a route for separated PW (review again while T720 is Hazoped).	*	*
Possibility of combustible gas mix and potential misuse of PSV on N4	18. Suggest the provision of manual vent for purging into service.	*	
	19. Consider the need to lock open or designate BV7102, 7110, 7118 and 7112/3 normally open.	*	
Tank pressure and level indication not indicated. LIT758 gives wrong reading as uses differential pressure.	20. Suggest associated valves BV7119 and BV7116 be locked open.		*
Tank level rises and P715A/B on P716 is starved of fluid	21. Suggest valve is designated normally open as quick response in an emergency to close may be required.	*	*
Tank T710 inadvertently drained	22. Suggest BV7108, BV7124 handle design should be fixed in place and, see action 11, rising stem gate valves used.		*
Tank level rises and if no action taken hydraulic pressure damages tank	23. Project to advise if adequate protection. Operations to advise on availability.	*	*
PCV710B continually opening and closing and tank pressure below normal	24. Confirm design philosophy and valve fit for purpose.	*	
T710 pressure low, positive pressure lost and air ingress possible	25. Confirm leak potential. Contact manufacturer with request to supply user list.	*	*

277

HAZARD IDENTIFICATION AND RISK ASSESSMENT

Hazop study action report form: water injection upgrade, Sheet no. 3

Line section: Produced water storage tank T710.
Design intention: Receives PW from storage surge facility and provides suction to P715A/B and P716.

PARAMETER	DEVIATION	POSSIBLE CAUSES	
Flow	More	Sweet fuel gas blanket PCV107A oversized, downstream control valve failed or set high	
	More	Tank leaks near base	
	Less	Undersized pressure vacuum valve on N4	
	Less	Undersized PCV710A sweet gas pressure control valve	
	Less	N14 sample point uses global valve	
	Reverse	Gas pressure higher at LP knock-out drum D910	
	Reverse	Tank T720 in operation with a high level T710 at a lower level	
	Reverse	Low tank pressure does not pressure lock BV7120 or NRV7116	
	Reverse	Gas flows back to sweet fuel gas blanket source when source of sweet gas ceases	
	Reverse	Reverse flow to N12 drain line if T720 is full	
Temperature	Higher	High feed temperature from upstream and solar radiation	
	Higher	Possible fire inside T710 from ignition of oil layer	
	Higher	Mechanical failure of shell fittings/instruments — for example, LSLL753 in a fire situation	
	Higher	Failure of pressure vacuum valve on N4	

278

CONSEQUENCES	ACTION REQUIRED	PROJ	OPS
Tank pressure rises (note PCVT108 is a fail closed valve)	26. Confirm sizing basis for PCV and establish upstream impact of excess gas returning to LP KO drum D910.	*	
Site waterlogged, oil contaminated and limited possibility of gas escape	27. Recommend constructing a bund.	*	*
Overpressure or underpressure damage to tank T710	28. Confirm design philosophy and have calculation independently audited.	*	
Partial vacuum in T710	29. See action 24.	*	
At tank level low there may be insufficient pressure to flow	30. Replace with a second ball valve.		*
Gas flows back to T710 and pressure differential across PDSL710 is reversed	31. Review downstream for protection against backflow — for example, check valve, emergency shutdown valve or trip.	*	
Produced water from T720 flows to T710 and also flows to pumps	32. Review if both tanks can be operated together and consider the use of a check valve in suction line to pumps.	*	*
Gas leaks to atmosphere via nozzle N10	33. Ensure BV7120 closed in normal operation and that handle cannot be dropped. Consider installing plug upstream of NRV7116.		*
It is possible that gas returned to system shutdown or under maintenance	34. Review the need for a check valve in line 1"–1321–A–GF–ELT.	*	*
T710 does not drain	35. Consider the operation of adjacent tank T720 when generating the drain down procedure.		*
Localized internal stresses	36. Check tank mechanical design.	*	
Tank shell and roof overstressed with loss of mechanical integrity	37. Consider a limited roof deluge or monitor (flowing to shell) facility.	*	
Possible loss of fluid from tank	38. Ensure instruments and fittings are fireproof.	*	
T710 suffers from over or underpressure	39. Suggest alarm on pressure vacuum valve for a fire condition.	*	

Hazop study action report form: water injection upgrade, Sheet no. 4

Line section: Produced water storage tank T710.
Design intention: Receives PW from storage surge facility and provides suction to P715A/B and P716.

PARAMETER	DEVIATION	POSSIBLE CAUSES	
Temperature	Higher	Mechanical failure of shell fittings/instruments — for example, LSLL753 in a fire situation	
	Higher	Failure of pressure vacuum valve on N4	
	Lower	Excessive cold if vessel is empty or on roof space during winter	
	Lower	Freezing of tank instrument legs and penetrations in general	
	Lower	Tank contents freeze if levels are low	
Pressure	Higher	External adjacent or local pool fire	
	Higher	PCV7108 fails closed in service	
	Lower	PVRV7101 seat freezes in cold weather	
Level	Higher	LAH756 and LSHH754 fail in a demand situation	
Composition	Impurities	Oil component separates into heavier layer and gas boils off	

APPENDIX 3

CONSEQUENCES	ACTION REQUIRED	PROJ	OPS
Possible loss of fluid from tank	40. Ensure instruments and fittings are fireproof.	*	
T710 suffers from over or underpressure	41. Suggest alarm on pressure vacuum valve for a fire condition.	*	
T710 liable to brittle fracture	42. Suggest check by independent audit that Charpy test values are adequate at minus 10°C.	*	
Loss of control and trip systems	43. Consider electric trace heating and insulation, preferably non-hygroscopic and fireproof.	*	
Possible tank damage and blockage of N11	44. Determine lowest freezing point and check design. Consider action as per action point 42.	*	
T710 pressure vent required in a demand situation	45. Ensure pressure/vacuum relief valve sized for fire condition. 46. Ensure pressure side sufficiently reliable.	*	
Total reliance on PVRV7101 to relieve high tank pressure	47. See action 45.	*	
Tank suffers pressure damage (vacuum) as level falls	48. Suggest valve selected can without high differential pressure break the effect of ice formation on seat. 49. Suggest consideration of two emergency vent designs to avoid common mode failure scenarios.	*	* *
Tank suffers from hydraulic damage/over stressing and possible loss of containment	50. Consider the use of a tank overflow — see action 6. 51. Determine requirement for trip reliability and compare with specification. 52. Consider frequency of operator checks and evaluate effect of improved vigilance.	*	*
Heavy materials settle forming residue and anaerobic conditions	53. Possible corrosion. Consider tank lining. If lining used watch for coating over shell base preparations.	*	*

Hazop study action report form: water injection upgrade, Sheet no. 5

Line section: Produced water storage tank T710.
Design intention: Receives PW from storage surge facility and provides suction to P715A/B and P716.

PARAMETER	DEVIATION	POSSIBLE CAUSES	
Composition	Impurities	Corrosion products or sludge block up shell and base penetrations	
	Impurities	Dirty gas carrying over hydrocarbons which stick on PVRV7101	
	Lost circulation	Oil settling on tank base	
Corrosion	Pitting	Sludge layer on tank base	
Radiation	Scale	LSA scale from reservoir water	
Foaming	Formation	Injection chemical inappropriate for service	
Impact	External	No protection from heavy plant	
	Internal	Dropped objects during construction	
Elevation	Lower	Tank base is flat with no slope from centre to shell	
Velocity	Lower	Size of PVRV not stated at N4	
Conductivity	Static	Lightning strike	
Sooner or later		PRVR7101 will fail in service	
Reliability	Less than adequate	Insufficient PVRV7101, LSLL753 and LSHH754	
Integrity	Civil	Tank pressure variations cause tank base to move	
Maintenance	Isolation	No isolation valves upstream of PVRV7101 and PCV710A	

APPENDIX 3

CONSEQUENCES	ACTION REQUIRED	PROJ	OPS
Loss of LSLL753 trip facility as BV7101 could block in service. Also blockage of N17.	54. Review effect on trip reliability. 55. Consider on-line maintenance to clear blockages.	*	*
PVRV7101 does not seal and gas lost to atmosphere	56. Consider maintenance of PVRV7101 and if achievable with single PVRV7101.	*	*
Carry-over to P715A/B and P716	57. Consider modified design of pumps and maintenance needs.	*	*
Possible pitting and general corrosion	58. See action 53.	*	*
Special precautions and not possible to handle/maintain contaminated equipment	59. Seek expert advice and test scale for radioactivity.	*	*
Possible stable oil foam production and separation of oil from water	60. Review T710 design philosophy to determine chemical needs.	*	
Possible T710 shell/base damage on impact	61. Consider installing a bund wall.	*	
Possible tank base damage	62. Review inspection requirements with final base check.	*	
Tank cannot be drained for certification, maintenance or repair	63. Consider installing a base with an adequate slope.	*	
High suction velocity causes maloperation	64. See action point 26 — PVRV7101 audit request.	*	
Damage to T710	65. Ensure construction provides for code requirements: earthing bosses and earth spike ohm checks.	*	*
Protection device not available and no other devices in service	66. Suggest use of multiple PVRV due to reliability issues and possible fire condition.	*	
Loss of T710 integrity or pump damage	67. Suggest reliability analysis be carried out in conjunction with a preventative maintenance review.	*	*
Variable load on foundations worst at high tank pressure	68. Review need for holding down bolts and effect on foundations.	*	
PVRV and PCV cannot be maintained on line	69. Strongly suggest that the use of isolation valves is considered (double isolation preferred).	*	*

Hazop study action report form: water injection upgrade, Sheet no. 6

Line section: Produced water storage tank T710.
Design intention: Receives PW from storage surge facility and provides suction to P715A/B and P716.

PARAMETER	DEVIATION	POSSIBLE CAUSES	
Maintenance	Isolation	Single valve isolation only on outlet line, N11	
Routine	Unforgiving design	Location of PIC710 on top of tank roof	
Labelling	Less than adequate	Sweet gas feed via PCV710A confused with primary pressure relief PCV7NB	
	Less than adequate	PVRV7101 not clearly marked	
	Less than adequate	LSLL753 located on tank shell	
Checking	Remote	PDSL710 on pressure let down valve PCV710B fails to operate	
	Less than adequate	Nameplate missing for T107	
Right equipment	Wrong object	Operator mistakes tank T720 for T710	
Time	More	Sludge settles onto tank base	
Start-up	Normal	Nozzle N10 purge inlet close to gas blanket valve	
	Normal	Purge gas has molecular weight similar to air	
Sampling	Less than adequate	Sample of air, gas, inert inadequate at start-up especially near centre/base of tank	
Utility services	Failure	SOVP710 fails to initiate PCV710B as required if air supply fails	

APPENDIX 3

CONSEQUENCES	ACTION REQUIRED	PROJ	OPS
Injection pumps require shutdown and suction lines drained before downstream spectacle blind swung	70. Strongly suggest that the use of isolation valves is considered (double isolation preferred).	*	*
Pressure indication not checked	71. Consider routing to control room and varying frequency of tank top pressure readings.		*
Action on wrong item possible with loss of sweet gas purge	72. Consider installing PCV710A at opposite side from 710B to provide maximum sweep from feed.	*	*
Unaware of importance of this critical safety device	73. Paint device red.		*
Unaware of device's importance and liable to impact damage	74. Paint device red. Consider installing a bund wall.	*	*
Over-reliance on differential pressure switch thus PCV710B not often checked	75. Suggest facility to perform local trip testing on PCV710B valve function via PDSL710 and SOVP710.		*
Operators may make error whilst logging data	76. Locate name plate and other important details adjacent to stairway.		*
Wrong details and routine checks inadequate	77. Suggest large TAG reference close to stairway light for clear identification.		*
Possible blocking of nozzle N15 on LSL753 and globe valve GLV7102	78. See action 53. 79. See action 54.	* 	 *
Purge gas does not sweep tank and is short-circuited by current design	80. Install N10 on opposite side to N2.	*	
Purge gas does not sink and tank T710 is not adequately purged	81. See action 76. 82. Consider installing N10 close to tank base or using sample point at N14; note no NRV on N14.	*	
Gas mixture within tank may be within explosive limit	83. Use gas measurement device with adequate length on sample/test line.		*
PCV710B slow or does not react and potential for reverse flow occurs to tank T710	84. Consider installation of NRV in line 4"–1313–G–GS.	*	

285

Hazop study action report form: water injection upgrade, Sheet no. 7

Line section: Produced water storage tank T710.
Design intention: Receives PW from storage surge facility and provides suction to P715A/B and P716.

PARAMETER	DEVIATION	POSSIBLE CAUSES	
Drawing	Omission	Tank dimensions not shown	
	Less than adequate	21.5" WG may be PVRV setting or T710 design pressure or PIC710 set point	
Shutdown	Emergency	Gas blanket vent valve PCV710B FC fails closed	
	Emergency	PVRV7101 valve and nozzle N4 too small	
	Emergency	Either side of PRRV7101 could be blocked as vent is exposed to atmosphere	
Emergency recovery	Less than adequate	Site emergency, fire and/or explosion on T710 or adjacent vessels	

APPENDIX 3

	CONSEQUENCES	ACTION REQUIRED	PROJ	OPS
	Tank details unknown without reference to further documents	85. Suggest tank diameter, height and design pressure are shown on drawing.	*	
	General confusion	86. Clarification required, see action 85.	*	*
	Over-reliance on PVRV7101 gas inventory or vaporized liquid discharged to atmosphere	87. Confirm shutdown philosophy particularly with regard to sub-systems connected to T710 and atmospheric emissions.	*	
	T710 is overpressurized and tank potentially suffers mechanical damage	88. Size of PVRV to be detailed as well as nozzle N4's nominal bore diameter.	*	
	Reduced flow capacity and potential over or underpressure extremes which could damage tank	89. Consider fitting wire mesh or equivalent over suction/discharge sides: ensure mesh cannot freeze or retain rain or moisture in all situations.	*	*
	Unknown unless emergency scenarios reviewed and possible outcomes established	90. Suggest site emergency procedures manual be produced and reviewed.	*	*

APPENDIX 4 — FURTHER STUDIES

HAZARDS

It is essential to understand the relevant safety regulations and acts affecting your tasks. Here is a way of doing it:

(1) Identify the major safety legislation applicable in your country and compare it with legislation in Europe or the United States as appropriate.

(2) Various countries have regulations related to the control of substances hazardous to health. Obtain details of two of these and compare them. What is the feature they all have in common and what does this suggest as a means of reducing hazards?

(3) The identification and control of major hazard sites is an important regulatory measure. Identify the different regulations which apply in three different countries.

(4) Control is the set of measures by which exposure to substances hazardous to health are reduced to, and maintained at, an acceptable level. This can be achieved in the workplace by appropriate process selection and engineering. It also requires personal action and appropriate management. Identify how this is achieved, noting the overlap between activities.

(5) Identify the hazards faced by an individual driving down a motorway in good weather. Note that the width of the hard shoulder is an important criterion affecting hazards.

HAZARD IDENTIFICATION

It is important to understand the terminology which abounds in this area:

(6) Compare the symbols used in this book for piping and instrumentation (P&I) diagrams with those used in your company. They will be different. The most widely-used terminology is American. Draw up an appropriate list.

(7) The terminology used in this text for Concept Hazard Analysis (CHA) and Preliminary Process Hazard Analysis (PPHA) differs from that in common use. This is because other terminology is considered confusing. Identify this other terminology and see how it compares with that used here.

(8) Use a CHA to identify any problems associated with the following proposed installation. Only the propane spheres need be considered in detail.

A 1200 m^3 sphere has been designed to store propane under pressure. It is to be located along with other materials as shown in Figure A4.1 (not to scale).

The main storage areas exists and is located south of a major road in a 145 m wide strip adjacent to a boundary fence with a motorway. The key units in the tank farm are to be as follows:

APPENDIX 4

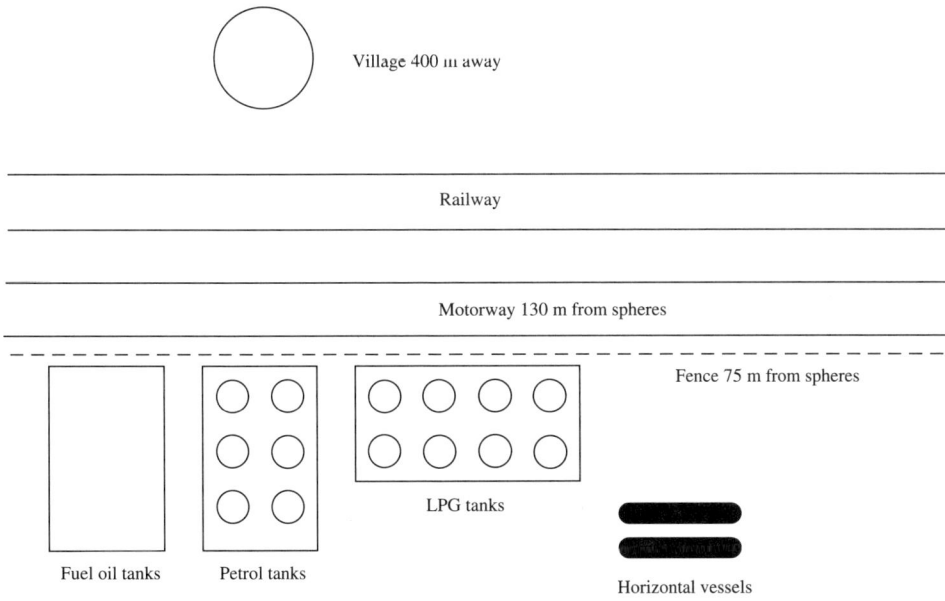

Figure A4.1 Approximate general site layout. (Adapted from IChemE video, *The Safe Handling of LPG*.)

- four spherical pressure vessels containing propane (1200 m^3);
- four spherical pressure vessels containing butane (2000 m^3);
- two existing horizontal pressure vessels used for propane or butane (150 m^3);
- ten existing floating roof tanks used for petrol and kerosene (2500 and 6500 m^3).

The location is about 450 m away from the nearest refinery unit and about 400 m from the nearest village. It is proposed to enclose the eight new spheres inside a 115 m by 55 m bund having a wall 0.5 m high with a central dividing wall 0.25 m high. Each square half bund area contains 2 propane and 2 butane spheres. The spacing between spheres will vary from 11.3 m to 17.2 m. The bund floor is to slope down to catch pits located under the centre of each sphere.

On top of each sphere is to be a three-way valve beneath two identical relief valves so one is always in service and one is isolated. Each sphere will have fixed water sprays both at top and mid height. In addition, a single spray will be directed towards the bottom connections. All the spheres will have fire-proof steel structures.

Samples of LPG are to be taken from each of the spheres on a routine basis for analysis. The LPG contains a certain amount of sodium hydroxide solution which separates out in the storage. Consequently this solution must be drained off prior to sampling the propane. Figure A4.2 (page 290) illustrates the arrangements for removing the caustic solution. The main bottom flange of each sphere is to be 1.2 m above the bund floor with at its centre a 50 mm connection on which were attached two plug valves. The valves are to be mounted underneath and close to the bottom of the sphere and terminated in a vertical pipe having its open end close to the bund catch pit. The short spool piece between the valves carries a side connection of 20 mm to a valve and sample connection. These draw-off arrangements are heated by small bore steam tracing beneath the lagging.

HAZARD IDENTIFICATION AND RISK ASSESSMENT

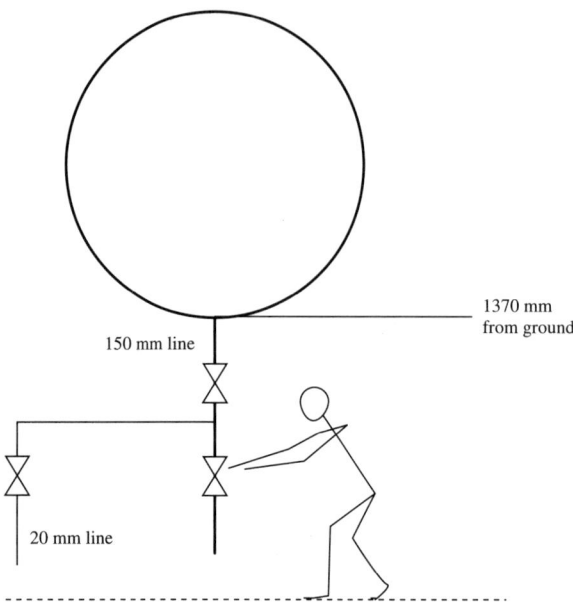

Figure A4.2 Illustration of the weak caustic draw-off activity.

Valve handles will not be permanently mounted or placed near the valves to prevent unauthorized tampering with the drainage system.

(9) A major study should be undertaken to test your knowledge. The waste heat boiler system (page 68) or the batch plant (page 177) can be used for this purpose. The steps involved are as follows:
- specify appropriate chemicals and reactions where appropriate;
- carry out a CHA;
- develop the P&I diagram to a level which you believe is appropriate;
- carry out a PPHA;
- develop the fault tree for a given incident and evaluate the occurrence of the top event;
- assume an incident severity category and evaluate the risk. Modify as appropriate;
- carry out a Hazop study.

(10) Carry out a Task Analysis on a simple item of office equipment — for instance, a drinks vending machine or an overhead projector.

(11) Carry out a Task Analysis on the valve used to add ingredient M into the batch reactor used in the ethylene derivative process (page 177). Evaluate the reliability of this operation using uncorrected data from the data bank given for Short Cut Risk Assessment.

(12) The batch plant problem (page 177) becomes more complicated with multiple autoclaves. For example, consider the changes brought about by having two autoclaves R101 and R102 (Figure A4.3). Instrumentation/control is as previously identified on page 177, together with appropriate interlocks. The study should include the following:

APPENDIX 4

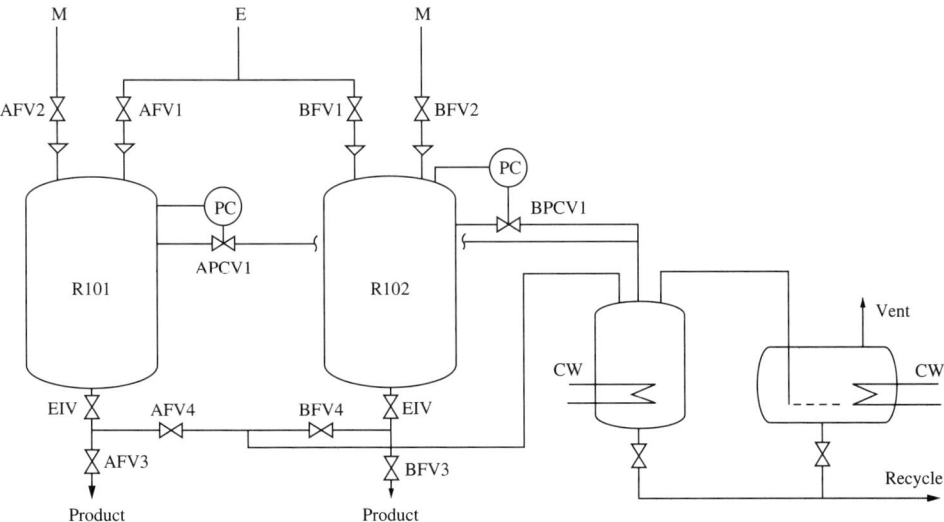

Figure A4.3 Main control valves on multiple autoclave system.

(a) Identify all likely sneak paths (ignore the process relief system except as a source of contamination in the dump tank).
(b) Carry out a combined Task Analysis/Hazop deviation study for valve BFV2.

FAULT TREES

(13) A system to pump acetic acid from a supply tank is illustrated in Figure A4.4, page 292. Construct a fault tree which has as its top event: 'No flow to the process'. The system functions automatically. When the regulator is energized, one of the pumps is started and acid passes through the feed pipes. If no acid is detected in the feed pipe the second pump is started. In order to simplify the problem assume only the faults mentioned in Table A4.1 on page 292 can occur. The failure of the pumps to start allows for the failure of the switching device.

The problem is best tackled by grouping together general problems and individual pump problems. The answer is 0.26.

(14) Evaluate the probability of the top event which results from an OR gate into which feed three OR gates each having a probability of 0.1. The co-products of probabilities must be taken into account.

(15) Use Boolean Algebra to simplify the following expressions:

$$Z = A + A \bullet B + B \bullet C + C \bullet D + B \bullet D$$

$$Z = A \bullet B + A \bullet B \bullet C + A \bullet E \bullet B + C \bullet D + C \bullet D \bullet E$$

$$Z = (G + X + Y) \bullet (E + X)$$

$$Z = (A \bullet C + A \bullet D)(E + C) + D$$

291

Answers:

$$Z = A + B \cdot C + C \cdot D + B \cdot D$$
$$Z = A \cdot B + C \cdot D$$
$$Z = G + X$$
$$Z = A \cdot C + D$$

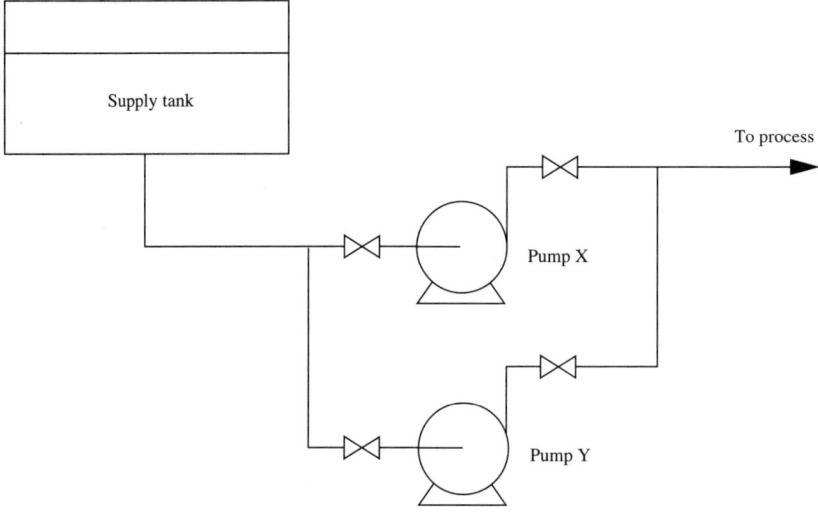

Figure A4.4 Pumping system.

TABLE A4.1
Faults assumed in acetic acid pumping system

Component	Symbol	Probability of failure	Failure mode
Cables on either pump	E and M	0.0001	Short circuit
Electricity supply	D	0.01	Power cut
Feed pipe	L	0.0001	Pipe rupture
Manifold	M	0.001	Rupture
Pumps	C and G	0.05	Fail to start
Supply tank	B	0.01	Level too low
Pipe outlets	A and F	0.0001	Rupture

(16) Evaluate the fault tree (Figure A4.5) assuming that the probabilities of events are as follows:

$A = 0.1, B = 0.2, C = 0.1, D = 0.1$

Answer:

$TOP = A(B + D) + C = (0.028 + 0.1 - 0.0028) = 0.13$

(17) For the system in question (15) are there any other features to consider? Is a general solution sufficiently accurate without developing the tree for the individual pumps? Use the following data to find out.

Failure of pumps and switch	0.19
Failure of regulator	0.05
Manifold failure	0.001
Process supply failure	0.01
Electricity supply failure	0.01
Blockage	0.01
Pump isolated in error	0.01
Inadequate priming	0.001

(18) What other equipment might be introduced to help the overall system in question (15) to function better? Develop a P&I diagram for the system. Hazop your diagram.

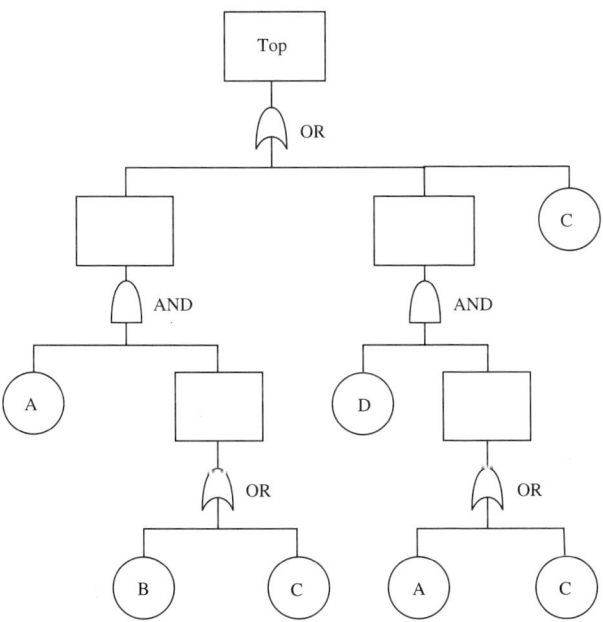

Figure A4.5 Fault tree for study (16).

EVENT TREES

(19) Draw an event tree to show what might happen if a driver in a car travelling along the road comes across a fixed obstruction. This occurs at a frequency of 1×10^{-6} per year. The driver sees this obstacle in sufficient time to avoid a collision ninety nine times out of a hundred if the foot brakes work with their normal efficiency. However these are not often in the best of condition and have a failure probability of 0.01. The driver knows to use the handbrake if the foot brake fails. This would bring the car to a safe stop with a probability of 0.95 each time it is used. Road conditions can be ignored and swerving around the obstruction or doing a handbrake turn means a collision as the road is narrow. Show that the probability of a collision based on this information is 1.0005×10^{-8} per year.

(20) The objective of this problem is to demonstrate that an event tree and a fault tree can be used to describe the same incident. The selection of the descriptions at junctions of an event tree can effectively be used to introduce gates.

An instrument cubicle is protected from a possible leak of flammable material into it by a purge stream of nitrogen. An alarm is fitted to signify the failure of this flow. It is possible for a flammable mix to be ignited in the cubicle if a fault develops in the electrical system.
(a) Construct an event tree to describe the sequence of events.
(b) Construct a fault tree to describe the sequence of faults.

RISK CRITERIA

The report of the Royal Society Study Group, *Risk: Analysis, Perception and Management*[21], is essential background reading on risk analysis. Note that there is by no means unanimous agreement with the conclusions of the report.

(21) Public risk management is said to involve the combination of organizational types, the combination of regulatory instruments and the combination of institutional rules to produce outcomes of the least-cost mix of Type I, Type II and Type III errors when traded off against other social values.

Type I errors traditionally means rejection of a true hypothesis and is often extended to mean sins of omission.

Type II errors traditionally means acceptance of a false hypothesis and may be extended to mean sins of commission.

Type III errors have been added in order to describe errors that arise from faulty specification of the problem, leading to real solutions to the wrong problem.

Consider these factors and how they affect the onus of proof. For example, compare the actions in product testing of drugs with those in the evacuation of the public in the face of a terrorist warning.

The cultural theory school holds, contrary to ideas that stress the effect of early conditioning, that cultural bias and the lifestyle that goes with them are to some degree chosen rather than predetermined. An individual's cultural bias is linked with the extent to which the individual is incorporated into bounded groups ('group') and the extent to which the interactions of social life are conducted according to rules rather than negotiated ad hoc ('grid'). By linking grid and group it is possible to generate four groups of people:

- Hierarchists (high grid, high group) are willing to accept a defined tolerable risk specified at a high level, as long as decisions are made by experts or in other socially approved ways.
- Sectarians or egalitarians (low grid, high group) accentuate the risks of technological development and economic growth so as to defend their own way of life and attribute blame to those who hold to other cosmologies.
- Fatalists (high grid, low group) do not knowingly take risks but accept what is in store for them.
- Individualists (low grid, low group) see risk and opportunity as going hand in hand.

To which group do you belong and do you agree with the definitions?

(22) Numerous studies have been carried out, particularly in the United States, which suggest that risk orderings of fatality judgements by lay people correspond quite well to the rank orderings implied by available statistics. Do your own views coincide with this belief?

(23) Draw up a suitable statement from the chief executive on a company's attitude to risk which might be promulgated as company policy. Indicate appropriate guidelines for risk criteria as might be issued for internal company decision-making.

(24) A societal risk study has been carried out to assess the frequency of people suffering at least harm from a release of chlorine. Eight directions of wind were identified. The initial data is given in Table A4.2. Plot the data as an F-N table.

TABLE A4.2

Direction	Weather	Frequency	Number of people
1	D5	2.8×10^{-6}	40
	F2	3.5×10^{-7}	1300
2	D5	2.2×10^{-6}	150
	F2	2.8×10^{-7}	2700
3	D5	2.0×10^{-6}	250
	F2	2.7×10^{-7}	4200
4	D5	3.5×10^{-6}	370
	F2	4.5×10^{-7}	4700
5	D5	5.7×10^{-6}	280
	F2	7.4×10^{-7}	3700
6	D5	4.9×10^{-6}	35
	F2	6.9×10^{-7}	1300
7	D5	2.8×10^{-6}	9
	F2	3.5×10^{-7}	1200
8	D5	2.9×10^{-6}	20
	F2	3.9×10^{-7}	200

APPENDIX 5 — LIST OF ACRONYMS

ALARA — as low as reasonably achievable
ALARP — as low as reasonably practicable
API — American Petroleum Institute

BATNEEC — best available techniques not entailing excessive cost
BLEVE — boiling liquid expanding vapour explosion

CESS — Critical Examination of System Safety
CHA — Concept Hazard Analysis
CHAIR — Concept Hazard Analysis Initial Review
CIMAH — Control of Industrial Major Accident Hazards
COSHH — Control of Substances Hazardous to Health
CSR — Concept Safety Review

EPA — Environmental Protection Agency (USA)

FAR — fatal accident rate
FDT — fractional dead time
FMEA — Failure Mode and Effect Analysis
FTA — Fault Tree Analysis

Hazop — hazard and operability study
HSE — Health and Safety Executive (UK)

LPG — liquefied petroleum gas

NII — Nuclear Installations Inspectorate (UK)

OSHA — Occupational Safety and Health Administration (USA)

P&ID — piping and instrumentation diagram
PES — programmable electronic system
PFD — process flow diagram
PPHA — Preliminary Process Hazard Analysis

QRA — Quantified Risk Assessment

SCRAM — Short-Cut Risk Assessment Method

REFERENCES

1. Turney, R.D., 1990, Designing plants for 1990 and beyond, *Trans IChemE*, 68 (B1): 12–16.
2. Bretherick, L., 1990, *Bretherick's Handbook of Reactive Chemical Hazards* (Butterworths, London, UK).
3. Elliott, D.M. and Owen, J.M., 1968, Critical examination in process design, *The Chemical Engineer*, November 1968, 377.
4. Kletz, T.A., 1988, *Learning from Accidents in Industry* (Butterworths, London, UK).
5. Kletz, T.A., 1991, *An Engineer's View of Human Error*, 2nd edition (IChemE, Rugby, UK).
6. US Dept of Labor, April 1990, *The Phillips 66 Company Houston Chemical Complex Explosion and Fire* (US Dept of Labor, Washington DC, USA).
7. Amalberti, R., 1992, Safety in process control: an operator-centred point of view, *Reliability Engineering and System Safety*, 38: 99–108.
8. Bergroth, K., 1993, Explosion in a sulphonation process, *Loss Prevention Bulletin*, No. 109, 1–5.
9. Ujita, H., 1992, Human characteristics of plant operation and man-machine interface, *Reliability Engineering and System Safety*, 38: 119–124.
10. Embrey, D.E., 1992, Module 6 of the MSc in Process Safety and Loss Prevention of The University of Sheffield, UK.
11. Rasmussen, B., 1987, Unwanted chemical reactions in the chemical process industry, *Risø–M–2631* (Risø National Laboratory, Finland).
12. Taylor, J.R., 1992, The sneak analysis procedure, *Proceedings of The Sneak Analysis Workshop, ESA-WPPO33, European Space Agency, Noordwijk, The Netherlands*.
13. Center for Chemical Process Safety of the American Institute of Chemical Engineers, 1989, *Guidelines for Process Equipment Reliability Data* (AIChE, New York, USA).
14. Center for Chemical Process Safety of the American Institute of Chemical Engineers, 1993, *Guidelines for Safe Automation of Chemical Processes* (AIChE, New York, USA).
15. Health and Safety Executive, 1992, *The Tolerability of Risk from Nuclear Power Stations* (HMSO, UK).
16. Health and Safety Executive, 1991, *Major Hazard Aspects of the Transport of Dangerous Substances* (HMSO, UK).
17. Health and Safety Executive, 1989, *Risk Criteria for Land Use Planning in the Vicinity of Major Industrial Hazards* (HMSO, UK).
18. CONCAWE, 1988, *Quantified Risk Assessment Report No 88/56* (CONCAWE, The Netherlands).
19. Allum, S. and Wells, G.L., 1993, Short-cut risk assessment, *Trans IChemE*, 71 (B3): 161–168.
20. Williams, J.C., 1988, A data based method for assessing and reducing human error to improve operating experience, *Proc IEEE 4th Conference on Human Factors in Power Plants, Monterey, California, USA, 6–9 June 1988*.
21. Royal Society Study Group, 1992, *Risk: Analysis, Perception and Management* (Royal Society, London, UK).

INDEX

A
acceptable risk 203–204, 206
accidents 3
ALARP (as low as reasonably
   practicable) 204, 205, 207

B
batch processes 176–189
BATNEEC (see Best Available Techniques
   not Entailing Excessive Cost)
benzene plants 96, 131–132
Best Available Techniques not Entailing
   Excessive Cost (BATNEEC) 241
Boolean algebra 14, 139–145
Boolean reduction 144
brainstorming 77

C
case studies,
   a lifting problem 24–25
   CHA of methanator 34–45
   Consequence Analysis of methanator 58–62
   Critical Examination 183
      of distillation column 77
      of ethylene derivative batch plant 177, 179
      of gas export riser 84–89
      of methanator 83–84
   FTA of gas leak 126, 149
      of high pressure in tank 126–127
      of high pressure on methanator 131
      of high temperature on benzene
         plant 131–132
      of simple pump system 138–139, 143
   Hazop of batch plant 187–188
      of benzene plant 96
      of methanator 112–120
      of pipeline 91–94
      of water injection plant 267–287

methanator information 30–36
methanator event tree 62
P&ID of compressor system 253–256
   of emergency shutdown system 249–252
PPHA of methanator 51–58
   of waste heat boiler 65–73
QRA of helipads in the jungle 224
   of pipeline 219–222
   of ship collision with oil rig 224–225
Task Analysis of batch plant 183–186
   of boil-over in reactor 165–166
   of diluting caustic soda 170
   of filling a batch tank 171–173
causes of death 199–201
CHA (see Concept Hazard Analysis)
CHAIR (see Concept Hazard Analysis Initial
   Review)
CIMAH (see Control of Industrial Major
   Accident Hazards Regulations)
common mode failures 144–146
compressors
   equipment knowledge base 255
   safety analysis table 256
computer control 196–197
Concept Hazard Analysis (CHA) 7–9, 18,
   19–45, 48, 254
   keywords 9
   of methanator 34–45
   methodology 24
   methods 19
Concept Hazard Analysis Initial
   Review (CHAIR) 19, 21, 34
   keywords 22–23
Concept Safety Review (CSR) 20
Consequence Analysis 7, 9–11, 15
   case study of methanator 58–62
control 3, 70, 198
   emergency 4, 50–57, 72–73,
      122–123, 245–256, 261
   normal 4, 50–57, 72–73, 122–123
      242–244, 252, 261

298

Control of Industrial Major Accident Hazards (CIMAH) Regulations	20, 21, 26, 121, 241	event trees	11, 62, 148–151, 174, 195, 294
		exothermic reactions	27, 31, 38, 83, 248, 251, 252
control (on alarm)	4	explosions	22, 34, 38
Control of Substances Hazardous to Health (COSHH) Regulations	20, 26	exposure to chemicals	28
		external error modes	153
control systems	248–256	external threats	23, 36
COSHH (see Control of Substances Hazardous to Health Regulations)			
Critical Examination	12, 16, 18, 19, 74–89, 177–179	**F**	
		Failure Mode and Effect Analysis (FMEA)	18, 153–154, 192–193
case studies		failure rates	229–231
distillation column	77	failure to control the situation	261
ethylene derivative batch plant	177, 179	FAR (see fatal accident rate)	
gas export riser	84–89	fatal accident rate (FAR)	201
methanator	83–84	Fault Tree Analysis (FTA)	14–15, 46, 124–151, 195
keyword dictionary	79–81		
methodology	81	case studies	
record sheet	75	gas leak	126, 149
critical tasks	180	high pressure in tank	126–127
CSR (see Concept Safety Review)		high pressure on methanator	131
		high temperature on benzene plant	131–132
		simple pump system	138–139, 143
D		evaluation of gates	133–134
dangerous disturbances	4, 9–10, 13, 22, 38, 40, 47, 49, 63–64, 262	event symbols	125
		gate symbols	124–125
dangerous dose	208	generic fault tree for a release	127
dangerous substances	12	overall methodology	146
death, causes of	199–201	fault trees	2, 3, 15, 58, 59, 71, 220, 221, 291–293
decision-making	237–238	fires	22, 34, 38, 60, 61
design	245	flammables	9, 34, 38
discharge	34	FMEA (see Failure Mode and Effect Analysis)	
documentation	103–106	frequency analysis	136, 227–228
		FTA (see Fault Tree Analysis)	
E			
effluents	34, 265	**G**	
emergency control systems	4, 50–57, 72–73, 122–123, 245–256, 261	gas leaks	126, 149
		gates, combination of	138
emissions	29	gate symbols	124–125
equipment failure modes	153–154	guide words/keywords	9, 22–23, 79–81, 91, 95–98, 100–102
equipment knowledge base			
compressor	255		
reactor	250		
separator	254	**H**	
event symbols	125	hazard analysis	6
Event Tree Analysis	14–15, 148–151	hazard controls	3, 242–245

299

hazard identification 288–291
hazard and operability studies (Hazop) 12–14, 18, 71, 74, 90–123, 160, 254, 267–287
   case studies
      batch plant 187–188
      benzene plant 96
      methanator 112–120
      pipeline 91–94
      water injection plant 267–287
   conduct of a meeting 96–103
   defining objectives and scope 99
   documentation 103–106
   flowchart for application of 102
   guide words 95–96, 100–102
   how to target effort 107–108
   procedure 91–92
   report 103–106
   team members 98–99
   use, strengths and limitations 108–111
hazardous deviations 259–260
hazardous disturbances 4
hazardous reactions 22, 27
hazardous substances 9, 22, 25, 28
hazard reduction 6
hazards 34, 90, 288
   introduction 1–18
Hazcheck knowledge base 46, 251, 254, 257–266
Hazop (see hazard and operability studies)
health and safety policy 213
human action error analysis 195
human error 153, 161–162
human reliability 163, 193–194, 232

I

immediate causes 3, 10, 13, 36, 47, 70, 127, 258
incidents 3, 37, 154–157
   Bhopal 5, 240
   Chernobyl 155
   factors contributing to 164
   Guatemala City 120
   Kegworth 158
   Kokkola 158–160
   Texas polyethylene plant 156
   Three Mile Island 158

   Zeebrugge 157
incident factors 164
incident scenarios 1–5, 7, 10, 13, 29, 47–48, 121, 134, 157–160, 257
   fault tree 2
individual risk 16, 199, 201
ineffectiveness probability of mitigation 233
instrument letter codes 243
interlocks 247

K

keywords/guide words 9, 22–23, 79–81, 91, 95–98, 100–102
knowledge bases 250, 254, 255

M

methanators 112–120, 131
mistakes 152, 154
method study 74, 172–174

N

normal control 4, 50–57, 72–73, 122–123, 242–244, 252, 261

O

occupational risk 201–202
Occupational Safety and Health Administration (OSHA) 20, 21, 26, 241
offshore rigs 84
operability problems 109
operators 174, 252
OSHA (see Occupational Safety and Health Administration)
overpressure 35, 38, 40, 63, 64, 177
overtemperature 35, 38, 40, 63, 64

P

P&IDs (see piping and instrumentation diagrams)
parameters 95–96
PES (see programmable electronic system)

pipelines 91–94, 219–222
piping and instrumentation diagrams
    (P&IDs) 11, 13, 14, 19, 32–33, 36, 42–43, 52–53, 66–69, 99, 110–111, 118–119, 177, 242–256, 268–273
    instrument letter codes 243
    symbols 244
planned reactions 9
pollutants 34, 265
PPHA (see Preliminary Process Hazard Analysis)
Preliminary Process Hazard Analysis
    (PPHA) 7, 8–11, 14, 17, 18, 46–73, 121, 160, 198, 254
    case studies
        methanator 51–58
        waste heat boiler 65–73
    methodology 48
    pro forma 50–51
    risk evaluation sheet 70
    sheets 54 57
    structure 48–51
pressure relief 245–247
probability 136
process deviations 4, 50, 97–98
process safety reviews 6
programmable electronic system (PES) 198
pump systems 138–139, 143

Q

QRA (see Quantified Risk Assessment)
Quantified Risk Assessment (QRA) 7, 11, 17, 210–211
    application of 240–241
    case studies
        helipads in the jungle 224
        pipeline 219–222
        ship collision with oil rig 224–225
    decision-making 237–238
    legislation 214–215
    objectives and scope 225
    requirements for 222–224
    role and use 212
    uncertainties 225–227, 236–237

R

reactions 23, 35, 63, 177, 178
    cause of incidents 178
    exothermic 27, 31, 38, 83
    hazardous 27
    planned 9
    unintended 9
reactor, equipment knowledge base 250
reactor system, analysis 82–83
releases 63, 263–266
residual risk 5, 238–240
risk
    acceptable 203–204, 206
    individual 16, 199, 201
    introduction 1–18
    occupational 201–202
    residual 5, 238–240
    societal 16, 199, 202–203, 209
    tolerable 206
risk assessment 16–17, 62, 65, 210–241
    elements of 210
    frequency analysis 227–228
    general factors affecting risk 233–234
    prioritization and sensitivity studies 218
risk criteria 199–209, 294–295
    severity categories 216–218
risk estimates 215
risk evaluation 17, 62, 65, 70, 199
risk ranking 236
risk reduction 235
root causes 4, 29–30, 120

S

safety reviews 7
safety schedules 120–123
SCRAM (see Short-Cut Risk Assessment Method)
second-chance design 245
separator, equipment knowledge base 254
severity categories 216–218
Short Cut Risk Assessment Method
    (SCRAM) 210, 215–218, 228–232
significant events 4
slips 152, 155
Sneak Analysis 107, 189–192
    sneak paths 191–192

societal risk　　　　　　　　16, 199, 202–203, 209
successive questions　　　　　　　　　　　　　77
system-induced errors　　　　　　　　　　　　163
system-related causes　　　　　　　　161–163, 162

T
Task Analysis　　　　15–16, 106, 152–175, 176,
　　　　　　　　　　　　　　　　　　　180, 198
   case studies
      batch plant　　　　　　　　　　　　183–186
      boil-over in reactor　　　　　　　　165–166
      diluting caustic soda　　　　　　　　　　170
      filling a batch tank　　　　　　　　171–173
   data　　　　　　　　　　　　　　　　　　167
   deviations　　　　　　　　　　　　　　　202
   plans　　　　　　　　　　　　　　　169–172
   steps/methodology　　　　　　　　　167–169

team members for Hazop　　　　　　　　　98–99
tolerable risk　　　　　　　　　　　　　　　206
trips　　　　　　　　　　　　　　　　　　　198
trip systems　　　　　　　　　　　　　　35, 41

U
unintended reactions　　　　　　　　　　　　　9
utility problems　　　　　　　　　　　　　　35

V
Venn diagrams　　　　　　　　　　　　　　133
violations　　　　　　　　　　　　　　152, 157

W
working methods　　　　　　　　　　　　　　16